From Safety to Safety Science

How do accidents and disasters occur? How has knowledge of accident processes evolved? A significant improvement in safety has occurred during the past century, with the number of accidents falling spectacularly within industry, aviation and road traffic. This progress has been gradual in the context of a changing society. The improvements are partly due to a better understanding of the accident processes that ultimately lead to damage. This book shows how contemporary crises instigated the development of safety knowledge and how the safety sciences pieced their theories together by research, by experience and by taking ideas from other domains.

From Safety to Safety Science details 150 years of knowledge development in the safety sciences. The authors have rigorously extracted the essence of safety knowledge development from more than 2,500 articles to provide a unique overview and insight into the background and usability of safety theories, as well as modelling how they developed and how they are used today. Extensive appendices and references provide an additional dimension to support further scholarly work in this field.

The book is divided into clear time periods to make it an accessible piece of science history that will be invaluable to both new and experienced safety researchers, to safety courses and education, and to learned practitioners.

Paul Swuste is an associate professor of the Safety Science Group, Delft University of Technology, the Netherlands, with an MSc degree in Biochemistry (Leiden University, 1978) and a PhD thesis, 'Occupational Hazards and Solutions' (Delft University of Technology, 1996). He has conducted research on risk assessments in high-tech-high-hazard industries, on the history of knowledge developments in safety science and on various occupational hazards. He has published frequently on these topics and co-organised the postgraduate master course 'Management of Safety Health and Environment' from 1994 to 2008.

Jop Groeneweg graduated as a cognitive psychologist from Leiden University in the early 1980s. In a career spanning about four decennia, he was involved in many projects, in and outside the university, to improve safety, predominantly in industrial organisations. As a professor of Safety in Healthcare at Delft University of Technology and a human performance expert at Leiden University and the TNO research institute, in the Netherlands, he aims to transfer his knowledge to the medical domain to reduce preventable adverse events while at the same time getting new insights that might help to further improve safety in the industry.

Frank W. Guldenmund graduated from Leiden University with degrees in both cognitive psychology and methods and statistics. In February 1992 he joined the Safety Science Group at Delft University of Technology. In his research, he focusses on the management of safety in industrial organisations and on the behavior of people within those organisations. He has been teaching safety science for nearly 30 years to both graduate and undergraduate students as well as to safety practitioners. Since 2002 he has been a trainer in the safety culture program of the International Atomic Energy Agency (IAEA), providing lectures and workshops on this topic worldwide. Currently, he is on the board of the Dutch Society for Safety Professionals (NVVK) and responsible for embedding (more) science into the work of safety professionals. He is editor of the society's journal as well as associate editor of *Safety Science*.

Coen van Gulijk is a senior scientist at TNO Healthy Living, a vising professor at the University of Huddersfield and affiliate researcher of the Safety Science Group of the Delft University of Technology. He is investigating and accelerating the digital transformation of safety models and safety management. He has taught safety science on an academic level in four universities in the Netherlands, one university in Belgium and one in the UK, and actively engages in international networks and scientific dissemination.

Saul Lemkowitz was an associate professor of the Chemical Engineering department at Delft University of Technology. He studied chemical engineering at Rutgers University, in the United States, and at Delft. His PhD thesis (Delft, 1975) focused on 'Phase and corrosion studies of the ammonia–carbon dioxide water system'. Dust explosions and explosion safety in the process industries were his fields of research and education, together with sustainability, industrial ecology and technology and society. He frequently published on these topics. Regrettably, Saul passed away on 13 February 2020.

Yvette Oostendorp finished her master's studies at Wageningen University and Research in environmental and industrial hygiene in 1983 and worked as a researcher at Wageningen UR on agreement between qualitative estimates and quantitative exposure measurements. From 1986 until 2004 she worked as an industrial hygienist at an occupational health service. She is author or co-author of several handbooks on chemical risk assessment for professionals in occupational health services. In 2004 she started as a senior advisor at the former Hazardous Substances Council, the advisory council for the Dutch parliament. Since 2012 she has worked at the Dutch Council for Environment and Infrastructure (RLI).

Walter Zwaard studied chemistry at Leiden University and received his PhD in 1983. He worked at Leiden University as risk manager, radiation safety officer and lecturer on laboratory safety. From 2004 until 2012 he was a member of the former Hazardous Substances Council. He has published widely on safety issues such as hazardous substances, accident prevention and risk management. He has written a number of books and edited several textbooks on safety. Since 1992 he has worked as a safety practitioner and consultant in both public and private sectors. As an instructor and lecturer, he participates in many courses for risk professionals.

From Safety to Safety Science

The Evolution of Thinking and Practice

Paul Swuste, Jop Groeneweg, Frank W. Guldenmund, Coen van Gulijk, Saul Lemkowitz, Yvette Oostendorp and Walter Zwaard

Routledge
Taylor & Francis Group
LONDON AND NEW YORK

nvvk veilig heids kunde

TU Delft
Technische Universiteit Delft

First published 2022
by Routledge
2 Park Square, Milton Park, Abingdon, Oxon OX14 4RN

and by Routledge
605 Third Avenue, New York, NY 10158

Routledge is an imprint of the Taylor & Francis Group, an informa business

British Library Cataloguing-in-Publication Data
A catalogue record for this book is available from the British Library

Library of Congress Cataloging-in-Publication Data
A catalog record for this book has been requested

ISBN: 978-0-367-43122-8 (hbk)
ISBN: 978-0-367-55024-0 (pbk)
ISBN: 978-1-003-00137-9 (ebk)

DOI: 10.4324/9781003001379

Typeset in Times New Roman
by Apex CoVantage, LLC

To Saul Lemkowitz (1941–2020)

Contents

Preface

This book presents a history of Safety Science knowledge. Our search for the development of safety-related knowledge provides insight into the background and utility of the present-day theories, models and metaphors. History shows the time-bound understanding and limitations of knowledge about causes of occupational, industrial and major accidents and shows also why ideas arise, trickle down into the day-to-day approaches to improve safety, disappear or lead a dormant existence. The past illustrates how concerns about accidents have led to attention to safety and later to structured knowledge – that is, the field of Safety Science.

Many articles about safety and most safety handbooks and manuals start with a historical section or only one chapter on the history of the safety domain. In contrast, we describe the subject in depth and re-examine all pertinent original articles. We go back to the early period of occupational safety, halfway through the First Industrial Revolution. The impetus for carrying out our study was Walter Zwaard's book *Chronicle of Dutch Safety* (Zwaard 2007). We decided not to limit our study to the Netherlands. Until after the Second World War, the United Kingdom and the United States were front-runners in safety, and the Netherlands is presented as a case study showing how Anglo-Saxon ideas on safety were translated to the Dutch national level. A restriction is the language of publications; we provide an overview and discussion only of English and Dutch literature. Of this literature overview, in total, we selected and read more than 2,500 publications, most from scientific journals and the rest from professional journals and books. The number of publications from the initial period – the end of the nineteenth century – is manageable. From the beginning of the twentieth century, the United States emerges as an important source of information. This trend is probably a reflection of the increased industrialisation of the United States that occurred during the start of the twentieth century, and especially during the Second World War. After the Second World War, Great Britain and, in particular, the United States served as examples in many areas of development, including Safety Science.

Due to our sole focus on Dutch and English literatures, contributions from central, southern and eastern Europe are few, because most is written in the native language. For example, developments in Germany and France from the early days of the Industrial Revolution are dealt with only anecdotally. Although Germany has a long tradition in occupational safety, German professional literature was not popular for many years, just before and after the Second World War. We did not include German literature in our study. This also applies to French literature. However, after the Second World War, and especially after Safety Science became an academic discipline in the 1970s and 1980s, the perspective changed. Authors from Scandinavia and Australia also published much, and the total number of publications increased exponentially. The non-Western industrialized countries, incidentally, occupy a separate position. It is known that many (major) accidents occur in those countries, but these countries, such as China, have only recently been reviewed in the English-scientific literature. Their contribution to safety-related knowledge development is therefore difficult to determine and so falls outside the scope of this book.

The text of our book is based on articles we published in *Safety Science* from 2010 onwards and on the aforementioned *Chronicle of Dutch Safety* (Oostendorp et al. 2016; Swuste, Gulijk, Zwaard 2010; Swuste et al. 2014; Swuste, Gulijk et al. 2016; Swuste, Groeneweg, Gulijk, Zwaard et al. 2018; Swuste et al. 2019; Swuste, Gulijk et al 2020; Swuste, Gulijk, Groeneweg, Guldenmund et al. 2020). In these articles and in this book, a number of questions play a pivotal role, such as the question of which theories, models and metaphors have been developed over time to explain the causes of industrial accidents and other major accidents, and the question of to what extent general management movements and knowledge developments have influenced the management of safety in companies. Additionally, we explore the context in which these developments have taken place, and what the consequences have been for safety in the Netherlands.

In this book we study 'knowledge development'; that is, the development of theories, models and metaphors. We consider a theory an empirically or otherwise validated model. And, more importantly, hypotheses can be formulated with a theory to test their tenability. A model is a schematic but not yet validated presentation of reality, neither supported nor confirmed through research results. An example of a model is a 'metaphor', or metaphorical usage based on a comparison. For example, the well-known 'dominoes' of Herbert Heinrich are a metaphor in which an accident process is compared to falling dominoes (Heinrich 1941). Because of their simplicity and strong visualisation, models and metaphors carry great expressiveness. (In Dutch, the word *metaphor* is also translated as 'image talk'.)

We regularly discuss the difference between science and profession in our book: the distinction between the development of safety as a science and the practical application of safety knowledge. With the latter, we mean knowledge development within the professional group, the profession, a development which does not run a priori in parallel with discourses in the scientific domain.

This book is written with the assumption that developments in the management of safety, as reported in the literature, are fuelled by both knowledge about causes of major accidents and by more general ideas about managing production. This assumption does not mean that the relationship between general management approaches and safety management is always crystal clear.

Legislation sometimes has had a stimulating influence on knowledge development, because legislation can stimulate new research. Moreover, legislation for safety practice has been the guiding principle in many companies, especially with regard to technical measures and administrative procedures. The influence of safety science knowledge on both legislation and company policy is in itself an interesting topic to explore. But science has, of course, its own dynamic, independent of legislation. And legislation is partly determined by knowledge and not vice versa: it ideally anchors safety knowledge within the law and as such is a social-political translation of this knowledge development. This book focuses on scientific developments, so safety legislation will be discussed only occasionally. The choice not to treat regulation in this book does not render it unimportant; authors like Lees (1980, 1996); Perrow (1984); Rowe (1988); Kjellén (2000); Hopkins (2000a, 2012); Coglianese et al. (2009); Mannan (2005, 2012); Pasman (2015); and Kjellén and Albrechtsen (2017) to name a few, deal with regulation in various degrees.

Each chapter, with the exception of the first, describes a period of 20 years. All chapters start with a brief characterisation of that period. Chapter content is ordered by country: the United Kingdom, the United States, and the Netherlands. For each chapter, the sequence of discussing British and American developments depends on the relative influence of the country on the safety-related knowledge development of that period. During some periods the Netherlands has played a leading role, but its developments will be discussed in the final paragraph of each chapter. After the Second World War, safety theories, models and metaphors came increasingly from other countries, notably Nordic countries, Canada and Australia. Therefore, the paragraphs 'United States' and 'United Kingdom' are renamed as North America (Chapters 5 and 6), North America and Australia (Chapter 7, 8) and Western Europe and Nordic countries (Chapter 5–8). From Chapter 5 onwards, this structure is, however, used more loosely because knowledge development, with the emergence of international scientific journals in the safety-related domain, takes on a more cross-border character from about the 1970s. In addition, from Chapter 5 onwards, developments in occupational safety (Chapters 5 and 7) and high-tech-high-hazard safety (Chapters 6 and 8) are described separately. Because Chapters 5 and 6 deal with same period, Chapter 6 will not present a characterization of the period. The same is true for Chapters 7 and 8.

In this book the term 'high-tech-high-hazard' is used for industrial sectors with potentially major accidents. The classical interpretation of hazard is energy, already proposed in the second half of the 1920s by DeBlois, as is discussed in Chapter 2. The processes in these sectors can become dangerous when hazards become uncontrollable and turn into risks. The magnitude and exposure are directly related to the energy content. Hazards are real and risks are constructs, as will be argued in Chapter 8. For this reason, the term high-tech-high-hazard is preferred above high-tech-high-risk.

This book is written for a wide audience, anyone interested in the development of safety and safety science, including managers and policy makers in the safety domain. Another target group is safety experts, and those working in the fields of occupational safety or high-tech-high-hazard safety, but also in other domains, like patient safety, transport safety and social safety. This book is also a reference for teaching and further education of safety experts.

The authors, spring 2021

Time travel

'Rarely is a problem unique. You can usually find something in past experience to offer some clue to the future'. The quote is from Heinrich and his co-author, Lateiner, in a book published seven years after Heinrich's death (Lateiner and Heinrich 1969). To honour this quote, this paragraph starts with a journey of a hypothetical time traveller coming from the second part of the nineteenth century to today's world.

Safety promotion exhibitions began in European countries and in the United States in the late nineteenth century and the beginning of the twentieth century. Visitors to these exhibitions would have gazed at technical devices such as protective coverings on moving parts, like rotating cogwheels, and spinning fans, for extracting hazardous fumes, and then seen modern technical facilities that could have made work safer but were nevertheless not applied in practice. If a visitor had been able to travel to the present with a time travel machine, he would have experienced a remarkable change in the approach to safety. Perhaps our time traveller had *Hard Times* in his knapsack, the novel by Charles Dickens that dramatically described nineteenth-century British industry, full of worker exploitation, insecurity and danger (Dickens 1854). Fortunately, nowadays, at least in the Western world, the dreadful working conditions described by Dickens hardly exist anymore. Applications of safety facilities that were exhibited in those early days have since become commonplace.

Not surprisingly, the number of accidents in industry has since fallen dramatically. In the United States, for example, around the end of the nineteenth century, hundreds of large explosions took place in factories every year. American industry killed thousands of workers every year, and tens of thousands were seriously injured (see, e.g., Klein 2009). And outside America's factories the situation was not much better. In 1910, for example, more than 3,000 American railway workers were killed and almost 100,000 were seriously injured. Much has improved and one does not have to go back a century to see progress. Indeed, a comparison of the 1970s' statistics with today's shows a spectacular decrease in occupational mortality, in plane crashes and in traffic accidents.

The time traveller would be surprised that today a field has emerged, Safety Science, with inspiring models and fruitful theories. A shift has occurred in thinking about causes of accidents. A century ago, the worker was more or less always to blame for an accident; he or she was obviously 'accident prone'. At that time, the employee was a cheap instrument, a moving part, and was trashed just as easily as a defective part. It wasn't uncommon that factory doors were locked during working hours to prevent employees from fleeing the workplace. Many workers died in fires, but the risk of being locked in during a fire was never considered. Investigations of disasters at the time usually focussed entirely on the role of the failing employee, rarely on defective technology and almost never on the managers, who were virtually blame-free. The concept of 'just culture' was a great unknown. Even in evidently unsafe situations, such as working with steam boilers without pressure relief, more than a century passed before adequate safety measures were introduced.

Today, causes of accidents are viewed more systematically. Organisational, people-oriented and technical factors are now seen as essential and combined elements in accident processes. The fact that contemporary management of organisations shares responsibility for accidents is relatively new. This approach stimulates management to take measures early in the product life cycle, in early design, and in pre-production stages, even though such measures in the short run sometimes reduce turnover and/or profit (see, for instance, Chemical Industries Association 2008; Rademaeker et al. 2014).

The innovative steps that made safety and, more recently, accountability and sustainability, core values of companies would be unknown, even unthinkable, in the eyes of our time traveller. He would be astonished to see that in some leading national and international companies, health and safety of workers and process safety are as important as quality of production, financial success, productivity and cost reductions. Through decades of optimisation and implementation, managers today encourage reporting of accidents and near-accidents and demand in-depth analyses of all serious incidents in order to uncover the problems at a system level and to be able to take corrective measures.

The way in which government views safety issues would also be new to our time traveller. Government nowadays is much more aware of its crucial role in maintaining and improving safety and, for that purpose, stimulates initiatives and tightens regulations (Covello and Mumpower 1985). Our time traveller would be astonished at the widespread negative safety perception of today's general public, this in spite of the enormous improvements in safety (Pinker 2018; Rosling et al. 2018). Even if modern 'safety monitors' show that not all safety improvements are introduced in all companies, the situation has nevertheless much improved. This does not mean there are no concerns for our time traveller: indeed, today's media coverage of dramatic events, such as plane crashes, is easily perceived as an indicator that we might well be living in unsafe times and that organisations are still doing far too little to improve safety (Tversky and Kahneman 1974; Fischhoff 2019). Additionally, companies are often influenced by the public's growing concern with safety because the companies' safety performance is still, by no means, at the desired level of zero accidents. Yet the idealistic goal of achieving zero accidents is an ethically justified objective that even our nineteenth-century time traveller would relate to.

Not only has the level of safety increased, the nature of today's safety discourse has broadened. For example, our desire that children's playgrounds be 100% safe would seem strange, even ridiculous, to our time traveller. Their time traveller's personal concern for child safety would be much more closely related to his own reality. In his own time, children labored crawling on their bare hands and knees in perilous coal mines, or changed spools of yarn in dangerous, unprotected spinning machines in textile mills.

Under the influence of safety innovations, optimisations and implementations, safety terminology has changed completely since the 1900s. One of the intellectual crown jewels of the modern approach to safety is thinking in terms of risks. The introduction of the concept of risk has opened up a multitude of possibilities for improvement that simply did not exist before. Risks can now be mapped and measured with QRA techniques, of which Bayesian networks are more recent examples, but many, many more techniques exist (Lees 1996; Cooke 2009; Pasman 2015). Inventive prevention measures can be devised, assessed, tested and implemented, so that risks can be controlled and accidents prevented. To this end, various instruments and metaphors have been developed and optimised: Risk Inventory and Evaluation, Task Risk Analyses, Last Minute Risk Assessments, and Bowties, to name just a few.

The eight chapters of this book describe developments in safety over many decades. Almost none of these were sensational; often developments implied gradual changes in thinking about safety and in gaining greater insights into the nature of the improvement measures. This usually incremental approach may give the impression per chapter that the improvements were only marginal. The reader may then get the false idea that not much improvement has been made. Like the legendary frog in the kettle filled with water that is gradually becoming hotter and hotter, the frog does not realize that the water eventually becomes fatally hot. In the development of safety, however, the opposite is true: the water is boiling, yet the frog is saved. There has been a spectacular improvement in safety, as a comparison between Chapters 1 and 8 show. Finally, this book attempts to look into the future. What are potentially new methods and methodologies to improve safety even further and thus reduce to an absolute minimum the number of people hurt and damage to installations and the environment?

Abbreviations

AGREE	Advisory Group on Reliability of Electronic Equipment
AIChE	American Institute of Chemical Engineers
ALARA	As Low as Reasonably Achievable
ALARP	As Low as Reasonably Practical
ARAMIS	Accident Risk Assessment Methodology for Industries
AWACS	Airborne Warning and Control System
BBS	Behavioural-Based Safety
BEVI	Dutch: Decree on External Safety of Industrial Plants
BLEVE	Boiling Liquid Expanding Vapour Explosions
BP	British Petroleum
BR	British Rail
BRF	Basic Risk Factor
CA	Competent Authority
CCPS	Centre for Chemical Process Safety
CO	carbon monoxide
COVO	Dutch: Safety Committee of Residents Living in the Rijnmond Area
CPR	Dutch: Commission for Prevention of Disasters
CSR	Corporate Social Responsibility
DGA	Directorate General of Labour
DSM	Dutch State Mines
ECSC	European Coal and Steel Community
EEC	European Economic Community
EFQM	European Foundation for Quality Management
ENIAC	Electronic Numerical Integrator and Computer
ESRA	European Safety and Reliability Association
FAR	Fatal Accident Rate
FMEA	Failure Mode and Effects Analysis
FTA	Fault Tree Analysis
GR	Group Risk
Hazop	Hazard and operability studies
HCI	Hazardous Chemical Index
HF	hydrogen fluoride

HRI	Hazardous Reaction Index
HRO	High Reliability Organisation
HRT	High Reliability Theory
HSC	Health and Safety Commission
HSE	Health and Safety Executive
IAEA	International Atomic Energy Agency
IBFNV	Federation Dutch Trade Union Movement
IChemE	British Institute of Chemical Engineers
IJEM	Individual, Job, Environment and Materials
ILO	International Labour Office
INK	Dutch: Quality Institute
INSAG	International Nuclear Safety Advisory Group
ISD	Inherently Safe(r) Design
ISI	Inherent Safety Index
ISO	International Organization for Standardization
ISRS	International Safety Rating System
IT	Information Technology
JSA	Job Safety Analysis
JITSO	Job Instruction Training, Safety Observations
LNG	Liquified Natural Gas
LOC	Loss of Containment
LOPA	Layers of Protection
LPG	Liquid Petroleum Gas
LTIF	Lost Time Injury Frequency
MIC	Methyl isocyanate
MORT	Management Oversight and Risk Tree
MAC	Maximum Acceptable Concentrations
NAM	Dutch: Dutch Petroleum Company
NASA	National Aeronautics and Space Administration
NCB	National Coal Board
NIPG	Dutch: Institute of Preventive Medicine
NSC	National Safety Council
NVVK	Dutch: Society for Safety Science
OARU	Occupational Accident Research Unit
OECD	Organisation for Economic Co-operation and Development
ORM	Occupational Risk Model
P&ID	Piping and Instrumentation Diagram
PI	Process Intensification
PIIS	Prototype Index for Inherent Safety
PR	Personal risk at a given location, also known as individual risk
PRA	Probabilistic Risk Assessment
PTSD	Post-Traumatic Stress Disorder
RoSPA	Royal Society for the Prevention of Accidents
QRA	Quantitative Risk assessment
RI&E	Dutch: Risk Assessment and Evaluation

SMS	Safety Management Systems
SPC	Statistical Process Control
SRA	Society for Risk Analysis
SRS	Systems Reliability Service
SWOV	Dutch: Foundation for Road Safety Research
TCDD	2,3,7,8-tetrachlorodibenzo-dioxin
TCI	Total Chemical Index
THERP	Technique for Human Error Rate Prediction
TNO	Dutch: Organization for Applied Scientific Research
TQM	Total Quality Management
TOR	Technic of Operations Review
TtA	Dutch: Journal of Applied Occupational Sciences
UKAEA	United Kingdom Atomic Energy Authority
Ukrainian SSR	Ukrainian Socialist Soviet Republic
UN	United Nations
USSR	Union of Soviet Socialist Republics
VI	Dutch: Safety Institute
WCI	Worst Chemical Index
WEBA	Dutch: Well-being at work
WGD	Dutch: Expert Working Group
WORM	Workgroup Occupational Risk Model
WOS	Working-on-Safety
WRI	Worst Reaction Index

1 The birth of occupational safety, safety and social struggle

1800s–1910

Occupational safety is a topic of all times. The dangers of mining were described many centuries ago. This chapter shows that fundamental and structural interest in workers' safety and causes of occupational accidents arose at the end of the nineteenth/ beginning of the twentieth century. This was the century of steam engines, the first Industrial Revolution. Also, the first sociological and empirical investigations started on living and working conditions, leading to the first social legislation and labour laws, and marked the birth of occupational safety.

DOI: 10.4324/9781003001379-1

The 1900s marked the beginning of an unprecedented age of technical and other innovations.

The 1902 French film *Le Voyage dans la Lune* (A Trip to the Moon) was an early **silent movie**. Its international success influenced and inspired other film-makers and led to the development of narrative film as a whole.

Meanwhile in the United States, the Ford Motor Company developed the **Ford Model T** in 1908. Later named the most influential car of the twentieth century, the Model T was successful because it provided inexpensive transportation on a massive scale. Moreover, it signified innovation for the rising middle class and became a powerful symbol of the United States' age of modernisation.

At the beginning of the twentieth century, **child labour** was widespread. In 1910, more than 2 million children under the age of 15 were employed in the United States. Factories and mines were not the only places where child labour was prevalent. Home-based manufacturing across the United States and Europe employed children as well.

Another cultural milestone was the birth of **cubism**, an influential art movement that revolutionized painting and sculpture in Europe. Pablo Picasso and Georges Braque were pioneers of the movement. Their paintings reflected interest in geometry and simultaneous perspective. Cubism inspired related movements in architecture, music and literature.

Thomas Edison patented the first useful rechargeable battery in 1901. The Edison battery was superior to batteries using lead plates and acid. Edison's batteries had a significantly higher energy density than the lead–acid batteries in use at the time and could be charged in half the time.

The first flight of the **Zeppelin** took place in 1900. After the outstanding success of the Zeppelin design, the word *zeppelin* came to be commonly used to refer to all rigid airships. Zeppelins were first flown commercially in 1910.

Halfway through the first decade of the twentieth century, **Einstein** developed the theory of relativity. Einstein had his 'amazing year' in 1905, when he published four groundbreaking papers, on the photoelectric effect, Brownian motion, relativity and the equivalence of mass and energy. It brought him to the notice of the academic world at the age of 26.

The invention of the internal **combustion engine** led to a true revolution. Combustion engines gradually replaced steam engines in many applications. In 1902, automobiles with these engines were put into production by Daimler Motoren Gesellschaft (Daimler-Mercedes).

silent movie

Ford Model T

child labour

cubism

Thomas Edison

zappelin

Einstein

combustion engine

Sources:

a. https://en.wikipedia.org/wiki/A_Trip_to_the_Moon.
b. https://en.wikipedia.org/wiki/Ford_Model_T#/media/.
c. https://nl.wikipedia.org/wiki/Kinderarbeid#/media/
d. https://en.wikipedia.org/wiki/Cubism#/ media/File:Pablo_Picasso,_1910,_Girl_ with_a_Mandolin_(Fanny_Tellier),_oil_on_ canvas,_100.3_x_73.6_cm,_Museum_of_Modern_Art_New_York.jpg.
e. https://commons.wikimedia.org/wiki/File:Edison-ni-fe.jpg.
f. https://en.wikipedia.org/wiki/Zeppelin#/media/.
g. https://en.wikipedia.org/wiki/Albert_Einstein#/media/File:Einstein_1921_by_F_Schmutzer_-_restoration.jpg.
h. https://nl.wikipedia.org/wiki/Bestand:PSM_V18_D500_An_american_internal_combustion_otto_engine.jpg

References to occupational safety appeared in ancient literature, but only anecdotally. The legal code of King Hammurabi (1810–1750 BCE) was probably the earliest document that dealt with the consequences of accidents. His 'eye for an eye, tooth for a tooth' approach stipulated that if someone erred with negative consequences, he or she should be punished with equivalent results. The oath of Hippocrates (460–390 BCE) is another example of an attempt to regulate the behaviour of, in this case, health care professions: 'Do no harm' is a precursor of the zero harm movement from more recent times (Figure 1.1).

The encyclopaedic work of Pliny the Elder, *Naturalis Historia*, mentioned accidents occurring amongst miners. In one of the methods of gold mining, called *arrugiae*, tunnels were dug over large distances. When shafts were insufficiently supported with bows, rocks could split, and tunnels could collapse and crush miners. According to Pliny, this type of gold mining was very risky, even riskier than diving for pearls at the bottom of the sea (Pliny 77). Another early reference to the hazards and risks of mining comes from the Renaissance period, almost 15 centuries later (Figure 1.2). *De Re Metallica* by Agricola was an early reference book on geology, mineralogy and mining, and devoted a few pages to mining accidents (Agricola 1556):

> *Sometimes, workmen slipping from the ladders into the shafts break their arms, legs or necks, or fall into the sumps and are drowned. Often, indeed, the negligence of the foreman is to blame, for it is his special work both to fix the ladders so firmly to the timber that they cannot break away, and to cover so securely with planks the sumps, at the bottom of the shafts, that the planks cannot be moved nor the men fall into the water. Moreover he (the foreman) must not set the entrance of the shaft house towards the north wind, lest in winter the ladders freeze with cold, for when this happens the men's hands become stiff and slippery with cold and cannot perform their office of holding.*

As the quote shows, the foreman has major responsibilities in terms of safety. Finally, the role of the individual in the accident causation process was mentioned in Proverbs 27:12: 'A prudent person foresees danger and takes precautions. The simpleton goes blindly on and suffers the consequences' (Holy Scripture 1996).

Like Pliny, Agricola also warned of mountain slides. As an example, he discussed a major mining accident which killed 400 miners at Goslar, in the Harz Mountains, a range in present-day Germany. To prevent these slides, tunnels should be supported sufficiently.

United Kingdom

The century of steam

Occupational safety slowly attracted some attention during the nineteenth century, a period during which the United Kingdom was the leading industrialising

Figure 1.1 Oath of Hippocrates

Source: A fragment of the oath on the third-century Papyrus Oxyrhynchus 254

Figure 1.2 Sixteenth-century mine shafts

Source: Agricola (1556)

nation, with its innovative technical discoveries. This period is called the Belle Époque in Europe, from the late nineteenth century until World War I, marked by its growing prosperity and the development of arts and science. The textile industry had a prominent economic position during the first period of the Industrial Revolution, which started in England in the eighteenth century. Step by step, European countries were changing from a feudal agricultural society to what was called the 'first modern society', an industrial state with an upcoming middle class (Pietersen 1981; Lintsen 1995a). Mechanisation of industrial production accelerated around 1830, and the iron and later the steel industry replaced textiles as the major sectors of the economy. The era of railways and other heavy industry started.

In spite of these historical developments, it was cholera that dominated the nineteenth century. Several pandemics swept across Europe and North America, leading to high mortality rates in densely populated regions, such as working-class areas in cities and around factories (Johnson 2006). In this period, religion played a pivotal role in public life, and diseases were often seen as a punishment by God. But slowly the link between hygiene, living and working conditions and cholera emerged, meaning something could be done to banish the disease. Banishing diseases like cholera was also stimulated by the rise of medical statistics: the registration and analysis of causes of deaths, and physician-hygienists who linked illness and mortality to substandard living and working conditions. Strictly speaking, the hygienists' statistics were a kind of risk analysis, even though the numbers obtained were not formulated and interpreted as future risks.

Charles Turner Thackrah was the first British physician and hygienist, and in 1832 he produced an overview of all trades and all occupational diseases in the town of Leeds (Thackrah 1832; Meiklejohn 1957). Thackrah followed an early tradition amongst physicians, namely visits and observations at workplaces, which was started in Italy by Ramazzini (1700). He showed that no more than 10% of the city's population enjoyed good health, and the city's mortality was in fact 150% higher than in the countryside. Thackrah was surprisingly optimistic about safety-related issues. Moving parts of machinery were already guarded in the textile industry, and compared to health, safety was a minor issue. In this period, people generally had the view that disease and suffering were part and parcel of the plan of the Almighty. But if accidents and diseases among workers were caused not by God but by interactions with machines and dangerous production methods, then men could act, for instance by issuing laws.

The living and working conditions in big cities, factories and mines were documented in the nineteenth century. Various British commissions reported on working hours and child and female labour, leading to the start of social legislation in 1802. The Factory Act 1802, also known as the Health and Morals of Apprentices Act, limited child labour and hours of work in the textile industry. The law passed with almost no opposition, illustrating the consensus about the unacceptable abuse of children in the textile mills. As well as regulating the admission of fresh air in the workplace, the law imposed a limit of 12 working hours per day, prohibited apprentices from night work and made it mandatory that apprentices were taught valuable skills like reading, writing and arithmetic. The law was extended in 1833 to other industries. The installation of the British Factory Inspectorate, charged with the supervision of the law, dated from 1833. And with the Factories Act of 1844, this inspectorate also had the duty by law to monitor safety in factories, such as checking various forms of guarding machines and installations (Le Poole 1865; Hale 1978). Initially, these kinds of inspections were frowned upon by industrialists and politicians, therefore only four inspectors were appointed, responsible for about 3,000 textile mills. Yet, their impact on machine safety and working conditions cannot be underestimated, and by 1865 the number of inspectors had increased to 35. The *Priestley versus Fowler* court case of 1837 redefined the responsibilities an employer had in the prevention of harm during work. For the first time an employee sued his employer for a work-related injury, and he received a sizeable pay-off. This case introduced the 'duty of care' as a concept in the relation between employers and employees.

In 1844, Friedrich Engels published *The Conditions of the Working Class in England*, based on his survey conducted in Manchester (Engels 1844; Sheeman 1973). In the 1860s and 1880s, the results of the first extensive sociological surveys amongst the London working class became public, describing the effects of the so-called economic barbarism that occurred during the nineteenth century (see, for instance, Mayhew 1861; Anonymous 1889a; Booth 1889). In other European countries, the results of similar surveys were published (Rosen 1976). These studies presented a picture of life and work in the big city as an ingeniously balanced mechanism, where every social class survived on the remains and waste of the class above. Big parts of the city looked like giant anthills, or worse, places with an overbearing stench of decay and excreta. The large-scale pauperisation, in both cities and factories, confirmed the dreadful picture already presented by Thackrah and Engels.

Around 1875, England's front-runner position was taken over, first by Germany, and later, after World War I, by the United States. Increasingly, technical progress was based upon scientific insights and rationalisations in terms of the products produced, the organisation of the production, and the planning and management. Applications of scientific discoveries in chemistry and physics formed the basis of new and emerging industries in these countries, such as the organic-chemical and the electro-technical industries. Techniques applied for thousands of years and based largely on trial and error transformed into technology.

In the United Kingdom, only one reference was published on occupational accidents. Calder's *The Prevention of Factory Accidents* is an account of manufacturing industry and related accidents. It is also a practical guide to the law on safeguarding, safe working and safe construction of factory machinery, plant and premises (Figure 1.3) (Calder 1899).

Calder was an engineer and inspector of factories for the north of England and Scotland. In his introductory chapter, Calder commented on the complete absence of any literature dealing with the practical aspects of industrial accident prevention. The main part of the book dealt with legislation and the safeguarding of machines and installations in various industries and was illustrated with drawings. An example of safeguarding of a rolling mill engine is shown in Figure 1.3, reducing risks from falling and preventing workers from getting caught in moving parts of the machine. Calder gave no explanation regarding the choices of his control measures, but implicitly he assumed that moving parts of machines, and heights, played an important role in accidents. But as prime causes of accidents, Calder mentioned a combination of ignorance, carelessness and the unsuitable clothing of workers. He specifically addressed occupiers and foremen of factories, as well as workers who were unaware of the nature of the forces and mechanical 'arrangements' that they had to control and which, if uncontrolled, could result in risks and accidents with often serious consequences.

Figure 1.3 Safeguarding a rolling mill engine, piston rods, plateau and stairs

Source: Calder (1899)

United States

US Steel, road to happiness

Before the twentieth century, in the United States, little was published on occupational safety. This changed at the beginning of the past century, when the Safety First Movement started as a private initiative of the steel industry, and the results of the Pittsburgh Survey were published. Occupational accidents were no longer seen as acts of God but actually as man-made. This survey is discussed in the next section.

In the United States, mining, steel and the product industry, along with trade, were becoming large industries, dominated by big conglomerates. As far as occupational safety was concerned, this area was seldom the domain of managers. Managers were not concerned with issues related to shop floor workers. Here the foreman set the rules, hiring and dismissing workers. And, not surprisingly, unskilled labour showed a large turnover, and the accident rates in industry were high. At the beginning of the twentieth century, national accident figures in industry were available, and in 1907 the first international comparisons were made. These statistics showed that the American steel industry had an occupational mortality three to four times higher than that of the German steel industry, where the mortality rate was 0.2 per 10^6 man-hours. Two years later, Fredrick Hoffman, a statistician at an insurance company, presented an estimation for the Bureau of Labor Statistics of the annual mortality in industry: 30,000–35,000 deaths, 350,000 severely wounded and 2,000,000 who required medical treatments, numbers that exceeded those of the American Civil War, in ther period 1861–1865 (Hoffman 1909; Anonymous 1915, 1926a; Aldrich 1997). The reliability of the numbers mentioned was not clear. But the overall picture of a higher mortality rate in the American industry than in Europe was an argument appearing repeatedly in various publications. Only a rudimentary concept of the causes of these accidents existed; accidents were seen as an unavoidable part of the job – they were inevitable or were caused by workers' behaviour. The shop floor had become an extremely hazardous place, and by emphasising guilt questions of accidents, preventing occupational accidents was almost impossible. This approach is reflected in the originally French proverb 'You can't make an omelette without breaking eggs'. US Steel was one of the companies where a growing burden of accidents was jeopardising production and productivity. This company was the country's largest steel concern, with branches spread all over the United States, and was the initiator of the Safety First Movement in 1906 (Palmer 1926). An illustrative poster from 1913 showed a country path as a metaphor for occupational safety (Figure 1.4).

The text at the bottom of the poster explicitly focussed on workers' behaviour and attitude. It was the first time such a national safety campaign had been launched. One can see this figure as a first metaphor of causes of accidents, with personal qualities and behaviour being major determinants of accidents. Many companies accepted this initiative with open arms. DuPont, a manufacturer of explosives, was one of them (Klein 2009). Due to the inherently risky nature of its production of explosives, DuPont realised that safety precautions were essential to ensure the welfare of its employees (Figure 1.5), and its first safety procedures were issued as early as 1911, regulating behaviour such as alcohol intoxication or disorderly fun (Figure 1.5).

DuPont stressed the need for research by stating that 'we must seek the hazards we live in' (DuPont 1927) and work together with scientists to improve methods of manufacturing. Even at the design stage, safety was addressed, and employees were provided with personal protective clothing, including non-flammable clothes. Even though many precautions were taken, explosions with devastating consequences occurred in the Wilmington powder mills in 1863, Brandywine in 1818, 1890 and 1915, and the Haskell powder plant in 1917, strengthening the company's drive to improve safety (Klein 2009).

The Pittsburgh investigation, Eastman's conclusions

US Steel had always emphasized the humanitarian motives of the campaign, but this did not alter the fact that the company suffered from negative publicity. One of the common slogans at that time was 'Steel is war', and unambiguous titles of articles appearing in the Chicago weekly press, such as 'Making steel and killing men' and 'The law of the killed and wounded', increased public attention to the working conditions at the company (Hard 1907).

The Pittsburgh Survey was the first sociological survey in the US, investigating working and living conditions of workers in the Allegheny district in

Figure 1.4 The road to happiness, US Steel, fall 1913

Source: http://vbn.aau.dk/files/17382694/Afhandlingsfil.pdf

Pennsylvania, and was conducted between 1906 and 1907 (Kellogg 1909). This survey provided a detailed image of the raw side of capitalism in this industrial district, dominated by US Steel. *Work Accidents and the Law* by the sociologist Crystal Eastman was one of many publications of the Pittsburgh Survey (Eastman 1910). By analysing hundreds of occupational accidents, as well as the financial

Figure 1.5 DuPont safety poster c. 1900

Source: History of Our Safety Core Value | DuPont

consequences of these accidents for the families concerned, Eastman was one of the first to address the importance of industrial safety. At the start of the twentieth century, compensation for occupational accidents was not regulated by law, in contrast to western European countries. Most US states enforced compensation laws around the 1920s (Ashford 1976; Aldrich 1997). Eastman's survey showed the petty financial contribution paid by the company to families of victims. In the case of fatalities, it was hardly enough to even pay for the funeral costs. Figure 1.6 shows 'The Puddler', a statue by Constantine Meunier in 1884, which illustrates the amounts paid per injured body part in 1907 ($100 in 1907 was equal to about $2,800 in 2020).

Different from the general opinion, Eastman showed that accident causation should be understood as arising in terms of labour conditions: long working hours, high production speeds, high temperature and noise levels, young age and a low level of education and experience of workers. Eastman suggested that excessive attention paid to the behaviour of workers was the main cause of accidents (Eastman 1910):

> *If a hundred times a day a man is required to take necessary risks, it is not within reason to expect him to stop there and never take an unnecessary risk. Extreme caution is as unprofessional among the men in dangerous trades as fear would be in a soldier.*

Eastman's detailed research on occupational accidents in the steel district was based upon a survey of 526 occupational fatalities occurring between June 1906 and June 1907. The so-called death calendar is the first illustration of Eastman's report, pointing to the fact that more than one occupational death was occurring every day (Figure 1.7).

During the survey, several sources of information were used: the coroner's files, company visits and observations on the spot, interviews with families, fellow workmen, other witnesses of the accident and employers. Not all sources of information were open for inquiry. In the coroner's reports only general information could be found and interviews with foremen and fellow workers highlighted only one side of the story, with escaping blame being the main driver. From the last source, employers, information was particularly difficult to obtain. Only in 30% of cases was any information made available. Because of these limitations, observations at the location of the accidents became the main source of information. The findings were quite an eye-opener in those days:

1. *Occupational accidents occur as frequently amongst educated white Americans as amongst uneducated immigrants;*
2. *The common opinion of both workers and foremen that 95% of accidents are caused by their victims implies that every accident is seen as a unique event. As a consequence, there is no prevention;*
3. *The analysis of accidents shows the repeated occurrence of similar types of accidents. Many of these accidents are preventable;*
4. *The foremen and the superintendents are primarily responsible for safety. This is the result of a sharp distinction between the workers and those who are in a position of authority. Most foremen send inexperienced workers to*

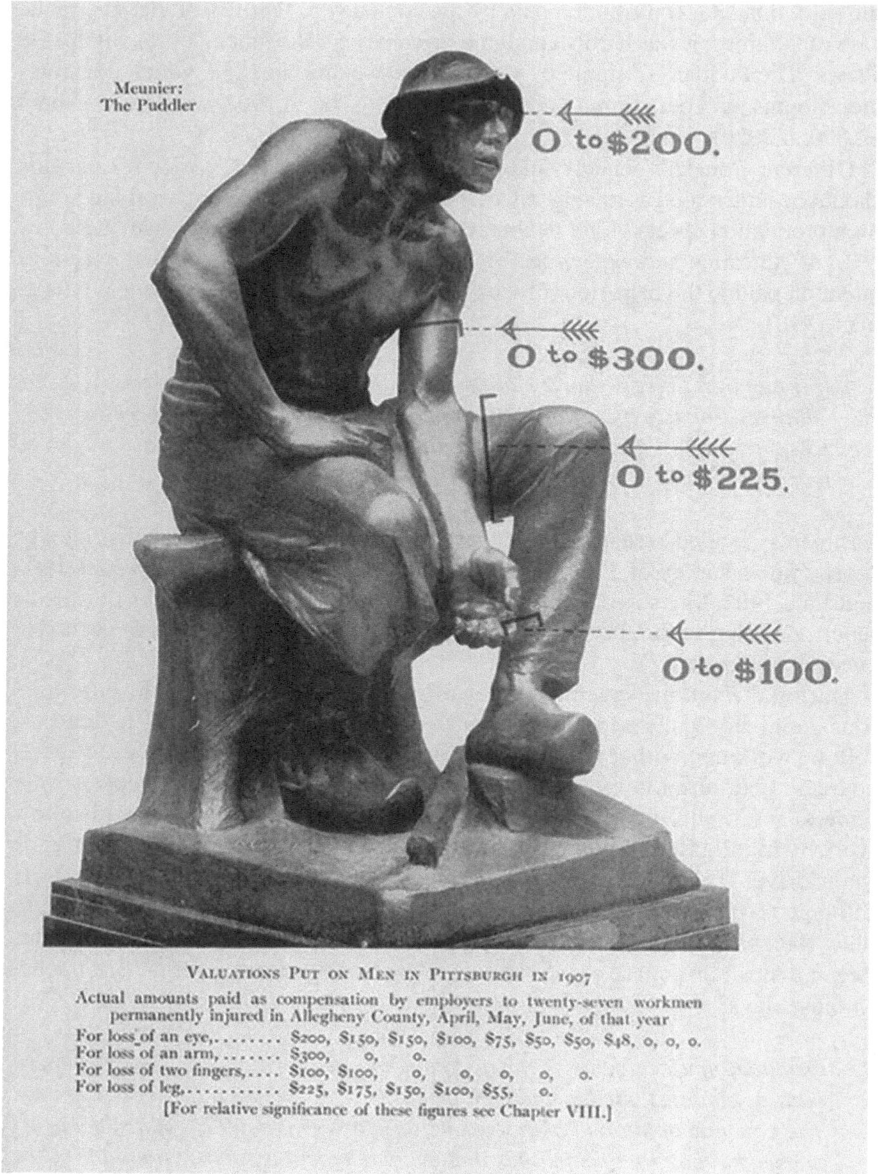

Figure 1.6 Compensation paid per body part, Pittsburgh Survey

Source: Eastman (1910)

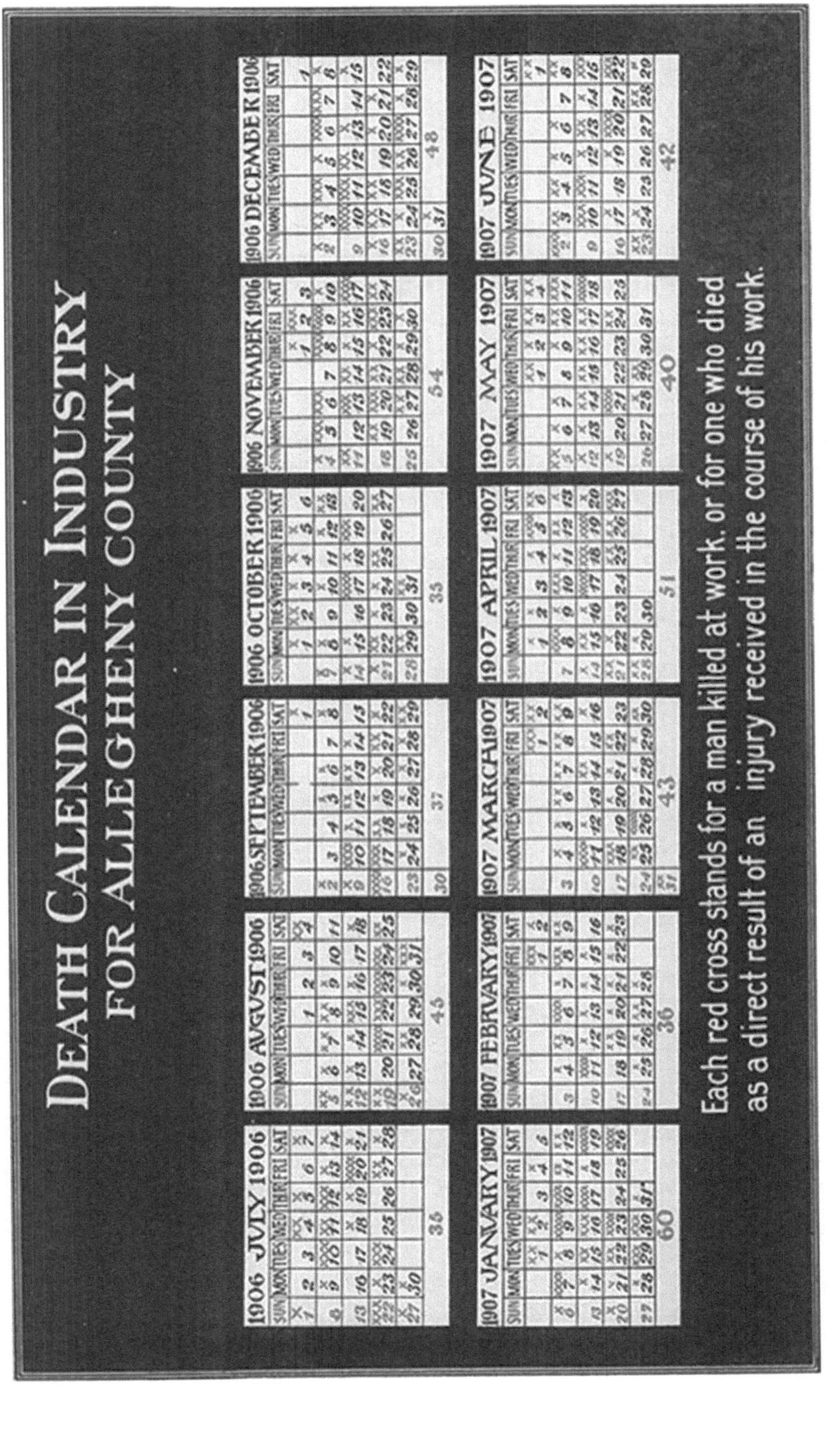

Figure 1.7 Death calendar, Pittsburgh Survey

Source: Eastman (1910)

dangerous places, because the foremen are cowards and will not go themselves, or because of high production demands or because of indifference;

5. *In the majority of cases, the financial burden of accidents is borne by the victims' families;*
6. *The main consequence of these accidents is huge social waste. From the perspective of social justice, legislation is needed to prevent accidents and to manage the financial consequences.*

The most important causes of occupational fatalities in the steel industry were also listed (Table 1.1). For the first time, a survey presented insight into the dominant accident scenarios, related to cranes, trains, heights and high temperatures, as was expected in the environment of a steelworks.

Responsibility for safety

Responsibility for occupational accidents, as reflected in conclusions 2 and 4, was classified by Eastman using five different categories (Table 1.2). Out of the 526 occupational fatalities, information could be obtained from 410 cases.

The first two were the victim and their fellow worker. Inattention played a dominant role. Later in the report this 'inattention' was put into perspective, as most of the victims were young children, or workers not aware of the dangers encountered. For those cases, the third category of foreman was used. Accidents with the fourth group, the employer, were the result of machine disturbances,

Table 1.1 Order of industrial fatalities in the steel industry in 1907 (after Eastman 1910)

Categories of accidents as % of total accidents (n = 195)		
1	Caught by crane activities	22
2	Caught by machines, trains	16
3	Falling from heights	12
4	Hit by explosion, hot metal	11
5	Contact with and loading of metal and steel products	9
6	Contact with electricity	4
7	Exposure to furnace gas (CO)	3
	Others	24

Table 1.2 Responsibility for industrial fatalities, Pittsburgh Survey (after Eastman 1910)

Indication of entire or partial responsibility as % of total accidents (n=410)	
Victim	26
Fellow worker	12
Foreman	10
Employer	29
None of the above	23

machine design, insufficient inspection and a high speed of production and working pressure. Also, accidents due to the organisation of the work and an insufficient guarding against moving parts of machinery were part of this group. The last group of accidents were those where no direct cause could be allotted. The results showed that the victim's responsibility in fatal occupational accidents was much less than commonly understood and advocated by the Safety First Movement. The categories in Table 1.1 and the responsibilities listed in Table 1.2 were starting points for prevention. Twenty years later these arguments were repeated in a review article written by of the head of the medical services of the International Labour Office (ILO) in Geneva (Carozzi and Stocker 1932).

Many scientific journals praised Eastman's publication, and in particular admired the chosen scientific approach (Anonymous 1911a). And in a later publication, Eastman advocated a thorough accident investigation method, an extensive safety supervision by the government and a liability system that would reward prevention activities and punish employers who did not take safety sufficiently seriously (Eastman 1911). It was not only the steel industry that was under attack in this period. Other branches of industry, including the meat, textile and automobile industries, were also subjects of novels and photo books. Well-known examples include *The Jungle* by Upton Sinclair, about the conditions in Chicago slaughterhouses (Sinclair 1906), and the 1908 photos by Lewis Hine for the National Child Labour Committee (Hine 1908) (Figure 1.8).

John Spargo's 1906 book *The Bitter Cry of Children*, about child welfare in general, and more specifically labour conditions in the coal-cutting industry (Spargo 1906), also highlighted the differences between classes in mortality rate in factories. In the poorest class, the probability of dying on the job was 350%

Figure 1.8 Child labour in the United States

Sources: Anonymous (1977a); Doherty (1981); Freedman (1994)

higher (350,000 per 10 million people) than in what he calls the 'well-to-do class'. He called for social reform to create better conditions for all. Between 1880 and 1910, states started to introduce legislation limiting the number of working hours of children. This signalled a shift in thinking about safety: it was not only the employers' but also the state's responsibility to ensure a safe working environment.

The Netherlands

The Netherlands during the century of steam

Industrialisation in the Netherlands started later than in neighbouring countries. Steam engines were introduced on a small scale only halfway through the nineteenth century (Lintsen 1995b). Also, surveys in the Netherlands on working conditions in workshops and factories were initiated somewhat later. Following the Socialist International Organisation, founded in London in 1864, national inquiries were used to document working and living conditions. Between 1870 and 1880, three such inquiries were conducted by the Workers Association Arnhem, by the General Dutch Workers Association and by Domela Nieuwenhuis. All three enquiries reported too-long working hours, on average 16–17 hours a day, low salaries and deplorable working conditions. The cost of living for working people in that period was higher than the salaries, thereby providing a financial incentive for child and female labour. The enquiries revealed a depressing picture of safety conditions. Supervision on boilers, steam engines and other dangerous installations was absent, and the conditions in factories worsened the pauperized condition of workers (Roland Holst 1902; Brugmans 1958; Welcker 1978).

In addition to these inquiries, Samuel Coronel, a physician and hygienist, published many articles on conditions in factories and workshops. One of his review articles on industrial hygiene concentrated on occupational accidents (Coronel 1876). According to the author, mechanisation had two side effects, apart from the mitigation of labour: firstly, the introduction of new hazards, such as steam, dangerous machines causing accidents and workshops crowded with machinery, leaving little space for workers to move. Secondly, Coronel noticed that adult workers were replaced by weaker forces, such as the rapid hands of children and women (Figure 1.9).

The first social legislation in the Netherlands was introduced much later than in the United Kingdom. The first Factory Act became law in 1874 and forbade child labour under the age of 12. This legislation was, however, largely symbolic in nature, due to a lack of supervision and enforcement. Thirteen years later, the eighth Parliamentary Enquête (enquiry) was conducted on the effectiveness of this law. Due to the fall of the government, the inquiry was limited to the regions of Maastricht, Tilburg, Amsterdam and the flax sector. Despite these restrictions, the results were a real bombshell, because the results of the interviews, which included many workers, were published very rapidly and confirmed in detail observations made by Coronel and also found in previous national inquiries (Giele 1981; Buitelaar and Vreeman 1985).

Figure 1.9 Child labour during the nineteenth century

Source: Binneveld (1991)

The Enquête revealed the working and living conditions of the Dutch working class were as poor as anywhere else, public opinion imagined only occurred in foreign countries, like England and Germany, but not in the Netherlands. The 1887 Parliamentary Enquête created a chain of initiatives. The Factory Inspectorate was installed in 1890, and social legislation slowly started, such as the Arbeidswet 1889 (Labour Act), prohibiting child and female labour, the Veiligheidswet 1895 (Safety Act), with regulations on machine guarding, and the Ongevallenwet 1901 (Accident Act), an act on industrial injuries, regulating financial compensation (Anonymous 1897a; Lochem 1943; Binneveld 1991; Schwitters 1991; Kerklaan 2006). Another initiative was the 1890 national exhibition 'To promote safety and health in factories and workplaces', organized in Amsterdam at the Paleis van Volksvlijt (Palace of Popular Diligence) (Zwaard 2007). This exhibition was a huge success, with more than a million visitors. Numerous examples of practical safety techniques were shown from industry, agriculture and mining. The predominant approach towards safety improvement was: fix the technology, fix the safety problem. The success of this exhibition led to the Safety Museum in Amsterdam (Anonymous 1891). This museum opened its doors in 1893 with a permanent exposition of safety techniques and started around 20 years earlier than similar initiatives in the United States and the United Kingdom. Partly financed by the state, provinces and the local authority of Amsterdam, safety promotion was one of the first activities established, followed by training programmes for employers, employees and civil servants from the Labour Inspectorate. The legal requirements of the 1895 Veiligheidswet (Safety Act) were one of the topics of these courses, which included advice, the rationale behind adequate safety equipment and procedures. Civil servants from the Labour Inspectorate regularly encountered resistance and difficulties with employers, who found regulations often impractical and needlessly expensive. The museum offered the opportunity to test the consequences of recommendations, and so enhanced the quality of inspections (Anonymous 1914a).

In 1892, five years after the Parliamentary Enquête, annual reports by the Dutch Factory Inspectorate again confirmed the deplorable conditions in factories, the overcrowding of machinery and the wretched conditions of most industrial installations. But for the first time, the carelessness, inattention, ineptitude and recklessness of victims were also mentioned as causes of accidents (Anonymous 1892). This attribution of causes can also be interpreted as a sign of the emancipation of the working class. While before, workers were seen simply as part of the indefinite poor, they were now seen as a group that needed attention, both from a health (e.g. preventing cholera) and a safety perspective. The upcoming liberals, with a strong focus on education, had organized a national discussion on the so-called *sociale questie* (social question), also called the *arbeidersvraagstuk* (problem of the working class). The results of national and parliamentary enquiries (Brugmans 1958; Romein and Romein 1973) made clear that inevitable class differences were not forced by 'God's will'. The liberal answer to the *sociale questie* was improved training and hygiene for workers. After all, healthy workers were more productive. According to liberal philosophy, however, these initiatives should not be initiated by government, but by private parties, for instance companies.

Around the turn of the century, two men played a dominant role in discussions on occupational safety: the engineer and liberal Frederik Westerouen van Meeteren (1851–1904) and the physician-hygienist and socialist Louis Heijermans (1873–1938). The following sections describe the contributions of both men.

Safety technique according to Westerouen van Meeteren

Westerouen van Meeteren viewed the Germans and their state-induced inspections and control of factories with some disgust. Westerouen appealed to all Dutch industrialists to start a safety society, before government did. This initiative, mirroring the French Association pour prévenir les accidents de fabrique (Association to Prevent Accidents in Factories) from Mülhausen, Elzas (Reid 1987), became the Nederlandse Vereeniging tot Voorkoming van Ongelukken in Fabrieken en Werkplaatsen (Dutch Society to Prevent Accidents in Factories and Workshops). The society was active between 1890 and 1901. Its aim was to prevent avoidable accidents, to initiate factory inspections, safety education and safety promotion, and to reward inventions of new safety equipment (Anonymous 1889b, 1890, 1893, 1896, 1897b; Krap 1890; Schwitters 1991). From 1893 onward, the association issued the professional periodical *De Veiligheid* (The Safety Journal) for 11 years. This journal contained many contributions from Westerouen van Meeteren. He was also the author of the first Dutch reference book on occupational safety and health, using predominantly German and French reference materials (Westerouen van Meeteren 1893; Bakker and Berkers 1995; Kerklaan et al. 2002).

Westerouwen van Meeteren's reference book was a combination of occupational medicine and occupational safety. He was an independent advisor on safety and health in factories and workshops for most of his career. In 1891, he was appointed technical advisor to the 2^{e} Afdeling van de Eerste Verzekeringsmaatschappij op het Leven en tegen Invaliditeit (Second Department of the First Insurance Company for Life and Against Invalidity) and was responsible for factory inspections. Being a technician himself, he justified the medical focus on the basis of efficiency. Eliminating health hazards was much more effective than combating safety hazards. Removing health hazards would affect many more workers than simply protecting against safety hazards, which would have consequences for one or a very limited number of workers, for instance only those close by and thus coming into contact with hazards. This point of view also reflected the position of occupational medicine, which at the end of the nineteenth century in western European countries was a more upcoming and stronger field of research than occupational safety.

Like Calder's reference book, Westerouwen van Meeteren's book is full of information on legislation and on statistics, reporting all registered fatal and nonfatal accidents in various trades, installations and machines during the period 1890–1891. Since the number of workers per trade was not given, accident rates were missing, and only absolute numbers were presented. Mechanical causes of accidents were also mentioned, such as moving parts of machines, and apart from general observations on lighting, working hours, ventilation and temperature, there were many illustrations of various examples of safety technique. Surprisingly, there were no references to particular workers' qualities, such as education level, or undesirable behaviour, like carelessness.

Heijermans' causes of occupational accidents

In the early twentieth century, occupational health and safety gained an important advocate with Louis Heijermans. At the invitation of the Sociaal-technische Vereniging van Democratische Ingenieurs en Architecten (Social-Technical Association of Democratic Engineers and Architects), as well as through his affiliation to the Sociaal Democratische Arbeiderspartij (Social-Democratic Labour Party), Heijermans began in 1907 a course in social and technical hygiene for students at Delft Technical High School, the forerunner of the Delft University of Technology. The course began with the lecture 'Education in Technical Hygiene'. Probably for the first time in the Netherlands, this new discipline was defined as a technical engineering discipline (Burdorf et al. 1997)

> *There is much to say about technical hygiene as a separate science, in addition to general health education. As far as medicine is concerned, the attention is seized by bacteriology, epidemiology, education, etc., so that a special study of technical hygiene is little or nothing* [. . .] *technical hygiene takes care of the working individual, who must be protected against the disadvantages that arise from the exercise of the profession and against the risk of accidents to which the workers are exposed. This discipline investigates what causes mine gas explosions, how machines are protected, how dust and poisonous gases can be removed, as well as the determination of normal working hours on physiological grounds.*
>
> (Heijermans 1907)

During his lecture Heijermans warned students and listeners not to have too much confidence in safety posters. Texts on posters were poorly understood by workers, who usually had only limited or even no education and had to worry about inconveniences usually related to these recommended measures. One year later Heijermans published his magnum opus *Handleiding tot de kennis der beroepsziekten* (Reference Book on Occupational Diseases) (Heijermans 1908) describing many different occupations, each with its own specific occupational health issues, illustrated by photos and descriptions of control measures to reduce adverse effects. Like Thackrah, Heijermans followed the example of Ramazzini. At the beginning of his career, Heijermans had written a few publications on occupational safety, espousing his opinions quite outspokenly, based on his frequent observations in factories and workshops. He cited three main causes for occupational accidents in his book *Gezondheidsleer voor arbeiders* (Hygiene for Workers) (Heijermans 1905). The first cause was the worker himself, which was the common opinion, referring to a victim's indifferent and careless attitude. Heijermans put this argument into perspective by pointing out that indifference was the consequence of daily confrontations with danger. The second cause was the tasks workers had to perform, which were often so stupid and monotonous that they killed all creative energy and transformed workers into machines. The last cause was the long and soul-destroying working hours, which made workers indifferent to danger. Based on the accident information from the Factory Inspectorate, Heijermans

Table 1.3 The most important categories of accidents from reported accidents (fatal and non-fatal), annual report of the Dutch Labour Inspectorate, 1902 (after Hasselt 1907; Valk 2007)

	Categories of accidents as % of total accidents (n = 7400)	
1	falling from heights	20
2	hit by falling objects	14
3	hit by tools for manipulating metals	6
4	hit by tools for manipulating wood	6
5	hit by several tools, like grindstones, presses, centrifuges	4
6	hit by lifts, winches, cranes	3
7	gripped by driving gear, shafts, belts, pulleys, transmissions	2
8	gripped by power tools and installations, steam engines, gas engines, etc.	1

presented a list of categories of major accidents (Table 1.3). The dominant role of power tools and industrial installations in accidents was confirmed by the Centrale Werkgevers Risico-Bank (Central Employers Risk Bank).

This private initiative of employers was one of the biggest insurers of workmen's compensation and approximately 50% of its payments had to be spent on this type of accident. The lack of attention during the design and construction of installations was also questioned (see also Anonymous 1909, 1911b; Vossnack 1913). And with the engineers of the Central Employers Risk Bank, the employers' role in preventing accidents was emphasised by designing or developing machinery that did not cause harm. Two decades later this focus on design was repeated by the International Labour Organization (ILO) (Carozzi and Stocker 1932) and again propagated at the Technical High School in Delft, at the foundation of the Safety Science Group (Goossens 1981).

As a result of sociological empirical surveys on living and working conditions, an irreversible trend started. The surveys revealed the desperate conditions of the working class and huge numbers of occupational deaths and injured. Governments in the United Kingdom, the United States and the Netherlands took action. Child labour was prohibited by law, working hours were reduced, factory inspectors were installed and safety techniques introduced. All three countries contributed substantially to the body of knowledge on safety. In the United States, the Pittsburgh Survey pointed at external factors as causes of accidents. Safety techniques, originating from the United Kingdom, had implicitly a similar message. This was also the argument in Dutch textbooks on safety and occupational health. In all three countries there was a discussion on the responsibility for safety. But there was also support for what was later called the 'individual hypothesis', where causes of accidents were the result of workers' behaviour: ignorance and carelessness. The American Safety First Movement was the most dominant representative of this hypothesis. In this period, doctors, technicians and sociologists shaped safety as a nascent occupational field. These men were the pioneers of a new period when the development of safety theories and metaphors started, and the first categorisations of accident scenarios were formulated.

2 Accident proneness, safety by inspection

1910–1930

This chapter discusses the second and third decades of the twentieth century. In the United States, attention is focused on managing safety and the first safety science manuals are published. In the late 1920s, an influential author, Heinrich, appeared on the safety stage. In the United Kingdom the first scientific safety theory was published: the accident proneness theory, later referred to as the 'individual hypothesis'. Also, arguments for the 'environmental hypothesis' were published. In the Netherlands the individual hypothesis remained popular and followed knowledge developments in safety from abroad.

DOI: 10.4324/9781003001379-2

As 1910 opened a new decade, drumbeats of war were heard across Europe.

The **First World War** (1914–1918) started in Europe and developed into a global war. It became one of the deadliest conflicts in history, with an estimated 9 million combatant and 7 million civilian deaths. The war shattered the social and demographic structure of Europe.

Meanwhile in Russia, the Bolshevik Party organized the October Revolution. **Lenin** did not have any direct role in the revolution, but he played a crucial role in the debate on the leadership of the party for a revolutionary insurrection as the party in the autumn of 1917 received a majority in the Soviets.

In 1912 the **sinking of the RMS Titanic**, a British passenger liner, took place in the North Atlantic Ocean after the ship struck an iceberg during her maiden voyage, from Southampton to New York City. More than 1,500 passengers and crew aboard died, making the sinking one of modern history's deadliest peacetime commercial marine disasters.

The worst was yet to come: the outbreak of the **Spanish flu** proved to be an unusually deadly influenza pandemic. Lasting from 1918 to 1920, it infected 500 million people, about a quarter of the world's population at the time. The death toll is estimated to have been anywhere from 17 million to 50 million, making it one of the deadliest pandemics in human history.

Discovery and progress continued in the 1910s. Krupp engineers patented **stainless steel**. This is used extensively in industry for corrosion resistance to aqueous, gaseous and high-temperature environments, and their mechanical properties at all temperatures.

Given the strategic location of Panama, building a canal connecting two oceans was seen as a potentially profitable enterprise. Construction was started in 1881 by the French. They failed and the US took over in 1902. It was by far the largest American engineering project to date. The **Panama Canal** opened in 1914 with the passage of the cargo ship SS *Ancon*.

In the 1910s, the medical use of **X-rays and radium therapy** was considered to be beneficial in different cases such as treatment of tuberculosis. Radium was generally used when a localised reaction was desired, and X-rays were applied when a large area needed to be treated. Radium was also believed to be bactericidal.

Surrealism, a cultural movement in Europe, developed in the aftermath of the First World War. Artists painted illogical scenes or creatures deriving from everyday objects. A well-known example is *The Elephant Celebes*, painted in 1921 by the German surrealist Max Ernst.

First World War

Lenin

Sinking of the RMS Titanic

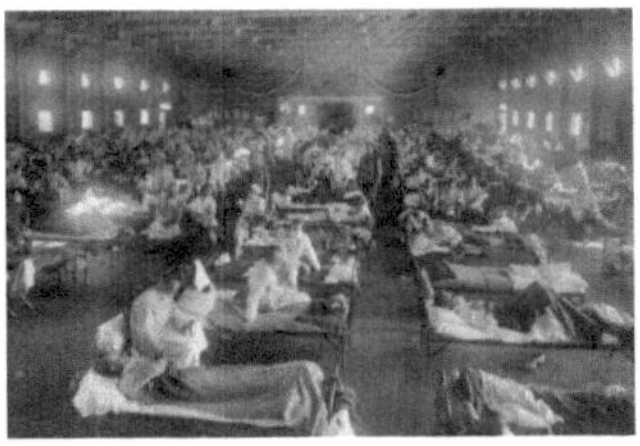

Spanish Flu

A NON-RUSTING STEEL.

Sheffield Invention Especially Good for Table Cutlery.

According to Consul John M. Savage, who is stationed at Sheffield, England, a firm in that city has introduced a stainless steel, which is claimed to be non-rusting, unstainable, and untarnishable. This steel is said to be especially adaptable for table cutlery, as the original polish is maintained after use, even when brought in contact with the most acid foods, and it requires only ordinary

Stainless steel

Panama Canal

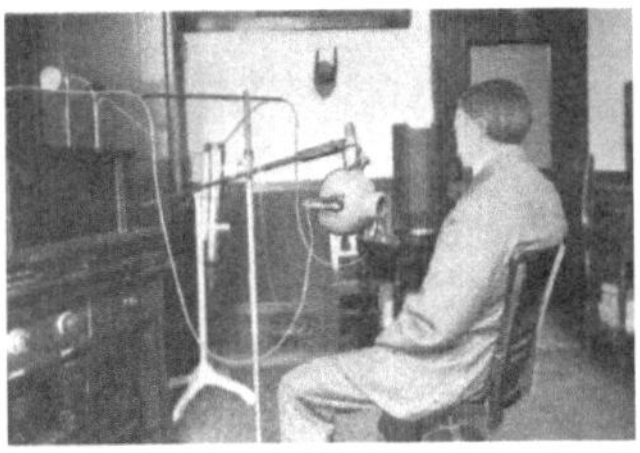

X-rays and radium therapy

Surrealism

Sources:

a. https://en.wikipedia.org/wiki/World_War_I#/media/File:Cheshire_Regiment_trench_Somme_1916.jpg
b. https://en.wikipedia.org/wiki/Vladimir_Lenin#/ media/File:Lenin_1919–03–18.jpg
c. https://en.wikipedia.org/wiki/Sinking_of_the_Titanic#/media/File:Stöwer_Titanic.jpg
d. https://en.wikipedia.org/wiki/Spanish_flu#/media/ File:Emergency_hospital_during_Influenza_epidemic,_Camp_Funston,_Kansas_-_NCP_1603.jpg
e. https://en.wikipedia.org/wiki/Harry_Brearley#/media/File:Stainless_steel_nyt_1-31-1915.jpg
f. https://en.wikipedia.org/wiki/Panama_Canal#/media/ File:SS_Ancon_entering_west_chamber_cph.3b17471u.jpg
g. https://commons.wikimedia.org/wiki/File:X-ray_ treatment_of_tuberculosis_1910.jpg
h. https://en.wikipedia.org/wiki/File:The_Elephant_Celebes.jpg

The 1920s was a decade of transition in Western society and Western culture.

The spirit of the **Roaring Twenties** was marked by a general feeling of novelty associated with modernity and a break with tradition. Formal decorative frills were shed in favour of practicality in both daily life and architecture. At the same time, jazz and dancing rose in popularity. Josephine Baker was a popular entertainer.

In the 1920s, **women's suffrage**, the right of women to vote in elections, gradually became accepted. National and international organisations formed to coordinate efforts towards that objective, especially the International Women's Suffrage Alliance, founded in 1904 in Berlin, Germany, as well as for equal civil rights for women.

The 1922 discovery by Howard Carter of the nearly intact **tomb of Tutankhamun** during excavations funded by Lord Carnarvon received worldwide press coverage. With more than 5,000 artefacts, it sparked a renewed public interest in ancient Egypt, for which Tutankhamun's mask remains a popular symbol.

The Jazz Singer marked the end of the silent movie era. The 1927 American musical drama film was the first talking movie with not only a synchronized recorded music score but also lip-synchronous singing and speech in several isolated sequences. Its release heralded the commercial ascendance of sound films.

In 1925, in London, Scottish inventor John Logie Baird gave the first public demonstration of televised silhouette images in motion. A year later he demonstrated the transmission of the image of a face in motion. It was the first **television** demonstration. It took several years before the new technology was marketed to consumers.

The **Wall Street stock market crash**, also known as the 'Great Crash', occurred in 1929. It started in September and ended late in October, when share prices on the New York Stock Exchange collapsed. It was the most devastating stock market crash in the history of the United States. The crash signalled the beginning of the Great Depression.

The **League of Nations** was the first worldwide intergovernmental organisation whose principal mission was to maintain world peace. Founded in 1920, it followed the Paris Peace Conference that ended the First World War. In 1919, US President Wilson received the Nobel Peace Prize for his role in the league.

In 1928, **penicillin** was discovered by Scottish scientist Alexander Fleming. His findings were not initially given much attention because Fleming was a famously poor communicator and orator. He presented a paper to the Medical Research Club of London, which was met with little enthusiasm by his peers. Only much later was the importance of his work recognised.

Roaring Twenties

Women's suffrage

Tomb of Tutankhamun

The Jazz Singer

Television

Wall Street stock market crash

Penicillin

League of Nations

Sources:

a. https://en.wikipedia.org/wiki/Roaring_Twenties#/ media/File:Baker_Charleston.jpg
b. Women_voter_outreach_1935_English_Yiddish.jpg
c. https://en.wikipedia.org/wiki/Tutankhamun#/ media/File:CairoEgMuseumTaaMaskMostlyPhotographed.jpg
d. https://en.wikipedia.org/wiki/The_Jazz_Singer#/ media/File:The_Jazz_Singer_1927_Poster.jpg
e. https://commons.wikimedia.org/wiki/File:Philips_ TX_400.JPG
f. https://en.wikipedia.org/wiki/Great_Depression#/media/File:American_union_bank.gif
g. https://en.wikipedia.org/wiki/History_of_penicillin#/media/File:Professor_Alexander_Fleming_at_work_in_his_laboratory_at_St_Mary's_ Hospital,_London,_during_the_Second_World_War._D17801.jpg
h. https://en.wikipedia.org/wiki/League_of_Nations#/media/File:No-nb_bldsa_5c006.jpg

The classical management school began in the late nineteenth century and placed top company managers at the centre of decision-making, which was then a revolutionary concept, as managers at that time had no interactions with shop floors. This was then the domain of the supervisor or foreman, who determined the course of production.

United States

The American management approach

There were two different schools in classical management: scientific management and general administrative management. Scientific management centred on ways of improving industrial and labour productivity by redesigning tasks and working methods. By contrast, administrative management theory examined organisations as total entities and focussed on ways of making organisations more effective and efficient (Pindur et al. 1995). The Americans Fredrick Taylor and Frank and Lillian Gilbreth were well-known pioneers of scientific management. The Franco-Turkish Henri Fayol and the German Max Weber were pioneers of administrative management. Apart from Weber, these authors were engineers and their technological background determined many characteristics of their management schools. For instance, they considered an organisation to be a mechanical entity and every person in the organisation was expected to think and act rationally. One strong stimulus of scientific management was the speech given by the US President Theodore Roosevelt in his 1908 address to state governors at the White House, when he prophetically remarked:

> *The conservation of our national resources was only preliminary to the larger question of national efficiency. The whole country at once recognized the importance of conserving our material resources and a large movement has been started which will be effective in accomplishing this object. As yet, however, we have but vaguely appreciated the importance of 'the larger question of increasing our national efficiency'.*
>
> (Cited in Taylor 1911)

This view was reflected in Taylor's publication *The Principles of Scientific Management* and more commonly referred to as Taylorism. In the early twentieth century, business flourished in America and capital was plentiful, but labour remained the limiting factor. In the late nineteenth century, Taylor experimented with different working methods at the Midvale Steel Company, where he worked, in Philadelphia, Pennsylvania. In particular, he saw that the participation of employees in production decisions was increasing production, as long as there was evidence of a standardized workflow. Complex processes were divided into simple subprocesses. These subprocesses were documented and linked to a reward system. Speed of work, cost reduction and a far-reaching division between head and hand labour in staff and line were important characteristics of this change. This led to significant changes in American industry, and conveyor belts and 'modern mass production' of goods were introduced. The ideal scenario was to provide an environment in which employees would not have to think about their work (Figure 2.1).

According to Taylor's view, employers and employees had similar goals. After all, higher production increased profits for employers and led to higher wages for employees. Scientific management constituted a first attempt to influence the behaviour of workers through reward systems. However, in most organisations, management was barely aware of what labour at their factories actually entailed and could not therefore guide the system. According to Taylor, this ignorance was the biggest obstacle to efficient production. His message was quite simple and at the time very radical: managers should make the decisions to implement scientific methods that could steer selection, training, task design, time studies and control. The active management approach was based on:

- *Observations, measurements, registration;*
- *Selection and training of workers;*
- *Development of standards and regulations;*
- *Close cooperation between management and employees.*

One limitation of this approach was its assumption that employees and employers were essentially economically driven beings. Workers were mainly seen as a means to achieving management ends. A little later, the renowned, and also infamous, 'time and motion studies' were introduced by Gilbreth and Gilbreth (1917). Classic management was characterised as using scientific methods – scientific, because this management involved introducing empiricism, together with measuring, monitoring and recording, all of which marked the start of the planning, organising, influencing and controlling of production.

Figure 2.1 Satirical and social-political presentations of conveyor belts.

Sources: (a) Dunn (1929); (b) *Modern Times* Chaplin (1936)

In line with scientific management, management of companies was concerned with appropriate selection and training of employees in order to prevent hiring so-called reckless and inattentive employees. Selection procedures had to ensure that the right man was put in the right place and was trained in the *safety idea*, as it was called at that time. Selection of employees for a particular job became possible with the rise of psychometrics. This form of applied relatively quantitative psychology was developed in the First World War and widely employed to select American pilots, and subsequently found its way into industry. Psychometrics tested personal qualities and skills and was thus aligned with the then prevailing behavioural explanations of accidents (Lochem 1943; Anonymous 1973; Hoorn et al. 1980). Training as well as education in safety were topics already developed before the First World War. This 'soft side of safety', including safety commissions and ditto games, was replaced increasingly by a 'hard side'. This hard side was not actual hardware but pertained to the selection and control of employees and was supervised by managers.

Behavioural management

Prior to the Second World War, the rise of industrial psychology introduced a new school, behavioural management, in which human behaviour, motivation and leadership were key features. This management school was inspired by behaviourism, a then modern empirical approach within psychology, which explained human behaviour based on stimuli, conditioning and the context in which behaviour occurred (Watson 1913; Pavlov 1927). The human relations movement was part of this approach. Two well-known pioneers were the Americans Elton Mayo and Fritz Roethlisberger. They investigated the behaviour of workers in the late 1920s and early 1930s at the Hawthorne Works of the Western Electric Company, in Cicero, Illinois, just outside Chicago, where electrical cables for telephones and other consumer products were produced. Their results showed that productivity was determined more by psychological factors, group dynamics at work and attention received from supervisors and management, and much less by economic benefits or physical working conditions. The main problem confronting behavioural management was the complexity of human behaviour. Behaviour and changes in behaviour were simply too difficult to predict. Human motivation thus seemed to play no significant role. The psychoanalytical movement started by Freud (1901), although perhaps one of the first attempts to understand the background of human behaviour, including human error, elicited no response in the safety domain. What kept lingering, though, was the 'slip', an undesired act that occurred without any conscious planning. In the behaviourist climate of that time, no value was given to the possible drivers of behaviour.

Safety technique

Safety technique was the start of occupational safety. In the nineteenth century, spectacular improvements were realized in several industrial sectors: for instance, the introduction of safety valves, and thicker steel walls reduced boiler explosions dramatically. The introduction of railway signalling had a similar positive effect, in that it reduced train collisions (Roper 1899; Rolt 1955). Safety technique encouraged the enclosing of moving parts of installations and machines and the fencing of stairs and platforms. One of the early American pamphlets on occupational safety addressed this issue and included illustrations of safety devices for steam boilers, electrical apparatuses and elevators (Law and Newell 1909). Safety technique was the domain of the inspectors of the Labour Inspectorate and of engineers and technicians. In these publications, causes of accidents were not explicitly addressed, but implicitly it was assumed that human–machine and human–height combinations were both causes of accidents.

The Safety First Movement, the Pittsburgh Survey and media attention had prompted many safety-related initiatives, and the connection between safety and production efficiency was gaining ground. Slowly, the assumption of the inevitability of occupational accidents was losing ground. In 1910, a review appeared of the Carnegie Steel Company, with 54 references by American and some European publications to occupational accidents in industry, mining and railways, and also to fire prevention, safety technique and first-aid techniques (Anonymous 1910). With the start of the National Safety Council in 1913, which was established by private companies, professional associations of engineers and insurance companies, the Safety First Movement gained momentum. Unions, however, were not involved in this initiative. They were scarcely active in the domain of occupational safety, the United Mine Workers Union being the only exception (Berman 1978).

The council became a focal point for safety-related initiatives. Two years before, the American Museum of Safety had opened its doors, and this museum was also organising safety initiatives, such as the professional journal *Safety* and national safety congresses, and issuing safety medals to companies that had the best safety practices (Palmer 1926; Anonymous 1928a, 1928b). The council was comparable to the British Royal Society for the Prevention of Accidents (RoSPA), which started in 1916, and the Dutch Safety Museum, which had opened in Amsterdam in 1893 (Anonymous 1891, 1914a; Zwaard 2007; RoSPA 2008). Almost all publications on occupational safety in the United States were initiated by the National Safety Council, which was supported by insurance companies. These companies were active in promoting the work of standard safety code committees; they published and distributed gratis excellent educational literature and promoted the statistical reporting of accidents. Additionally, the insurance companies maintained a large corps of safety engineers for consultation work and, with these engineers, were largely responsible for the extension and improvement of accident prevention methods (Stone 1931; Busch 2018). Three topics dominated the discourse on safety. Aside from safety technique and organisation of safety, financial compensation for occupational accidents was also a major issue. Publications on financial compensation had a legal focus. Stimulated by Eastman's results and her call to manage the financial consequences of accidents, American safety engineers visited Europe, particularly the UK and Germany, to learn how their national compensation systems functioned. Germany in particular received much attention during these visits. In Germany, every occupational accident, irrespective of its cause, was compensated. The Americans were surprised by the impressive growth figures of German industry, in spite of the high collective costs of the compensation insurance. In Germany, reducing the inherent hazards of machines and tools was given greater priority than placing blame on victims or on fellow workers (Hoffman 1909; Page et al. 1910; Schwedtman and Emery 1911; Villard 1913; Blanchard 1917).

Safety publications

In America, compensation was addressed from workers', unions' and insurance companies' perspectives, and was illustrated by cases of financially hopeless conditions of families of workers who suffered from accidents or were even killed on the job (Dunn 1929; Eastman 1908; Hard 1910; Mitchell 1911). The huge costs of the frequent accidents were also a major financial issue for insurance companies. Beyer (1916), Blanchard (1917) and later Heinrich (1927) also explicitly pointed at the hidden costs of these individual accidents, including loss of production, replacement of workers, costs of caring for victims and the possible high legal costs of dealing with claimed legal employer liability. Accidents disrupted production, which reduced efficiency. Companies with a high safety level were perceived as producing efficiently and maintained a steady production of high-quality products (Beyer 1917). Similar conclusions were published by the American Engineering Council in its 1928 report 'Safety and Production'. The survey, amongst 14,000 companies, showed a strong correlation between occupational safety and the efficiency of industrial production (Aldrich 1997; American Engineering Council 1928; Tolman 1928).

Another group of publications dealt with safety techniques. These were professional publications from specific industry branches. The state of the art of safety technique was described and illustrated with photos. The engineering approach was dominant in these publications (Anonymous 1913, 1914b; Beyer 1916; Ashe 1917; Schaack 1917; Williams 1927) (Figure 2.2). The last group of publications were reference books, with both a general overview of safety technique, but not restricted to a particular branch of industry, and an extended discourse on organising safety in companies. This last topic was new. Nowadays we would use the term 'safety management', but in these publications that term was not used as such. Two reference books were published in the period covered by this chapter

Figure 2.2 (a) Grinder, complete protection; (b) stairways; (c) safeguarding of a cutting machine.

Sources: Ashe (1917); Beyer (1916); Schaack (1917)

(Cowee 1916; DeBlois 1926). The publication of these reference books was a sign of a growing professionalisation of the safety domain and will be discussed in the next section.

A common topic in all publications was the need for adequate safety procedures, including control and, more generally, education of workers in occupational safety. The National Safety Council translated this into a three-E slogan, referring to *Engineering, Education and Enforcement*. Engineering meant the design of mechanical safety devices. Education included training of workers to work safely. Enforcement implied a documentation of safety rules and procedures by the employer, while compliance was monitored through supervision. General safety rules for all workers were announced through posters, safety boards and bulletins. These rules and posters stressed behavioural aspects of safety. In addition to general rules, task- and machine-specific safety procedures were promulgated as safety codes. This focus on safety codes and procedures was in line with the documentation of tasks and working procedures initiated by Taylorism. Compliance with the safety rules was the job of the foreman, and not of the safety officer of the company. The special position of the foreman in safety was emphasised in all publications; he was the direct contact with the workers. If a foreman was not convinced of a particular safety measure, getting it into practice was like banging one's head against a brick wall (Greenwood 1934).

Professionalisation of occupational safety

Professionalisation of safety was visible not only in institutions dealing with occupational safety but also in the rise of a professional journal, in reference books and in a series of articles that presented an integrated vision of safety, including safety technique as well as organisational and individual factors influencing safety.

The first reference book on occupational safety, by Cowee (1887–1975), *Practical Safety, Methods and Devices, Manufacturing and Engineering*, was published a few years after the foundation of the National Safety Council (Cowee 1916). Cowee was manager of the safety department of the Utica Mutual Compensation Insurance Corporation. Apart from the topic of occupational accidents, Cowee also published on insurance issues as well as on stock markets (Cowee 1911, 1931, 1938, 1942, 1960). The reference book was an extensive review on the status of safety technique in hazardous branches, such as the construction sector, the metal and steel industries and mining, together with a limited review of a number of hazards, such as fire, explosives and electricity. Detailed information on causes of accidents was not provided. Indeed, Cowee believed that workers' behaviour was the dominant factor in accident causation, stating that only about 30% of all occupational accidents were preventable by guarding dangerous parts of machines, while for about 60%, safety education was the route to prevention. The origin of these percentages and this division was unclear, but it was nearly identical to that of the National Safety Council, focusing on the reckless behaviour of workers, who had to be educated to work safely in a mechanical environment. Safe work, it was believed, should be a habit, a message resembling today's slogan of 'Do it safely or not at all' (Anonymous 1914b; Beyer 1916; Ashe 1917; Schaack 1917; Willems 2004). Cowee addressed safety management by extensively explaining both the role and the organisation of safety committees. These committees would meet without the presence of a superior and rotate the chairmanship on a monthly basis. The aim of the commission was to spot unsafe conditions in factories and workplaces and report these to the plant's safety committee.

The reckless behaviour of workers remained a recurring topic in quite a few publications. The first American pamphlet on occupational and public safety stressed that care and faithfulness were the main defences against accidents (West 1908). Some publications adopted the view that recklessness was just a part of man's nature. Requesting workers to work safely was ineffective, as were repeated warnings with similar messages. Instead, experiences from accidents should quickly be translated into 'lessons learned', to teach the worker he was the loser in the case of an accident (Anonymous 1913). But there was also some opposition to this view: workers' reckless behaviour was not seen as the main cause of accidents, but rather a consequence of the increased speed of production, monotonous work and long working hours of 70–80 hours a week or more. Workers' 'reckless behaviour' was thus nothing more than a state of chronic exhaustion (Bogardus 1911a, 1911b; Myers 1915; Lee 1919; Whitney 1925; Muntz 1932). Employers were never accused of reckless behaviour, although they were responsible for the choice of machinery, the selection of raw materials and the frequency of maintenance (Eastman 1908).

Safety management according to DeBlois

A book by Lewis DeBlois (1878–1967), *Industrial Safety Organisation for Executives and Engineers* (1926), was the second industrial safety textbook. DeBlois worked as chairman of DuPont de Nemours' safety commission of the department of high explosives, and later as chairman of the central safety commission of the company. As Cowee had been, DeBlois was also then president of the National Safety Council. In 1923, he was elected as the ninth president. DeBlois was rather outspoken on the role of management in occupational safety, on the reliability of accident figures and on causes of accidents. Unlike most other publications on industrial safety, DeBlois stressed the primary and central role of the company's head manager and his prime responsibility for effecting a high level of occupational safety. Frequently percentages like 75–90 % were cited as the contribution of victims' behaviour to causing accidents. That might be the case, according to DeBlois, but when accidents were occurring repeatedly, apparently management did not take the topic seriously, and no action was taken.

> *Process disturbances, for instance products jumping out of presses, occur unexpectedly, and should be regarded as accidents, for many times these conditions do cause injuries of workers. Whether this happens or not, accidents and efficiency were each other's opposites. You cannot have them both in*

> *your company. A company with unexpected process disturbances is by definition not efficient.*
>
> (Citation of Williams in DeBlois 1926)

Also, the employer should propagate the safety message through his attitude and behaviour and become the leader of safety, just as he was the leader in other company's issues. He had to show a strong personal commitment to safety and regularly had to repeat the message of the central importance of safety.

DeBlois did not use the term 'Safety First Movement' in his book. Taken literally, it would have stimulated cowardly behaviour, according to him. At that time, it was thought that risks and risk taking were an essential part of a man's upbringing and allowed them to learn from their mistakes. Consequently, the word 'first' disappeared, and thus only the term 'safety movement' was used. The safety movement had then reached its adolescence. Due to increasing mechanisation and mass production, there was still an apparent raise in accident rates. Eye-catching was the word 'apparent', and a substantial part of DeBlois' work dealt with methods to measure safety, and to show its value in increasing profits. National accident figures and figures per state seemed to fluctuate with the economy. Compared to the beginning of the nineteenth century, the number of fatal industrial accidents (see Chapter 1) in the twentieth century seemed to drop. The 1926 figures, presented by the US Secretary of Labor, estimated an industrial mortality of at least 23,000 per annum, and a yearly number of non-fatal industrial accidents of 2.5 million (Anonymous 1926a). Economic prosperity regulated employment, which in turn determined exposure to hazards. The accident figures per state and the national figures were, however, very unreliable and did not show an effect of increased attention to occupational safety. It seemed that America was uninterested in occupational fatalities and wounded workers, and forgot to count them. If there was any registration, medical categories were used, which only contained very limited information on causes (DeBlois 1927). Sometimes only general causes were given, such as accident from falling, or specific employment sectors were mentioned, like accident in a mine, or physiological classifications such as fracture were used with no recorded connection to accident causes. The situation might be different at individual companies and branches, which could show spectacular reductions, both in absolute numbers of accidents and in rates per 1,000 workers. Iron and steel companies were examples, but these accident reductions were not general to industry as a whole. For example, accident figures were still high and unchanging over long periods in mining, wood processing and textile industries. There was obviously something wrong with the management of safety when accident figures remained at high. In order to change, safety had to become second nature for everyone.

The important contributions of DeBlois were his ideas on accident causation and his general rules for prevention. Accidents had to be understood as a consequence of a sequence of events which either directly or indirectly caused harm and damage. Hazard was the starting point of an accident, and hazard was equivalent to the potential or kinetic energy, or could be of a mechanical, electrical or chemical nature. The chain of events was further subdivided into the release of energy and, in causes of harm, the direct cause of received harm. Examples of causes having long-term effects were decisions made by management and foremen which addressed safety insufficiently. DeBlois further introduced the terms 'probability' and 'exposure'. Probability, according to DeBlois, was nothing more than common sense reduced to a calculation. It determined with exactness what a well-balanced mind understands by a kind of instinct, without providing any details on the accident process involved. Exposure was familiarity with a hazard. The probability of an accident might be low, but if the exposure was high, the chance of an accident occurring increased. Also, rules for prevention were quite simple. Prevention was affected primarily by reducing hazards, such as speed, mass, pressure or temperature, replacing chemicals or other means to prevent the release of energy. A second method of prevention was the reduction of exposure, such as the guarding of moving parts of machinery. Prevention by control measures to reduce harm was in general less effective than the other two strategies.

Finally, DeBlois commented on employing psychological tests used to select workers for specific jobs and tasks. Such tests were routinely being applied by the US Army, and they might also possibly be used to identify accident-prone workers. Without referring to the British research on accident proneness, the topic of the next chapter, DeBlois mentioned that the experimental state of psychological tests meant that they were not yet suitable for application in an industrial setting.

Heinrich's influence

Next to Cowee and DeBlois, a third influential author was Herbert Heinrich, originally employed as a mechanic and later as safety engineer at the Travelers Insurance Company (Busch 2018). Like Cowee and DeBlois, Heinrich emphasised the organisation and management of safety through safety services in companies and independent safety commissions manned only by employees. Heinrich, in his message to foremen (Heinrich 1929), even went so far as to see the role of safety management as primary:

> *Accident prevention was a management task. Management supports safety initiatives, analyses causes of accidents, develops and addresses solutions* (DeBlois 1926). *Good management was better than good tools.*
>
> (Heinrich 1929)

Heinrich played an important role, not least because he regularly published his ideas in papers in *National Safety News*, the professional journal of the National Safety Council, with titles such as:

1927 The 'Incidental' Cost of Accidents
1928 The Origin of Accidents
1929 The Foundation of Major Injury
1929 A Message to Foremen

Within the Safety First Movement, Heinrich advocated the importance of data collection and applied mechanised principles for the prevention of accidents. In 1927, he published the results of an analysis of tens of thousands of cases of compensation for industrial accidents, which later became known as the 'Rule of Four'. Firstly, Heinrich estimated that the average cost of an occupational accident for an employer was approximately four times higher than the amounts actually paid for compensation due to all associated indirect costs (Heinrich 1927). A second ratio proposed by Heinrich related to causes of accidents. His percentages deviated from those of Cowee, though according to Heinrich, 98% of all accidents were preventable, 10% through improved technique, while 88% were the result of unsafe acts by workers, and to a much lesser extent by the foreman. The other 2% of accidents were not preventable and should be regarded as 'acts of God' (Heinrich 1928). In his third rule Heinrich, like DeBlois, introduced an accident as an event occurring as a process:

> *If a person is injured, then the sequence of events is: first the cause, then the accident followed by the injury.*

Finally, Heinrich proposed a ratio for the seriousness of accidents, which he called 'accident mechanism': 1 (major injury): 29 (minor injury): 300 (no injury). In later publications, this ratio was transformed into the well-known 'iceberg' or 'pyramid' metaphor (Heinrich 1929). He provided no justification for his ratios, nor any other information on the quality and origin of the data behind these ratios. He quoted the following, actual dialogue from a company:

Workman: *Can't stand over each individual workman continuously.*
Safety engineer: *Don't need to. Workmen violate the same safe-practice rule on average 300 times before they are hurt. You should be able to see at least a few of these violations and take executive remedial measures at that time.*

In Heinrich's view, attempts at prevention of damage did not make much sense. Eliminating the causes of accidents was a logical preventive activity, and this 'moral' is strikingly expressed in the last sentence of Figure 2.3. Just like the 1:4 rule for the cost of accidents and iceberg ratios, Heinrich came up with a simple set of numbers to indicate the relative importance of behaviour and supervision over physical causes.

Safety propaganda

The Safety First Movement message also became visible through many forms of persuasive safety information, in which the unsafe behaviour of employees remained the main concern. A series of initiatives was set up to influence behaviour through persuasion, with 'education', 'training' and 'information' being keywords. On a large scale, safety posters were developed (Figure 2.4), safety films

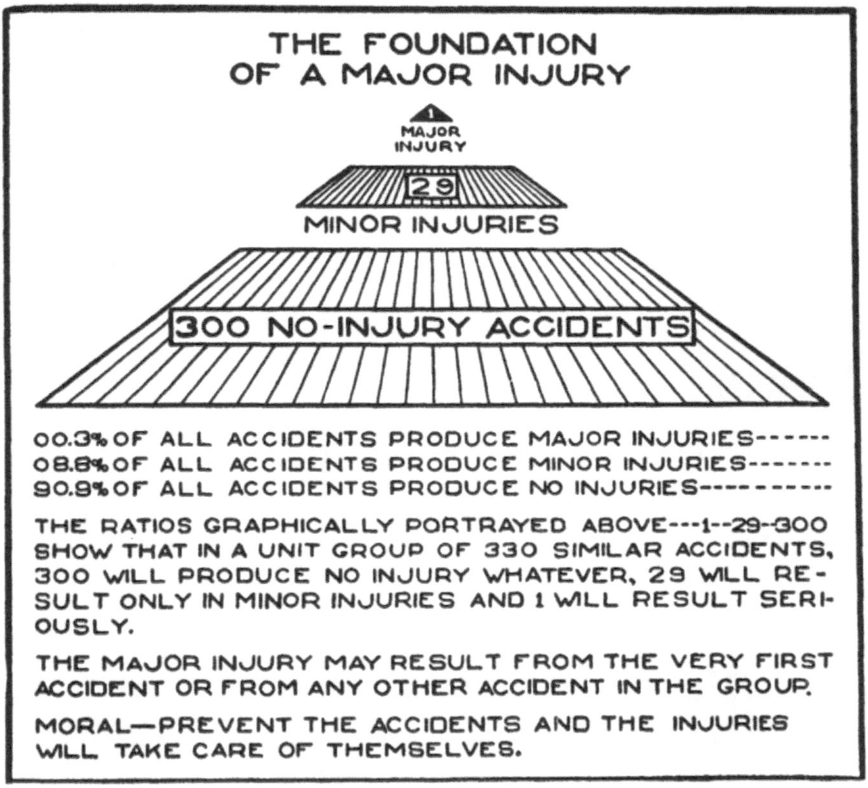

Figure 2.3 The accident 'iceberg' or 'pyramid'

Source: Heinrich (1929)

were made, and the Safety First message even appeared on the back of postal envelopes.

In companies, suggestion boxes arrived with prizes for the best proposals, safety games were organised, bulletins illustrating safety improvements were shown, and at entrances to factories, safety noticeboards and safety scoreboards were clearly displayed (Figure 2.5).

The safety message was also spread through safety committees, which became more and more prevalent in companies. In these committees, only employees (no foremen or managers) were allowed to take part; this was to ensure an open atmosphere among the members. The assumption was that the presence of a foreman at these committees would create a major barrier to idea formation. Such

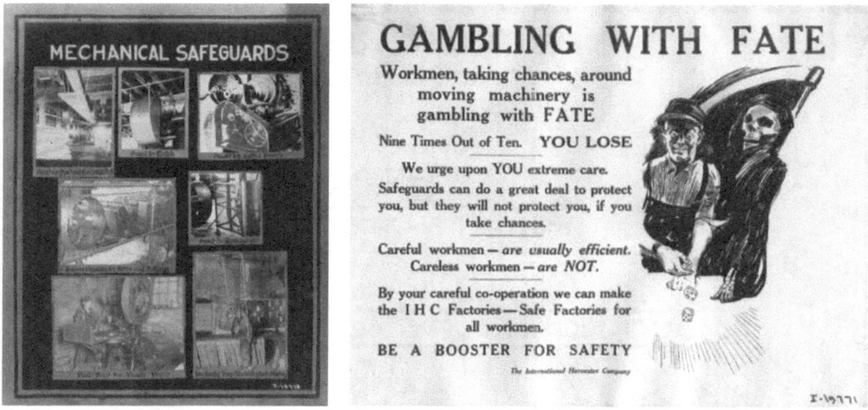

Figure 2.4 Examples of American safety posters. (a) 1916, (b) 1919

Source: Arxiu Nacional de Catalunya (1987)

Figure 2.5 (a) Pennsylvania Dixie-Cement Corporation, (b) DuPont's safety bulletin, (c) the Chicago Great Western railroad

Source: DeBlois (1926)

committees were also set up in the Netherlands, where there was a strong preference for committees to consist exclusively of employees. Gaining the trust of workers, knowledge of workplaces and insight into practical safety at workplaces were the arguments for this approach (Patijn 1903). The committees appointed a safety officer to report unsafe situations to a company safety service, many of which were then being created. Professionalism was further taking shape as the American Society of Safety Engineers, established in 1911 from the insurance branch, merged with the National Safety Council in 1925. As a result, differentiation of safety-relevant information was taking place. Following examples of German insurance companies, the US Bureau of Statistics and American insurance companies were developing risk measures. Lost-time accidents, for instance, were defined as accidents with a minimum of at least one lost day. Injury frequency rates were obtained by dividing the total number of accidents by the total number of hours of exposure. These measures of safety became accepted generally and facilitated a relatively simple method for benchmarking safety among companies.

A second development was the classification of causes and consequences. This classification can be considered the start of a scenario-based analysis in a developing knowledge domain. For instance, falling was separated into two categories, namely falling from heights and falling at ground level. Furthermore, distinctions were made between falling from scaffolding, from ladders and through openings (Anonymous 1920, 1921, 1926b). At this time, safety codes were also undergoing quite a number of changes, because the First World War had further emphasised the importance of standardisation. In 1919, the National Safety Council together with the Bureau for Standardisation started this development, and they subsequently developed 40 different codes for various machines and equipment, for instance grinders and paper mills. Public hearings were used to help establish these codes (Williams 1927).

Slowly safety became a commercial market, as apart from films, posters and pamphlets on safety information, there was a demand for personal protective equipment and mechanically engineered adjustments of machines and installations. The occupation of the safety engineer in companies developed along with this safety market. But the biggest impact of occupational safety was its capitalisation (its money value), which was seen clearly by focusing on the costs of a lack of safety. Together with control and supervision, occupational safety was placed at the centre of management activities. This became the main message of the Safety First Movement: laws and regulations would have only a limited effect on occupational safety as long as active support from management was lacking. DuPont even went further: any injury was an expression of a failing operational management (DeBlois 1919; Aldrich 1997).

United Kingdom

Safety research

In contrast to the United States, in the United Kingdom the topic of occupational safety was not initiated by market parties, such as insurance companies and large industries, but by the government. The British publications were not only limited in number but also different in nature due to the UK's long tradition of social legislation, dating from the start of the nineteenth century. Halfway through that century, specific safety regulations were already in force for the textile industries (Hale 1978). The research focus was a typical British approach to occupational safety in that period. This scientific approach contrasted with American efforts, which had a more applied and managerial interpretation of occupational safety. The Pittsburgh Survey, of course, was a notable exception. The scientific focus of the UK was initiated by a governmental committee investigating industrial fatigue. Already in 1904, such a committee had reported to government on the large number of rejections of recruits to the army on physical grounds during the Second Boer War (1899–1901). The conclusion pointed to physical deterioration caused by several factors, such as the nature and conditions of industrial work, and the general living conditions (McIvor 1987). The committee recommended a scientific enquiry into the physiological causation and effects of over-fatigue of workers. A 1911 survey of the departmental committee on accidents in factories and workshops showed an alarming rise in fatal industrial accidents, due to a lack of safeguarding machines, fatigued workers, and young workers and children employed at dangerous machines (Home Office 1911). But it would take another war before other inquiries would start. In 1915, the British government had formed the Health of Munitions Workers Committee, a commission of the Ministry of Munitions, to investigate the productivity and efficiency of munition production. The second task of this committee was to investigate the health and safety of women, children and the elderly, who replaced men in these factories during World War I. This committee combined representatives of the medical profession, academics, labourers, employers, the Factory Inspectorate and government. This committee had been active until 1918, and was transformed into the Industrial Fatigue Board, a permanent governmental research organisation, to continue the work started by the Health of Munitions Workers Committee. In 1929, the board was renamed the Industrial Health Research Board. The goal of these committees and boards was to study the influence of the organisation of work, influenced by Taylorism. The effects of working hours, lighting, temperature and ventilation, and short cyclical work were investigated, not only on productivity, but also on the incidence of accidents.

Accident proneness

As in the United States, in the United Kingdom national accident registrations had also shown a shocking rise in industrial accidents during World War I. A few explanations circulated for this alarming increase, such as levels of mechanisation, and consequently the high speed of production, the introduction of new hazards and long working hours. Another argument was the conditions in industry due to the war. Because most men had been recruited by the army, a massive influx of female, young and older workers in the industry occurred (Figure 2.6).

Figure 2.6 (a) Poster, 1918, (b) Women in munitions factories, Scotland, 1918

Source: Ministry of Munitions 1916 IWM PST 8184

These groups of workers were considered inherently less competent than men. Several studies were conducted to investigate the perceived causes of this lower level of competence. But the Industrial Fatigue Board believed the distribution of accidents amongst workers in hazardous professions should be the prime point of focus. The first question to be answered was thus whether or not industrial accidents were equally distributed amongst workers or were limited to a specific group of workers. This question had been the starting point for the epidemiologist Major Greenwood and his assistant, the physical statistician Hilda Woods, of the British Ministry of Munitions (Greenwood and Woods 1919; Farewell and Johnson 2010). This was the first British published research on occupational accidents. It had been conducted amongst women working at munitions factories during the last two years of World War I. Their publication was titled 'A report on the incidence of industrial accidents on individuals with special reference to multiple accidents'. The term 'accident proneness', which later became commonplace, was not yet mentioned in this publication. Instead, 'susceptible workers' was used. The population of the survey consisted of 3,889 women from ten different factories. Only in two factories were all women included; at all other factories a random sample of women participated in the survey. The period of the survey varied between two and five months, and aside from the number of accidents, information on locations and tasks of the women working was known for only a limited number of factories: those with either heavy lathe operations or profiling operations. Using statistical analysis, three different hypotheses were tested:

1. *The distribution of accidents was based on chance, i.e., a population exposed to equal risks had an equal risk of industrial accidents;*

2. *All members of an exposed population had an equal risk of accidents. However, after the first accident, the probability of any following accidents occurring would change. The probability would either increase (accidents were contagious) or decrease (workers became more careful);*
3. *The probability of accidents and repeated accidents was not equally distributed in an exposed population. There were clumsy, awkward workers, and there were prudent workers.*

These different hypotheses had different distributions of accidents over the exposed population. A Poisson distribution would describe the first hypothesis, while a negative binominal distribution would describe the third. The second hypothesis generated a distribution different from the first and the third. Analysis suggested that the contagious view of accidents gave the best description (Figure 2.7).

The authors defended their research by pointing out possibilities for prevention. If accidents were governed by chance, then prevention should be directed towards workstations and to the improvement of general conditions in factories. If the second hypothesis proved to be true, the prevention should minimize accident contagiousness or improve watchfulness of careful workers after a first accident occurred. If the third hypothesis were correct, then selection of workers with a low 'accident sensitivity' was the solution in the case of proven susceptible workers, as well as a quick removal of workers causing multiple accidents. The distribution of repeated accidents with women had given a good fit for the second and third hypothesis, once only a single period was considered. The preference of the authors was for the third

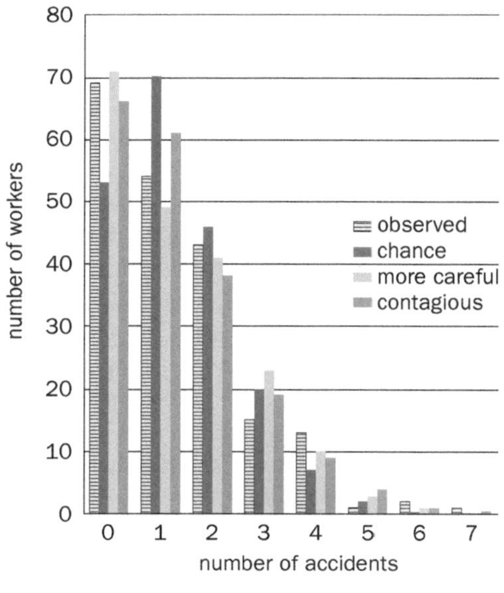

Figure 2.7 Distribution of accidents according to hypotheses

Source: Greenwood and Woods (1919)

hypothesis, due to the correlation of accident figures between two periods of time. However, this proof was based on a population of only 65 women. According to the authors, the consequences of the accidents, the physical harm that occurred, were determined by chance. This implied that accidents with light or with severe consequences would have the same or similar causes, an assumption also underlying the accident iceberg metaphor later developed by Heinrich. Furthermore, the number of accidents had no effect on productivity, and causes of accidents had no relation to external factors, like speed of production. In line with the third hypothesis, these causes had their origin in workers' personal traits. The survey resulted in a rather vague definition of 'susceptible workers', only defined as a stable trait belonging to a person. It was also not clear whether all accident-prone workers had a common personal background. Nevertheless, from the 1920s onwards, the 1919 article of Greenwood and Woods became a standard publication, referred to for decades by many researchers in the field.

The individual hypothesis

The individual hypothesis was based on the nineteenth century assumption about the hereditary skills and talents of an individual. This relationship, which scientifically was considered generally valid during that period, was based on Darwin's ideas. Darwin believed that spontaneous genetic variations could develop in either a forward or backward direction. In safety research, Greenwood and Woods' theory cited such ideas. The widely proclaimed message of the Safety First Movement that genetic predisposition and personal characteristics were major factors in accident causation was believed to be confirmed by Darwinian theory.

From the 1920s onwards both hygienists, including occupational physicians, and psychoanalysts became interested in accidents. Traditionally, reference books on occupational medicine, appearing around the turn of the century, paid little attention to occupational safety (see, for instance, Arlidge 1892; Oliver 1902). This, however, was changing in the 1920s. Hygienists made a comparison between occupational accidents and two other major causes of death: cancer and phthisis (then a term for tuberculosis). The incidence of accidents in those professions investigated was 3.2 per 1,000 workers (0.3–3.5), while cancer had an incidence of 2.0 (0.4–2.3), and phthisis 4.6 (1.0–5.6). Greenwood, together with the physician Collis, wrote an extensive work, *The Health of the Industrial Worker*, and the physicians Hope, Hanna and Stallybrass published a similar work, *Industrial Health and Medicine* (Collis and Greenwood 1921; Hope et al. 1923). Both works contained extensive chapters on occupational accidents, financial compensation of victims and prevention. The state of the art of safety technique was presented and illustrated with photographs, and ample attention was paid to determinants of accidents, both external and individual predispositions to accidents. The classification of accidents by Collis and Greenwood is shown in Table 2.1.

Although this list of causes did not directly relate to careless workers (i.e. the individual hypothesis), the authors declared that a lack of control of workers caused 60–80 % of these accidents. Failure to enclose moving parts of machinery would explain only 10% of the accidents, which mirrored American research

Table 2.1 Reported industrial fatalities, annual report by British Labour Inspectorate, 1913

Categories of accidents as % of total accidents (n = 1310)		
1	Falling from heights	34
2	Caught by machines	32
3	Hit by falling objects	10
4	Contact with escaping gas or steam	3
5	Hit by an explosion	3
6	Contact with hot metal, liquid	2
7	Contact with electricity	1
8	Hit by crane or crane load	1
9	Hit by tools	0.5
	Others	13

findings (Bellhouse 1920; Anonymous 1930). Collis and Greenwood were outspoken about the use of safety legislation. Experience thus far had shown that legislation rarely, if ever, led to reforms of working conditions. Legislation generally followed the state of safety technique, which was already being introduced by more enlightened employers. These employers requested protection by law, arguing that the safety improvements they made were a financial limitation to their trade, and therefore such costs should be shared by industry in general.

Between thinking and doing

During the period between the two world wars, both psychologists and psychoanalysts investigated individual factors influencing the occurrence of accidents. In addition to the often mentioned inattention of workers, stress factors were causes of accidents, as sensorimotor variables and subconscious drifts. These stress factors caused workers to adapt insufficiently to work, resulting in accidents. Research on sensorimotor variables basically assumed that susceptible workers were *doing, instead of thinking*, and this impulsive behaviour was a major cause of accidents. The psychoanalytical approach to accidents was radically different from that of psychologists. Unconscious or semi-conscious impulses, driven by experiences of fear, aggression and childhood feelings of guilt, brought the employee into a particular mental state, which made accidents unavoidable. There could even be a self-punishment factor (Armstrong 1949; Hale and Hale 1972).

In addition to the explanations mentioned above, tests were developed to detect susceptible workers, and to exclude them from dangerous work. In 1925, the psychologist Eric Farmer introduced the term 'accident proneness' in his paper presenting differential tests for these workers. From then on, accident proneness became the indication of the safety theory that explained accidents by the individual hypothesis. Simultaneously, independently the same concept of *Unfallneigung* (accident proneness) and the *Unfäller* (accident prone worker) was developed in Germany by Karl Marbe. Marbe was a psychologist from Würzburg University who used accident reports from the Berufsgenossenschaften, the professional organisation of the German accident insurance cooperatives (Marbe 1925; Farmer 1925; Hale and Hale 1972; Burnham 2008). Both Marbe and Farmer developed the 'aesteto-kinetic' battery of tests, which measured the capacity for coordination and concentration of workers, to select the *Unfäller* (Farmer and Chambers 1926, 1929, 1939; Farmer et al. 1933; Farmer 1940, 1942). They tried these tests in various populations, including apprentices in the marines and the army, London bus drivers, shipwrights, mechanics and electricians. With an exposed population, varying between 650 and 1,843 workers, and a sample time of between 12 weeks and five years, the results showed only a rather low correlation of 0.2–0.4 between the test results and accidents. Researchers could not conclude otherwise than that accident proneness was dependent on many different factors, which they did not yet quite understand.

Environmental hypothesis

A study by Osborne and colleagues (Osborne et al. 1922) was research in the tradition of the British Industrial Fatigue Board, using the environmental hypothesis. During World War I, the influence of temperature on cutting accidents among ammunition workers was investigated. The study population was large, 1,800–6,000 workers from several factories. The lowest incidence occurred at 21°C and 23°C. Higher temperatures, and temperatures below 10 °C, showed a sharp rise in accidents. Classifying accidents according to the hours of the day indicated a strong relation between the energy use of the plant and output to production. These results were further confirmed by laboratory research, which was part of the same study. It showed that the accuracy of movements decreased sharply with higher production speeds, thereby suggesting that production speed was an important determinant of accidents.

The Netherlands

Individual factors

In the Netherlands, the Safety Museum, started as an initiative of industry, remained the leading institute on safety. The number of visitors grew from 10,000 in 1915 to 30,000 in 1927 (Zwaard 2007) and the museum developed more and more activities. *Nieuws van het Veiligheidsinstituut* (News from the Safety Institute) was launched in 1924, and changed in 1927 to a monthly magazine, *De Veiligheid* (The Safety Journal), which became a platform for information exchange between safety practitioners. Industrialists were interested in safety, efficiency and the applicability of scientific results. Partly because of this interest and a growing demand for highly educated staff, the Nederlandse Organisatie voor Toegepast Natuurwetenschappelijk Onderzoek (TNO; Dutch Organisation for Applied Scientific Research) was established in 1932. The task of the institute, method and organisation had been stipulated in 1930 in a special law. The mission of the TNO was to ensure that society, and in particular companies, took full advantage of the results of scientific research.

Following the examples from abroad, the Safety Museum started to develop other safety initiatives in the 1920s, such as the organisation of national safety conferences, the design of safety posters and weekly radio presentations on safety-related issues. These radio programmes, titled *Doe het veilig* (Do It Safely), were broadcast on Sunday mornings, giving practical information on the hazards of tools, safe use of ladders and the importance of safety goggles. Safety posters were also part of a large societal offensive aimed at improving public health. The posters feature not only industrial accidents but also tuberculosis and venereal diseases. Well-known artists were asked to produce safety posters, and these posters became very successful, as indicated by the growing demand (Anonymous 1940a). The metaphors in these posters referred to fear, shame and guilt, and vividly described to the public the consequences of negligent, reckless and immoral behaviour (Hermans 2007) (Figure 2.8). With diligence, allusions to social class differences were avoided. Aside from the aforementioned diseases, safety posters depicted known situations on shop floors, in factories and at home, places where safe behaviour on the part of workers prevented accidents. References to the shortcomings of employers were also avoided, under the assumption that this avoidance prevented an otherwise negative impact on employers' safety efforts. Just like the American posters (Figure 2.4), the early Dutch posters contained quite a lot of text. These texts were later reduced to general safety-related remarks, with more emphasis given to pictorial presentation.

All posters stressed the importance of behavioural aspects of accidents. This was in line with the assumption that occupational accidents were relatively simple events, with only one important cause, such as not following instructions or insufficient use of personal protective equipment. These two causes were starting points for prevention. Human causes were getting a face through the introduction of 'Jan Ongeluk' (John Accident), a careless worker who frequently created dangerous situations for both himself and surrounding workers.

> . . . *after all the protection, insofar as it was absent or insufficient, was applied to machinery and manufacturing facilities, respectively improved, large numbers of accidents remain not preventable. Chance, fate if you want to call it, indifference, insufficiency, boldness, etc. will always contribute to a high percentage of these accidents.*
>
> (Correspondence between Werkspoor and Hoogovens 1927 in Desta and Heuvel 1987)

In the 1920s, the National Safety Congress organized a sequence of conferences, modelled after the 1890 exhibition. The second National Safety Congress took place in 1926, the third and fourth in 1927 and 1929, respectively. The 1926 congress dealt mainly with safety at home and road safety, which were topics of major public interest. Meanwhile, household comfort had greatly increased with the introduction of gas, electricity and all sorts of inventions. For these, however, there was no legislation concerning safety requirements, in spite of the fact that the public had at that time no knowledge of how to handle these new acquisitions

Figure 2.8 Examples of Dutch safety posters

Source: University museum University of Amsterdam

1922: Why did you not wear a cap like me? Loose hair in proximity of machines and gears is dangerous.

1922: Because he was careful, Grandpa reached 70 without having any accident.

1926: That is what happens when the emergency exit is blocked.

1927: Distraction leads to accidents.

1927: Stacking incorrectly is perilous.

safely. Not surprisingly, the use of faulty devices, defective geysers, wrong fuses and broken cables claimed many victims every year. At the conference in 1926, the chairman of the Algemene Nederlandse Wielrijders-Bond (General Dutch Cyclist Federation, or ANWB) painted a gloomy picture of road safety. Here, too, it appeared that safety lagged behind technical developments:

> *Days after days all sorts of cars are driving, even bashed motorbuses with defective brakes, with . . . on days driving cars, even driving bashed motorbuses with defective brakes, with a steering wheel that cannot be controlled because it is worn or clogged, with emergency doors, stuck of barred by luggage, with wheels not properly protected against runoffs, with doors stuck or luggage barred, with wheels, properly protected against runoffs, with drivers serving 16 hours a day, who were children or silent drinkers, with staff*

lighting matches over leaking reservoirs, with drivers pushing the cyclists from their legitimate path . . .

Beginning in 1927, the National Safety Congress returned to occupational safety, which then received much more attention in terms of legislation and government policies than did road safety and home safety. This did not mean there was no interest in safety in the home or on public roads. The Safety Museum regularly dealt with both of these themes in *De Veiligheid* (The Safety Journal). However, the government was still leaving road safety to private initiatives, especially the ANWB, and local authorities. In the eyes of the government, traffic accidents were mainly due to the recklessness of road users.

This chapter has shown that occupational safety was an increasingly important theme in the 1920s. Safety museums and safety posters appeared in the countries discussed. In the United States, safety scoreboards were introduced outside factories, reflecting the importance of safety management. Here the behavioural management approach focused on carelessness and unsafe acts of workers, as shown in the safety textbooks of authors like Cowee and Heinrich, supporting the 'individual hypothesis', as it was later called. DeBlois' publication supported the 'environmental hypothesis' and postulated an accident as being a process starting from a hazard, or energy. DeBlois also introduced the concept of exposure to hazards and the probability of accidents, concepts that after World War II played a role in discussions on the concept of risk. In the United Kingdom, unsafe acts were grounded in science, a notion that was published, and later became known as the 'accident proneness theory'. In the Netherlands, the Safety Museum and the National Safety Congresses promoted safety and mainly followed safety developments from abroad. These developments increasingly set the tone in the following decades. In addition, the most important safety concepts of the accident process were becoming commonplace, as will become apparent in the following chapter.

3 Dominoes, safety by technique – prevention

1930–1950

Chapter 3 discusses developments in safety in the 1930s and 1940s. In this period, accident models were published for the first time. In the United States, Heinrich's visualisation of the accident process as a series of falling dominoes played a decisive role. But also, medical professionals were entering the safety domain, questioning the 'individual hypothesis', as was the case in the United Kingdom. The Dutch contribution to safety knowledge at the time was still rather limited. In the interbellum, the field of safety science started to mature, as manifested by the publication of several influential handbooks.

DOI: 10.4324/9781003001379-3

In the 1930s, industrial production increased sharply with a shift from the United Kingdom to the United States.

Due to rapid industrialisation, the 'typical American' in the 1930s was no longer a farmer but an industrial worker. The **industrialisation** affected urbanisation (concentration of labour), immigration (accommodation of incoming workers) and foreign affairs (industrial superpower).

During the 1930s, radio was considered an intimate and credible medium. The public used it as a news source and expected it to provide factual information. **Radio as mass communication device** reached millions of people instantly and altered social attitudes, family relationships, and the ways people relate to their environment.

Meanwhile, the **Great Depression** struck the world. It illustrated how rapidly the global economy could decline. The depression had devastating effects in both rich and poor countries. Personal income, tax revenue and profits dropped. International trade fell by more than 50% and unemployment in some countries rose to as high as 33%.

In the midst of the Great Depression, **Adolf Hitler** and his Nazi Party restored economic stability in Germany and ended mass unemployment using heavy military spending. The regime undertook extensive public works. Returning economic stability boosted the regime's popularity. The Nazi regime dominated neighbour countries through military threats in the years leading up to war. In Italy **Mussolini** decided to cooperate with Nazi Germany.

The **first practical radar systems** were developed in the 1930s. The first use of radar (an acronym from 'radio detection and ranging') was for military purposes: to locate air, ground and sea targets. This evolved, in the civilian field, into applications for aircraft, ships and roads. Later it became the primary tool for weather forecasting.

In 1935 the French scientist Irène Joliot-Curie, together with her husband, won the Nobel Prize in chemistry for their discovery of **artificial radioactivity**. (Her parents, Marie and Pierre Curie, had earlier received Nobel Prizes for their pioneering research on radioactivity.)

Hitler wanted a cheap, simple car to be mass-produced for his country's new road network. It took lead engineer Ferdinand Porsche and his team until 1938 to finalize the design. The **Volkswagen Beetle**, a two-door, rear-engine economy car, was manufactured and marketed by German automaker Volkswagen.

Gone with the Wind is a 1939 American epic historical romance film adapted from the 1936 novel by Margaret Mitchell. The film was immensely popular when first released. It became the highest-earning film up to that point and held the record for more than a quarter of a century. Although the film has been criticized as historical negationism glorifying slavery and the Lost Cause of the Confederacy myth, it has been credited with triggering changes in the way in which African Americans are depicted cinematically. The film is regarded as one of the greatest films of all time.

Industrialisation

Radio as mas communication device

Great Depression

Adolf Hitler and Mussolini

First practical radar systems

Artificial radioactivity

Volkswagen Beetle

Gone with the Wind

Sources:

a. https://nl.wikipedia.org/wiki/Fabriek#/media/Bestand:Airacobra_P39_Assembly_LOC_02902u.jpg
b. https://en.wikipedia.org/wiki/The_War_ of_the_Worlds_(1938_radio_drama)#/ media/File:Orson_ Welles_War_of_the_ Worlds_1938.jpg
c. https://en.wikipedia.org/wiki/Great_Depression#/media/File:UnemployedMarch.jpg
d. https://upload.wikimedia.org/wikipedia/en/archive/b/b1/20121113100103%21Hitlermusso2_edit.jpg
e. https://commons.wikimedia.org/wiki/File:FuMG_64_Mannheim_41_T_radars_at_Grove_ DK_1945.jpg
f. https://en.wikipedia.org/wiki/Discovery_ of_nuclear_fission#/media/File:Irène_et_ Frédéric_Jol iot-Curie_1935.jpg
g. https://en.wikipedia.org/wiki/Volkswagen_Beetle#/ media/File:Posche-Typ32.jpg
h. https://en.wikipedia.org/wiki/Gone_with_the_Wind_(film)#/media/File:Clark_Gable_and_ Vivien_Leigh_-_Wind.jpg

The 1940s were dominated by the Second World War and its aftermath.

The **Second World War** (1939–1945) was the deadliest conflict in human history, with 70 million to 85 million fatalities. It comprised massacres, genocides including the Holocaust, strategic bombing, premeditated death from starvation and disease, and the only use of nuclear weapons in war. The war changed the political alignment and social structure of the globe.

At the **Bretton Woods Conference,** held at the Mount Washington Hotel, in New Hampshire, delegates from all Allied nations gathered in 1944 to regulate the international monetary and financial order after the war. Two institutions created at Bretton Woods, the IMF and the World Bank, were to play an important role in globalisation. They were created to restore and sustain the benefits of global integration and international economic cooperation.

During the war the US brought together a scientific team for the goal of producing fission-based explosive devices before Germany did. The use of **nuclear weapons** ended the Second World War. In 1945, a uranium-based weapon was detonated above the Japanese city of Hiroshima, and a plutonium-based one above Nagasaki, Japan.

With the **Cold War**, beginning in 1947, a period of geopolitical tension began between the Soviet Union and the United States and their respective allies after the Second World War. The conflict was an ideological and geopolitical struggle for global influence by the two powers, following their victory against Nazi Germany. The doctrine of mutually assured destruction discouraged a pre-emptive attack.

The **United Nations** (UN) was established after the Second World War with the aim of preventing future wars, succeeding the ineffective League of Nations. The UN objectives included maintaining international peace and security, protecting human rights, delivering humanitarian aid, promoting sustainable development and upholding international law.

The de Havilland DH 106 Comet was the world's **first commercial jet airliner**. Its prototype first flew in 1949. For the era, it offered a relatively quiet, comfortable passenger cabin. Additionally, a large number of the control surfaces were equipped with a complex gearing system as a safeguard against accidentally overstressing the surfaces or airframe at higher speed ranges.

In 1943 the **Colossus** was the first programmable, electronic, digital computer. It was developed by British codebreakers to help in the cryptanalysis of secret German messages during the Second World War. This secret computer was operational two years earlier than ENIAC, the first electronic general-purpose digital computer.

Bambi is a 1942 American animated film that has been recognized for its eloquent message on nature conservation. Conservation is so central to *Bambi* that the film is credited with having inspired many 1960s environmental activists at an early age.

Second World War

Bretton Woods Conference

Nuclear weapons

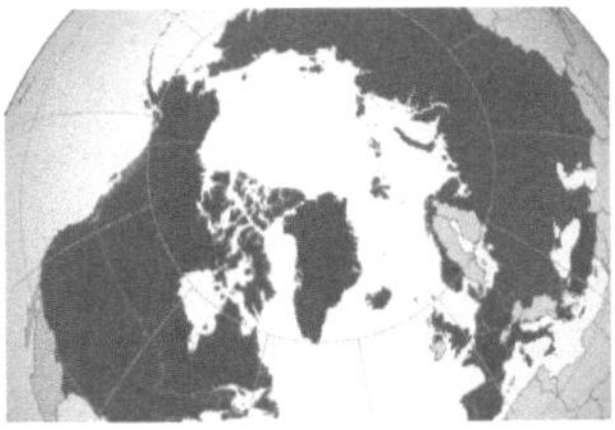

Cold War

United Nations

First commercial jet airliner

Colossus

Bambi

Sources:

a. https://en.wikipedia.org/wiki/World_War_II#/media/File:German_troops_in_Russia,_1941_-_NARA_-_540155.jpg
b. https://commons.wikimedia.org/wiki/File:Image-Mount_Washington_Hotel.jpg
c. https://en.wikipedia.org/wiki/Atomic_bombings_of_Hiroshima_and_Nagasaki#/media/File:Atomic_bombing_of_Japan.jpg
d. https://en.wikipedia.org/wiki/Cold_War#/media/File:NATO_vs._Warsaw_(1949–1990).png
e. https://en.wikipedia.org/wiki/Headquarters_of_the_United_Nations#/media/File:North_by_ North west_movie_trailer_screenshot_(13).jpg
f. https://en.wikipedia.org/wiki/Jet_aircraft#/media/File:BOAC_Comet_1952_Entebbe.jpg
g. https://en.wikipedia.org/wiki/Colossus_computer#/media/File:Colossus.jpg
h. https://en.wikipedia.org/wiki/Bambi,_a_Life_in_ the_Woods#/media/File:Bambi_book_cover.jpg

Following the school of general and behavioural management, discussed in Chapter 2, quantitative management had already been developed before the Second World War, between 1930 and 1940. This management school was based on mathematics, especially statistics, resulting in rational decision-making based on what were believed to be reliable quantitative models. The approach was first developed for military problems and was originally known as *operational research.* Later its name changed to *operations research.* The approach provided important data for quantifying risks, and for supporting management decisions during planning and project monitoring (Moore 1968). In the early years, operational research was mainly applied to assessing the effectiveness of military operations, minimising costs and carrying out risk analysis. In 1936, the Bawdsey Research Station used operational research to assess the reliability of the Royal Air Force radar installations in preparations for a possible war against Germany. Later, other military operations were scrutinised, such as the massive carpet bombing of German cities at the end of the Second World War, based on statistical data analysis and probabilistic risk assessment of alternative operations. After the war, these techniques were applied in the private sector. Quantification and mathematical models obviously had both strengths and weaknesses. In terms of the latter, often not all relevant data were available and/or reliable. Also, policy decisions, which often involved politics and ethics, also consider factors that were not quantifiable.

In this period, four reference books on occupational safety, management of safety and prevention were published in the United States (Heinrich 1931, 1941, 1950a; Armstrong et al. 1945), while only one reference book came from the United Kingdom (Vernon 1936), and one from the Netherlands (Mesritz and Ree 1937). These publications presented the background of the quantitative management school.

United States

Heinrich's contribution

Heinrich (1886–1962), introduced in Chapter 2, was a leading figure in this period. He was the only author who remained active in the safety field until after World War II. Heinrich was very productive. In addition to reference books, he also produced nine influential papers during this period (Table 3.1). Heinrich contributed to safety in production from a position within a governmental organisation, the War Advisory Board, and the National Safety Council. Additionally, he created a number of numerical models and metaphors that were, and still are, popular in managing and understanding safety in many companies.

Heinrich's first reference book on occupational safety, *Industrial Accident Prevention: A Scientific Approach*, used the term 'scientific', as Taylor had previously done (Heinrich 1931). The term 'scientific' should not be taken too literally: 'scientific' referred to an approach based not so much on the sciences, but rather upon facts instead of beliefs. Heinrich worked as an assistant superintendent at

Table 3.1 Publications of Heinrich, 1931–1950

Reference Books, Year	*Articles, Year*	*Titles of Articles, Reference Books*
1931		*Industrial Accident Prevention: A Scientific Approach* (1st edition)
	1932	'The Safety Engineer Aids the Life Underwriter'
	1938	'Accident Cost in the Construction Industry'
	1938	'It's up to the Foreman!'
1941		*Industrial Accident Prevention: A Scientific Approach* (2nd edition)
	1942	'Men in Motion'
	1942	'The Foreman's Place in the Safety Program'
	1945	'Key Men in Industry: Part 1'
	1945	'Key Men in Industry: Part 2'
	1945	'Key Men in Industry: Part 3'
1950		*Industrial Accident Prevention: A Scientific Approach* (3rd edition)
	1950	'The Human Element in the Cause and Control of Industrial Accidents'

the Engineering and Inspection Division of Travelers Insurance Company. His reference book was basically a summary of previous publications and textbooks by other authors. The state of the art of safety technique was presented, and, like DeBlois, Heinrich supported a management approach. Furthermore, he made a clear distinction between causes and consequences of accidents. One of Heinrich's main contributions was his transformation of complex safety issues into quite straightforward numerical presentations, thereby making occupational safety more predictable. For instance, the ratios of accident costs in 1927 (4:1), the causes of accidents in 1928 (88:10:2) and the accident consequences in 1929 (300:29:1) have already been mentioned in Chapter 2.

Heinrich clearly explained the 4:1 proportion in accident costs in his 1931 reference book using cases and examples. Nevertheless, the background and therefore the justification of this ratio remain unclear. Heinrich did mention other ratios, for instance that of the construction industry, based on one case study, namely 4400:1 (Heinrich 1938a). On average, he argued, the ratio in construction was higher than 4:1, more like 1:6. The 4:1 proportionality, Heinrich stated, is a rule of thumb, proven to be approximately true in many cases in practice. He mentioned this ratio in each of his following reference books (Heinrich 1941, 1950a), indicating a strong belief in its validity. Insights were added by another insurance company, Lumbermens Mutual Casualty Co. While supporting the concept of a ratio for accident costs, Keefer's (1945) contribution to Armstrong's reference book argued that accidents should be reported in terms of accident frequency rates, or the number of disabling injuries per million hours worked (Armstrong et al. 1945). Despite Heinrich's method being crude and of unclear origin, it provided safety practitioners with a clear communication tool that emphasised the relation of safety with the costs of 'unsafety' – accident costs.

Interestingly, Heinrich used the proportionality rule differently during the Second World War. In one of his later papers, he mentioned as adverse effects not only costs but also lost (production) time. He aimed at the soul of the American war industry; accidents delayed weaponry production, which was totally unacceptable. He used strong language to make his point:

> *Stop this senseless slaughter and maiming of workmen, stop this unintentional sabotage of your production effort, and thus help in a smashing big way to WIN THE WAR.*
>
> (Heinrich 1942a, capitals in original)

The domino metaphor

Heinrich mainly concentrated on the psychological conditions surrounding accidents. This became obvious in his 88:10:2 distribution, which focused on the behavioural aspects of workers. These behavioural aspects were so-called proximate or immediate causes. These causes produced directly, with no intermediate agency, an effect, or an immediate result. He stated:

> *The committee (of Standardization of industrial Accident Statistics) adopted the following definition of proximate cause: that the accident should be charged to that condition or circumstance the absence of which would have prevented the accident; but if there be more than one such condition or circumstance, then the one most easily prevented.*
>
> (Heinrich 1931)

Using this definition, it is easy to interpret human psychology as the one and only cause of accidents, but Heinrich was aware that multiple factors were involved. In the first edition of his reference book, the workman was not the only one at fault; so too was the supervisor (Figure 3.1).

This is in contrast to the accident proneness theory, popular in Europe at that time, which exclusively blamed the victim.

Terms like 'unsafe acts of person' and 'man failure' did not appear in Heinrich's work until the second edition (Heinrich 1941), suggesting that he might have been influenced by British scientific views on accident causation.

The 88:10:2 distribution was considered an aberration even in Heinrich's time. Keefer (1945) had already mentioned that unsafe acts and carelessness were not primary causes of accidents. He cited alternative results, pointing out that accidents were always combinations of unsafe acts and mechanical causes. The National Safety Council claimed that 61% of accidents had mechanical causes, while the State of Pennsylvania even mentioned 95%. Keefer blamed industrial

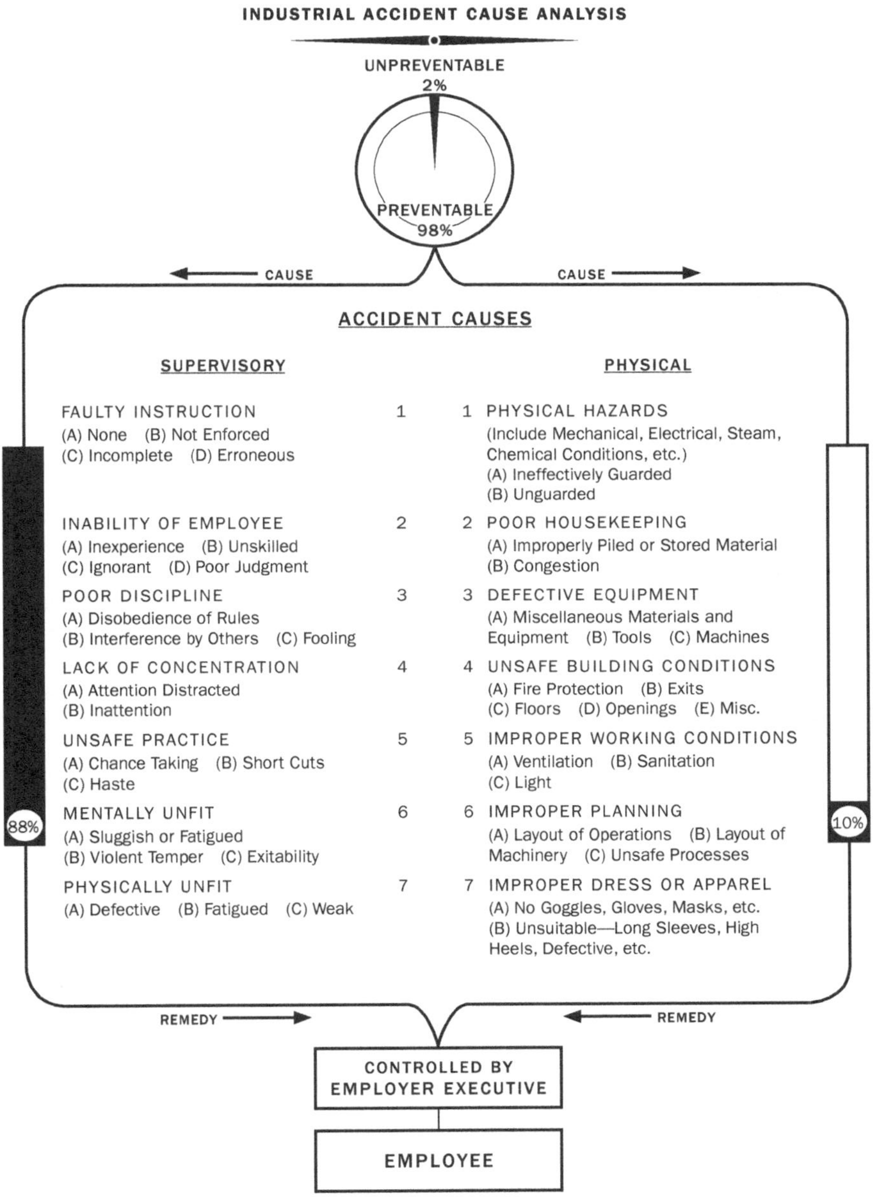

Figure 3.1 Causes of accidents

Source: Heinrich (1931)

executives and foremen for using Heinrich's distribution as a reason for blaming victims. Keefer considered interventions for controlling behavioural causes as secondary because the study of behaviour could not be considered an exact science. Confronting Keefer, Heinrich (1950b) proposed that psychology should be integrated into safety practices, because most errors were due to failures in supervision and human errors. He reasoned that a safety engineer could improve safety in four ways: (1) by applying technical solutions; (2) by using instruction and persuasion; (3) by improving the selection of workers and ergonomics; and (4) by employing disciplinary actions.

One of Heinrich's best-known metaphors is the falling dominoes, a visualisation of a generic accident process. The metaphor was first presented in the second edition of his book (published 1941), in which he developed the concept of a sequence of events into the dominoes in ten axioms of accidents:

1. *The occurrence of an injury invariably results from a completed sequence of factors – one being the accident itself;*
2. *An accident can occur only when preceded by or accompanied and directly caused by one or both of two circumstances – the unsafe act of a person and the existence of a mechanical or physical hazard;*
3. *The unsafe acts of persons are responsible for the majority of accidents;*
4. *The unsafe acts of a person do not invariably result immediately in an accident and an injury, nor does the single exposure of a person to a mechanical or physical hazard always result in accident and injury;*
5. *The motives or reasons that permit the occurrence of unsafe acts of persons provide a guide to the selection of appropriate corrective measures;*
6. *The severity of an injury is largely fortuitous – the occurrence of the accident that results in the injury is largely preventable;*
7. *The methods of most value in accident prevention are analogous to the methods required for the control of the quality, cost and quantity of production;*
8. *Management has the best opportunity and ability to prevent accident occurrence, and therefore should assume responsibility;*
9. *The foreman is the key man in industrial accident prevention;*
10. *The direct costs of injury, as commonly measured by compensation and liability claims and by medical and hospital expenses, are accompanied by incidental or indirect costs, which the employer must pay.*

The list of axioms was a mixture of findings from his earlier research and his personal insights into accident prevention. Axiom 10 represented findings from his earlier research on the hidden costs of accidents. Axiom 8 also contained important advice: management should take responsibility. This mixture of beliefs, advice and evidence made these axioms seem like personal opinions. Also note that at least four of the axioms related directly to the domino metaphor, namely axioms 1, 2, 4 and 6.

Figure 3.2 shows the dominoes in action: the sequence of events is stopped by removing one of the dominoes. The first domino is the social environment that workers live in and their hereditary traits.

FIG. 4. The unsafe act and mechanical hazard constitute the central factor in the accident sequence.

FIG. 5. The removal of the central factor makes the action of preceding factors ineffective.

Figure 3.2 Prevention of accidents by removing dominoes

Source: Heinrich (1941)

Heinrich thought that people with dubious hereditary or social backgrounds were more susceptible to accidents. The second domino is *fault of person*. By fault of person he meant character faults of individuals, such as recklessness, restlessness and contempt for safety, rather than an individual making a mistake. The third domino shows the direct cause of accidents: an unsafe act combined with mechanical or physical hazards. This does not always lead to an accident, but it is a necessary precondition. The fourth domino represents the actual accident. The fifth domino equals the fortuitous injury and it is impossible to predict how serious that will be. However, when the safety engineer reduced the number of accidents, the number of injuries was automatically reduced. The choice of removing the third domino, the direct cause of an accident, was a deliberate preventive action. This third domino was therefore the safety engineer's most effective way of lowering the number of accidents, as he had much less opportunity to influence the social environment, the workers' hereditary traits or poor personality traits. Heinrich published the dominoes only in his reference books and not in any of his papers. Apparently, he assumed that everyone knew his metaphor.

Heinrich's dominoes are still recognized by safety specialists today and referred to in present-day educational programs to illustrate the concept of consecutive events leading to accidents. This continued use attests to the importance of the work of early thinkers in the field of safety, with Heinrich at the forefront. There is no way of telling why Heinrich thought his ten axioms would cover the work area, nor of proving that his axioms cannot be broken down into smaller parts, a requirement for an axiom. Nor did he report his methodology for deriving the five distinct steps, and no data were provided supporting his metaphor. After DeBlois in 1926, Heinrich was the second author to postulate an accident as a sequence of events. This was the starting point for developing accident scenarios that later developed into fault trees and other linear accident models and metaphors, like the Swiss cheese and the bowtie metaphors (see chapter 8).

Most likely, Heinrich's use of the domino metaphor was not a random choice. In American politics, dominoes were popular from the 1930s until the 1950s (Ninkovich 1994). The metaphor was also used to describe the crash of the American banking system in the years 1931–1933. Also, in the arena of international politics with the acute strategic interdependence between allies, the rest of the coalition could fall as well if key strategic partners were to fail. The obvious utility of the domino metaphor is that it translates the *sequence of events*, a relatively abstract phrase, into an easily visualized and hence immediately understood picture.

The National Safety Council

Heinrich contributed to the war efforts for the Second World War and was appointed as chairman of the Safety Division for the War Advisory Board in 1942 (Manuele 2002). The appointment was channelled through the National Safety Council (NSC), the leading national occupational safety and road safety organisation. The council's reputation and achievements in safety prompted President Roosevelt to use it in the US war effort. He charged the NSC to:

> *[M]obilize its nationwide resources in leading a concerted and intensified campaign against accidents, and to call upon every citizen, in a public or private capacity, to enlist in this campaign and do their part in preventing wastage of human and material resources of the nation through accidents.*
>
> (National Safety Council 2008)

The role of the foreman

In the war years after 1942, Heinrich's papers emphasised the importance of the foreman as the key player in promoting safety and production in industry. These foremen were the heart of American industry and were perfectly positioned for promoting safety. Heinrich had already mentioned this in his paper in 1929 and in the first (1931) and later editions of *Industrial Accident Prevention*. In one of his papers he emphasised that 'it's up to the foreman', because the foreman is close enough to the work floor to understand the workmen and has time for coordinating safety activities (Heinrich 1938b). Safety activities included: collecting information from the workmen, creating safety awareness and enthusiasm, and coordinating safety meetings. In 1942, Heinrich published an article directly addressed to America's foremen (Heinrich 1942b). He explained that much was expected from the foreman, like meeting high production targets *and* promoting safety. A notable suggestion was that foremen should have safety meetings without the interference of higher management, so that they could speak freely amongst themselves.

In 1945, Heinrich produced three papers called 'Key Men in Industry' (Heinrich 1945a–c) in which he created a framework for supervisory industrial work, which was exclusively aimed at foremen. The framework also contained tasks for supervision. He stated his ambition to develop the framework in a similar way to what he had done for safety, but unfortunately the papers fell short of reaching that ambition.

Accident investigation, chance and effect

Only once did Heinrich make a link between safety and insurance. In his paper 'The SafetyEngineer Aids the Life Underwriter' (Heinrich 1932), he demonstrated how safety could benefit from cooperation with a causality engineer. The causality engineer knew which jobs were hazardous by using accident statistics. The safety engineer could use that knowledge to make those dangerous jobs safer. Heinrich gave an example of how statistical analysis could support safety interventions. A causality engineer identified a dry-cleaner's shop that was experiencing an exceptionally high number of fire-related accidents. The safety engineer stepped in and investigated, finding that cleaning was performed with petrol-like substances. These substances have low boiling points and correspondingly very low flashpoints, so combustible vapours spread easily. The safety engineer proposed using kerosene, which has a higher boiling point and a much higher flashpoint than ambient air and thus evaporates much less easily than commonly used solvents, such as petrol, whose flashpoint is much lower than ambient air. Subsequently, the number of fire accidents decreased at this shop. This is how Heinrich envisaged the cooperation between the causality engineer and the safety engineer. The safety engineer usually did not use statistics actively but could certainly learn from the statistical analyses of causality engineers.

This was an early call for the use of statistical analysis to improve safety. In 1926, DeBlois also wrote about accident probabilities and their relationship with hazardous situations. Although quantitative management propagated a probabilistic approach, the safety domain apparently was not yet ready to use this type of analysis in its mainstream safety activities. In future papers, Heinrich would no longer refer to statistical analysis. But through his axiom 6, about the preventability of accidents, he remained sensitive to probabilistic approaches.

The discussions on the probability of accidents by Heinrich and DeBlois fit logically in the quantitative management school, a fact that is often overlooked. But the fact that these early safety researchers recognized and acknowledged the importance of statistics for improving safety demonstrated their exceptional insight and their role in the development of safety principles and tools for accident prevention.

In 1950, Heinrich combined his knowledge of accident prevention into a framework for safe and efficient production. Heinrich's framework provided a method for analysing accident occurrence, thereby improving prevention. This framework combined all aspects of safety that had been developed in the post-war period and logically organized them into a practical safety management system for accident prevention (Figure 3.3).

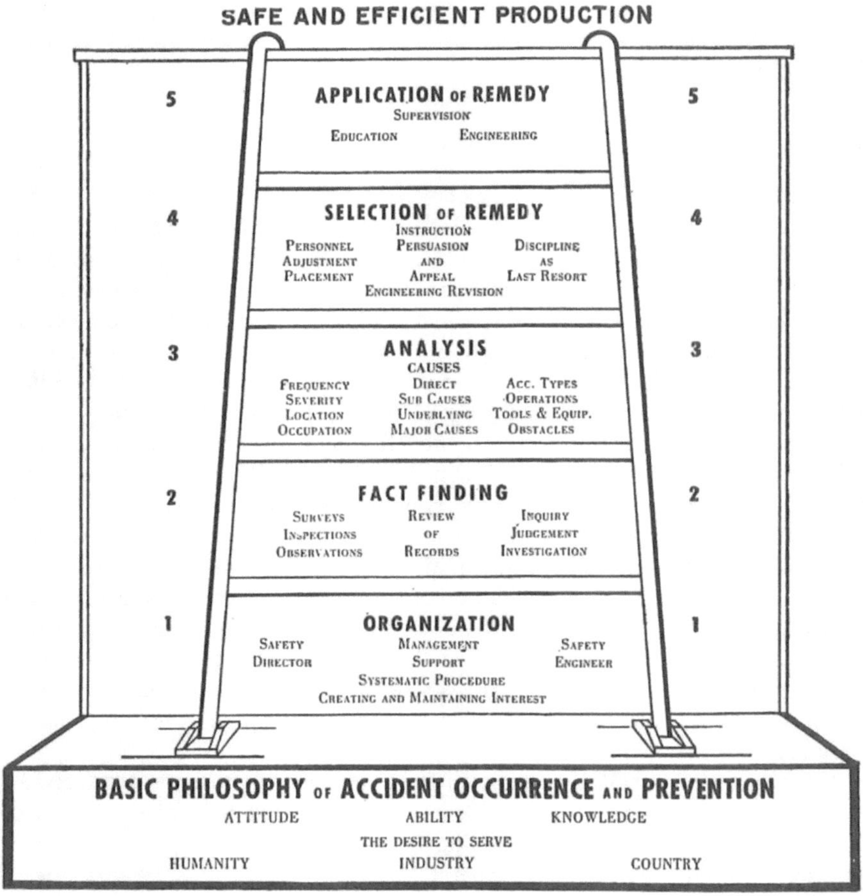

Figure 3.3 The 'safety ladder' metaphor of safety management

Source: Heinrich (1950a)

The framework has five steps: 1) organisation, 2) fact finding, 3) analysis, 4) selection of remedy, and 5) application of remedy. With the finalisation of this framework, Heinrich's reference book became a standard for the development of safety systems in any industry. In that respect, this framework was the pinnacle of his achievement. It was a concisely formulated, complete safety management system that contained many elements of modern safety systems.

However, the concept of a safety management system was not original. Other authors, such as DeBlois (1926) and Armstrong and colleagues (1945), had already described these concepts and the necessary steps to prevent the occurrence, or at least reduce the effects, of accidents in industrial settings. Heinrich provided his characteristic easy-to-understand pictorial representation. Again, it was impossible to validate the correctness of the contribution. It seems likely that Heinrich was in a position to develop and test this system while at the War Advisory Board. Unfortunately, he did not shed any light on the system's underpinning. Thus, lacking clear proof, his management model remained invalid in the eyes of science.

Criticism of Heinrich

Heinrich's strength was predominantly the practical applicability of his work. The Great War had just ended and left ruins all over Europe, the Spanish flu had killed more people than the Great War, the greatest financial crisis of modern times had taken place, accidents still killed thousands and another world war was upcoming. Also, accidents were no longer 'acts of God' but events that could be prevented, or at least dealt with, as has been shown in previous chapters.

These developments called for simple practical solutions that worked directly when applied. It seemed therefore logical that Heinrich wrote primarily for safety practitioners rather than for scientists and thus he refrained from scientific underpinnings. Heinrich gave safety researchers food for thought and tools for practice. He studied many concepts in regard to safety and presented them in recognisable ways, including the hidden costs of accidents, the foundation of a major injury, generic causes of accidents, the accident process expressed in terms of his domino metaphor and a framework for safe industrial production. Even today, Heinrich's dominoes are an essential part of safety instruction (see, for instance, Dutch MVK 2008).

Heinrich claimed to have used 75,000 accident reports from industry and insurance compensation cases to develop the 88:10:2 ratio and 50,000 accident reports for his 300:29:1 ratio (and an unknown number for the 4:1 ratio). Obviously, Heinrich had access to huge amounts of data, but he did not report his data in any of his publications. In Appendix 5 of the first edition of his book, some incident data were shown for fatalities and compensations in the late 1920s, but it is unclear whether he used these data for his analyses. It may have been impractical to publish large amounts of data. Data may also have been confidential, or perhaps Heinrich did not wish to present his work as science. Whatever the reason, the absence of his data and his lack of description of his methods makes it impossible to verify the findings leading to the ratios he developed, making his findings purely descriptive models and metaphors. This became clear in his explanation of his 300:29:1 ratio. In the first edition, Heinrich mentioned that the incidents had to have the same cause, in the second edition the incidents were similar and in the third they were of the same kind involving the same person. This ambiguity showed that this was the least tenable of his premises, but he still used it repeatedly in his texts to instruct foremen to look out for dangerous acts. The concept that accident consequences were fortuitous, that a universal distribution of consequences existed, and the lack of data made the absolute numerical values of Heinrich's ratios highly debatable.

The epidemiological triangle

After the Second World War, enthusiasm for the psychological approach to accidents began to fade. Accident-prone workers were difficult to define, and the long-term effects of interventions based on the psychological approach proved difficult to demonstrate (see Chapter 2). Despite the shortcomings of early accident proneness research, the concept of an accident-prone personality nevertheless became popular, as shown by the many references in the scientific literature (Anonymous 1931, 1931–1939). These early studies can be criticised, however, since much of the experimentation was devoid of personality measures, and studies that used personality measures were open to subjective interpretation. Many of these study designs can be considered a shotgun approach – that is, a large number of variables were measured without any theoretical basis for their inclusion in the study. Under these conditions it can be expected that significant results will occasionally be obtained simply on the basis of chance alone. It therefore seemed more effective to make machines and installations fail-safe, so that the mistakes of workers would not lead to accidents, rather than to educate workers in safety matters with the help of behavioural programs and safety posters (Figure 3.4).

When medical professionals became interested in the domain of safety, they introduced the epidemiological triangle, which gave a new impetus to safety research. This epidemiological model focused on interactions between a victim, the host and situational variables, these being the physical and socio-economic environment, and an agent, this in the form of energy (Figure 3.5). This model circumvented the conceptual and theoretical weaknesses of psychological research, which was inclined to be largely descriptive and failed to focus on the aetiology of accidents and the exposure to hazards.

The epidemiological triangle was a model primarily meant to control the spread of infectious diseases that had ravaged Europe since the nineteenth century. The model was based on the assumption that by stopping one of the vectors of the triangle, the spread of disease could be stopped (see, for example, Johnson 2006). The fight against cholera, in which drinking water was seen as the main vector, was the example given. During this period, the real agent, the cholerae bacteria, was still unknown, as was the underlying disease process (Pinwell 1866) (Figure 3.6).

The approach based on the epidemiological triangle model was successful in controlling cholera and was also applied to safety management problems. Gordon (1949) was the first to introduce this approach. In safety science the host was the victim. The agent was energy in all its forms, energy constituting the key agent for causing the injury. The surroundings were extrinsic factors, influencing the agent and the possibilities of exposure, including physical, biological and socio-economic factors.

United Kingdom

Accidents and their prevention

The title of Vernon's reference book *Accidents and Their Prevention* (Vernon 1936) refers to an engineering approach. But Vernon (1870–1951) was not an

Figure 3.4 Examples of American safety posters, 1936–1937

Source: Arxiu Nacional de Catalunya (1987)

engineer. He had academic qualifications in chemistry, physiology, biology and medicine. In 1915, he had worked in a munitions factory at Birmingham, which had prompted his interest in occupational safety. Vernon joined the Health of Munition Workers Committee, as well as its later successors. His work was remarkable, not only because industrial and mine safety was the topic of his book, but also because it covered the transport sector, including road transport, and additionally safety at home. Vernon supported his approach to safety with extensive

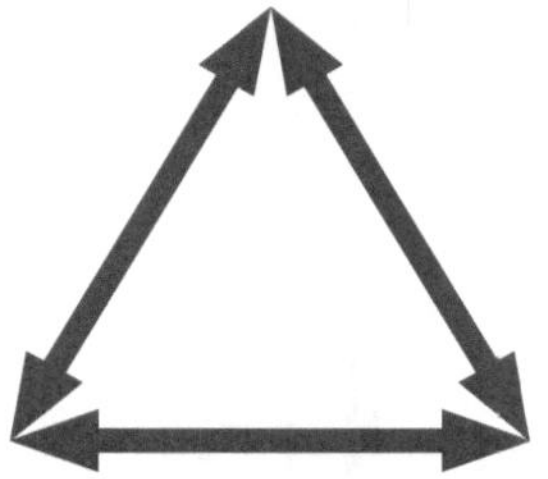

Figure 3.5 The epidemiological triangle

Source; After Gordon (1949)

data from his own research and with data from the British Labour Inspectorate. The contribution to fatal accidents of these last two sectors, transport and home, was much higher than that of industry. Of the 17,000 registered fatalities in 1932, 43% related to the transport sector, a third were domestic and only 9% related to industry. Table 3.2 summarizes major causes of industrial fatalities, according to statistics of the British Labour Inspectorate.

The environmental causes of accidents were also an important theme in Vernon's research, and his book showed the influence of temperature, worker fatigue, speed of production, insufficient ventilation and alcohol consumption (Osborne et al. 1922; Vernon 1919, 1920; Bedford 1951; Smith 1951).

Vernon referred to Marbe's and Heinrich's work, and to the accident proneness theory. But Vernon was rather critical of aesteto-kinetic tests developed by Farmer (see Chapter 2). For factory workers these tests were not reliable, while for road and rail workers they were. Furthermore, the individual hypothesis used the concept of homogeneity, that is, both for the groups studied and for exposure to hazards and risks. This concept was, however, disputed, and it was very unlikely that workers with the same job would encounter the same hazards or risks. This argument touched on a rather fundamental point. Only one factor was seen as an explanation for accident causality: personality. However, none of the research reports on accident-prone workers provided any contextual information (Vernon 1936). Vernon still believed that the human factor was a major contributor to accidents, but compared to the complexity of human factors, mechanical defects were relatively easy to address. Thus, Vernon's work provided detailed information on technical solutions for safety. As previously mentioned, the influence of human behaviour was much more complicated than the machinery used at the time, and the organisation of work should have been a starting point for prevention, more specifically the safety committees, as

Figure 3.6 Death's dispensary

Source: Pinwell (1866)

introduced in the United States. These safety committees should have kept the interest in safety alive amongst managers and workers, and should have trained workers in safe working methods, and should have conducted safety inspections on the shop floor. This last point was particularly important to Vernon, his

Table 3.2 Fatal accidents in industry in 1932 (after Vernon 1936)

Important groups of accidents (1932) in % of total (n = 602)		
1	Falling from heights	35
2	Hit by machine	14
3	Hit by falling object	11
4	Cranes	10
5	Hit by moving parts	7
6	Hot metal	3
7	Hit by object	7
8	Hit by tools	1
	Other causes	13

argument being that legislation without any form of inspection hardly made any sense at all.

The Netherlands

Limited knowledge development

While in the United Sates and the United Kingdom the development of knowledge of the safety domain flourished, the Netherlands was a different story. The only substantial contributions during this time frame were two dissertations. The physician Ter Borg obtained a doctorate at the University of Utrecht on research at Hoogovens (a Dutch steelworks) (Borg 1939). With the use of questionnaires, he obtained information on accident factors at Hoogovens and 15 other Dutch companies, including Philips (lightbulbs), Wilton (shipyard) and the City of Amsterdam (public works). He supported the accident proneness theory, referring to Greenwood and Woods' study as well as that of Marbe. The other doctoral dissertation was by the psychologist Herold, who defended his work at Nijmegen University. Herold investigated 66 miners working in the south of the Netherlands who had worked for at least five years at one of the Dutch mines (Herold 1945). He used the aesteto-kinetic test battery that was previously developed by the British research group of Farmer et al. The relationship between accidents and test results was not presented in this study, but the opportunities to positively influence the innate or acquired accident disposition through company training were. The notion that a lack of sufficient attention by workers was the main determinant of accident processes had also been presented in a number of previous publications, among them a Dutch textbook for secondary technical education (Mesritz and Ree 1937), articles in the trade press (Kraft 1950; Hart 1950) and in Ter Borg's dissertation.

Ample attention was given to accident proneness and to safety education for workers. The *Dutch Safety Journal* published a series of articles on this topic, referring to German and French safety literature until halfway through the 1930s, and from 1936 onwards referring mostly to British and American publications (Anonymous 1928a, 1928b, 1936, 1937a–c, 1940b; Copius Peereboom 1941;

Gorter 1935, 1946; Patijn 1945, 1946; Sievers 1941; Steiner 1939; Young 1946). But there was also a focus on design-related aspects of safety (Copius Peereboom 1941; Gorter 1946; Winkel 1936), and on the American slogan 'Safety Pays Off', which manifested in efficient and undisturbed industrial production (Anonymous 1939a, 1939b; Winkel 1936).

Safety Museum

The Safety Museum had a very practical approach to occupational safety. This was emphasised once again in 1933, when the director-general of Labour brought together all safety-related bodies: the Labour Inspectorate, public and large private insurers, employers' and employees' organisations, and the Labour councils. The Safety Museum played a central role and started registering occupational accidents, yielding an estimated 100 lost-time accidents per 1,000 workers occurring in the Netherlands every year. That number rose to 115 in 1939 and 157 in 1947 (Redactie 1999; Bus and Swuste 1999). These relatively detailed accident registration data provided information for improving safety posters (Figure 3.7).

Some posters presented general messages such as 'Work with care', linking this phrase either to family life or, more specifically, to your mother ('Think of Mother'). Other posters highlighted safety messages for specific machines or specific activities, like welding or working with lathes. The Labour Inspectorate was particularly concerned about accidents involving electricity, a subject of many safety posters. Industrial medical care was also developing. In the 1930s, the first occupational health services were created in larger companies. From the start, early tuberculosis detection was the core of the activities of these health services, which used slogans like 'The small cost improves the great benefit' (Burdorf et al. 1997). Safety committees in industry, on the other hand, were not mandatory and, in practice, did not flourish.

Safety inspectors

After the war, the Labour Inspectorate discussed a plan to close the gap in companies' safety provision by appointing safety inspectors in companies. In 1947, the Working Group of Safety Inspectors started, which led, after several name changes, to the current Nederlandse Vereniging voor Veiligheidskunde (NVVK; Dutch Society for Safety Science) (Desta and Heuvel 1987). This working group started with the first training of safety experts. After the war, the discussion about safety committees and safety services also started cautiously again. The opinions of employers and employees about safety commissions were still divided. The Dutch government, with the director-general of Labour as their mouthpiece, strongly supported these safety committees and services, which were meant to be managed by a safety inspector or engineer who:

> *must be the person who makes the work interesting for the committee and protects it from being overweight and from embarking on paths that are not*

Figure 3.7 Dutch safety posters 1931–1942

Source: University museum University of Amsterdam

1931 (There is a danger of such lampholders, and not of safe lampholders)

1939 (If your family is sacred, work safely)

1939 (Weld safely)

1940 (Unsafe stamping presses, hydraulic press stances, etc., 546 accidents per year

1942 (Work safely! Think of Mother)

its own. From that point of view, safety commissions would only be appropriate in companies where a safety service exists.

(Desta and Heuvel 1987)

This chapter has described the development of safety in the United States, the United Kingdom and the Netherlands in the 1930s and 1940s and discussed how several basic concepts were introduced into safety by technicians and physicians. The United States was dominant in this period. These concepts include, firstly, that the occurrence of accidents can be viewed as a process; secondly, the central role of the foreman; and thirdly, the development of an important model for safety management. As regards the third, Heinrich's domino metaphor for the accident process was a particularly influential model. The two different hypotheses on the occurrence of accidents, the environmental hypothesis and the individual hypothesis, remained opposites, both in the United States and in the United Kingdom. The Netherlands followed developments from the United States and the United Kingdom, and the emergence of safety committees in Dutch companies started slowly. At the end of the 1940s, in the aftermath of the Second World War, the times became especially ripe for prevention and technology, as the next chapter will show.

4 Prevention, behaviour and the makeable man

1950–1970

This chapter focuses on the 1950s and 1960s and also describes the scaling up of the process industry. The 'modern management approach', which originated in post-war Japan, also emerged in this period. The growing scope of the safety domain in the United States included the prevention of property damage, and developed a human factor approach. New models and safety techniques were developed, both in the United States and in the United Kingdom. In these and many other countries, and also in the Netherlands, a new approach, ergonomics, had a major impact on the development of safety science. A new safety theory, 'task dynamics', emerged in the Netherlands. Although the concept of risk was not used broadly, confidence in technology rose to great heights in this period, and in occupational safety more and more attention was being paid to prevention.

DOI: 10.4324/9781003001379-4

In the 1950s communism challenged the world.

The 1950s truly became the **golden age of television**. As the number of households with TVs multiplied, many of the genres that today's audiences are familiar with were developed: westerns, kids' shows, situation comedies, sketch comedies, game shows, dramas, news and sport programmes. In the 1950s, American parents were scared of the polio epidemics that threatened their children each summer. When a polio vaccine was licensed in 1955, the country celebrated and **children's polio vaccination campaigns** were launched. In the US, following a mass immunisation campaign, the annual number of polio cases fell from 35,000 in 1953 to 5,600 by 1957. The **European Union** (EU) traces its origins to the European Coal and Steel Community (ECSC) and the European Economic Community (EEC), established respectively in 1951 and 1957. EU policies ensure the free movement of people, goods, services and capital within the internal market, enact legislation in justice and home affairs, and maintain common policies on trade, agriculture, fisheries and regional development.

In 1957, the Soviet Union launched the first artificial earth satellite, **Sputnik 1**, into an elliptical low Earth orbit. The satellite's unanticipated success triggered the Space Race, a part of the Cold War. The launch was the beginning of a new era of political, military, technological and scientific developments.

Ben-Hur, a 1959 American epic historical drama film, had the largest budget as well as the largest sets built of any film produced at the time. The film was nominated for 12 Academy Awards and won an unprecedented 11. The implications of the story, the relationships among the characters and the feuding forces within the society of the time have many parallels in life today.

In 1955, in Montgomery, Alabama, Rosa Parks rejected a bus driver's order to relinquish her seat in the 'coloured section' to a white passenger, as the whites-only section was full. Parks became an international icon of resistance to **racial segregation**. It was the start of the civil rights movement.

The 1953 **North Sea flood** was caused by a combination of a high spring tide and heavy storm. It struck the Netherlands, Belgium and the United Kingdom. After the major flood, the Dutch governments conceived and constructed an ambitious flood defence system beginning in the 1960s called the Delta Works.

Many consider the **transistor** to be one of the greatest inventions of the twentieth century. The first mass-produced transistor radio was released in 1957, leading to the mass-market penetration of transistor radios. The success of transistor radios led to transistors replacing vacuum tubes as the dominant electronic technology.

Golden age of television

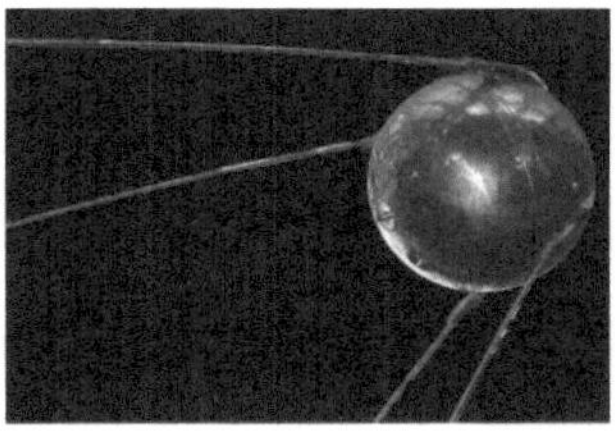

Sputnik 1

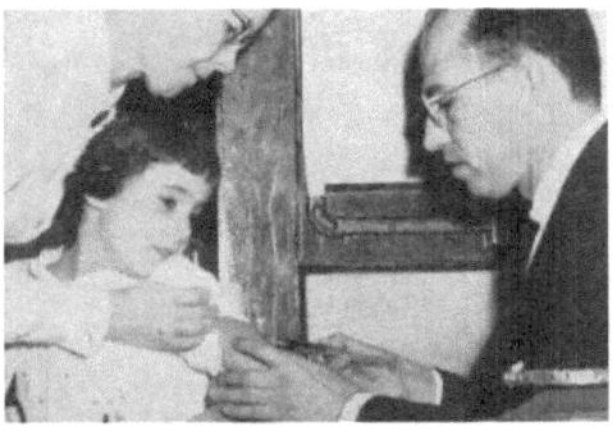

Children's polio vaccination campaigns

Ben-Hur

Civil rights movement

European Union

North Sea flood

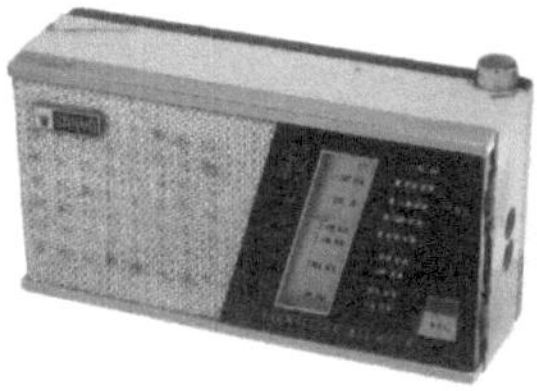

transistor radio

Sources:

a. https://en.wikipedia.org/wiki/Color_television#/media/ File:A_Colour_Television_Test.jpg
b. https://en.wikipedia.org/wiki/Sputnik_1#/media/ File:Sputnik_asm.jpg
c. https://en.wikipedia.org/wiki/Jonas_Salk#/media/ File:Salk-child-Karsh.jpg
d. https://commons.wikimedia.org/wiki/File:Ben_ hur_1959_poster.jpg
e. https://en.wikipedia.org/wiki/Rosa_Parks#/media/ File:Rosaparks_bus.jpg
f. https://nl.wikipedia.org/wiki/Marshallplan#/me¬dia/Bestand:Marshall_Plan_poster.JPG
g. https://en.wikipedia.org/wiki/North_Sea_flood_ of_1953#/media/File:Watersnoodramp_1953.jpg
h. https://en.wikipedia.org/wiki/Transistor_ra¬dio#/media/File:Sanyo_Transistor.jpg

The 1960s was a decade of crisis and innovations in which a new social order emerged.

The 1960s saw the height of the **civil rights movement**. In 1963, hundreds of thousands of participants converged in Washington, DC, for the March on Washington for Jobs and Freedom, demanding an end to racism and bigotry against African Americans. Martin Luther King delivered his iconic 'I have a dream' speech during the largest political rally for human rights ever seen in the United States.

Two years later, the **Vietnam War** escalated as the US increased its involvement in the war through bombing campaigns and by committing hundreds of thousands of US ground troops to the fight. US involvement in South-East Asia was driven by a domino theory: the fall of one country to communism would result in surrounding countries succumbing to communism.

The **Woodstock festival,** in 1969, was a pivotal moment in popular music history as well as a defining event for the counterculture generation. As one of the biggest rock festivals of all time, it proved a cultural touchstone for the late 1960s. The phrase 'the Woodstock generation' became part of the common lexicon.

The construction of the **Berlin Wall** by East Germany began in 1961. When it was complete, a guarded concrete barrier divided Berlin physically and ideologically. A well-known crossing point between the East and the West was called Checkpoint Charlie. The wall made West Berlin an isolated exclave in a hostile land. Western powers portrayed the wall as a symbol of communist tyranny, particularly after East German border guards shot and killed would-be defectors.

President John F. Kennedy proposed a national goal in 1961: landing a **man on the Moon** and returning him safely to Earth. In 1969, the Apollo 11 lunar module *Eagle* landed on the Moon. Armstrong's first step on the Moon was broadcast on live TV to a worldwide audience. The astronaut described it as 'One small step for man, one giant leap for mankind'. America had won the 'space race'.

In 1966, Mao Zedong launched the **Cultural Revolution** to preserve Chinese communism by purging the remnants of capitalist and traditional elements from Chinese society. The revolution damaged China's economy while tens of millions of people were persecuted and killed. Temples, churches and mosques were closed down or destroyed.

Kennedy did not witness this success as he was fatally shot in 1963. The **assassination of John F. Kennedy** evoked stunned reactions worldwide. As the incident took place during the Cold War, it was at first unclear whether the shooting might be part of an attack upon the United States. The news shocked Americans and left a lasting impression on people around the globe.

The Sound of Music is a 1965 American musical drama film. It became the highest-grossing film of all time and held that distinction for five years. The popularity of the film has not dwindled and it is still broadcast throughout the world.

civil rights movement

Vietnam War

Woodstock festival

Berlin Wall (checkpoint Charlie)

Man on the Moon

Cultural Revolution

Assassination of John F. Kennedy

The Sound of Music

Sources:

a. https://en.wikipedia.org/wiki/I_Have_a_Dream#/media/File:Civil_Rights_March_on_Washington,_D.C._(Dr._Martin_Luther_King,_Jr._and_Mathew_Ahmann_in_a_crowd.)_-_NARA_-_542015_-_Restoration.jpg

b. https://en.wikipedia.org/wiki/File:DakToVietnam1966.jpg

c. https://en.wikipedia.org/wiki/Woodstock#/ media/File:Swami_opening.jpg

d. https://nl.wikipedia.org/wiki/Berlijnse_Muur#/media/Bestand:Bundesarchiv_B_145_Bild-F079005–0021,_Berlin,_Grenzübergang_ Checkpoint_Charlie.jpg

e. https://en.wikipedia.org/wiki/Moon_landing#/ media/File:Buzz_salutes_the_U.S._Flag.jpg

f. https://en.wikipedia.org/wiki/Cultural_Revolution#/ media/File:1967–11_1967年_毛泽东接见红卫兵油画.jpg

g. https://en.wikipedia.org/wiki/Assassination_of_John_F._Kennedy#/media/File:JFK_limousine.png

h. https://en.wikipedia.org/wiki/The_Sound_of_Music_(film)#/media/File:The_Sound_of_Music_Christopher_Plummer_and_Julie_Andrews.jpg

United States

Modern management

The modern management approach focused on information processing and decision-making. This approach became increasingly important after the Second World War. Managers had to plan, organise, manage and supervise to ensure high production levels while maintaining high levels of quality and safety. Business organisations were open systems – no longer seen as independent entities but perceived to be interacting within a commercial environment with internal and external stakeholders. The American statistician William Deming and the businessman Joseph Juran were pioneers of this approach, the introduction of quality control being their main achievement, with employees and customers playing a major role (Pindur et al. 1995).

Quality control, product versus process

Just after the Second World War, the Americans Deming and Juran revolutionized the quality assessment of industrial products in American-occupied Japan by shifting assessment from the finished product to an earlier stage, the production process (Deming 1982; Juran 1951; Juran and Barish 1955). America supported Japanese industry in order to prevent the then economically weakened country from becoming vulnerable to communism. For that purpose, Deming worked in Japan during the Allied occupation. The Marshall Plan had a similar goal for Europe (Leitner 1999; Judt 2012). According to Deming, weak management was the pivotal problem causing the low quality of Japanese products. Deming used a simple but effective mantra for motivating Japanese managers:

> *If management focuses on quality, costs will drop; but if the focus is on cost, quality drops.*
>
> (Anonymous 2010)

Both Deming and Juran played an important role in rebuilding Japanese industry. Deming's work was based on earlier collaborations with Walter Shewhart at Bell Laboratories in New York. As early as 1939, Shewhart and Deming published their 'statistical process control', or SPC, to facilitate understanding and to minimise production process variations (Shewhart 1931; Shewhart and Deming 1939; Greisler 1999). Two kinds of process variations occurred: relatively small ones, occurring regularly due to common variations in the production conditions, like temperature variations; but large process variations could also occur, forcing a process to reach, or even exceed, its tolerable limits. These large variations rarely followed a normal distribution, as small variations did. Large variations occurred less frequently, but these could be potentially highly disruptive to the process, such as faulty machine settings or extremely low-quality raw materials. With statistical process control, process variations could be better predicted and therefore corrected before actual products were produced, marking a major step forward in the quality control of products. SPC was first introduced in manufacturing, and later applied in processes with quantifiable outputs. Deming's and Juran's contribution was the initiation of Total Quality Management (TQM), which became a vital part of the modern management school. TQM would much later, in the 1980s, become an important management school in the Western world. These principles of quality control in industry had a major impact on safety management (Burggraaf et al. 2020).

Four major schools emerged. Some have been discussed in previous chapters: classical and scientific management, behavioural management, quantitative management and finally modern management, including Total Quality Management. These four schools were not mutually exclusive for a certain period of time, and coexisted (Figure 4.1).

The latter days of Heinrich

After 1950, Heinrich published only three more papers, all in the *National Safety News* (Table 4.1). He produced no new insights into industrial safety but discussed home safety (Heinrich 1951). Heinrich was, however, drawn into a dispute with Blake about the 88:10:2 ratios (Heinrich and Blake 1956). Blake challenged this ratio and argued that fixed numbers reduced the vigilance of safety workers as it inhibited their initiative for starting safety investigations, which could lead to ineffective behaviour-correcting programmes. In his reply, Heinrich stated defensively that no one else had repeated his investigation, thus no one had corrected the numbers. Heinrich concluded that his numbers remained valid. That year, Heinrich published his last paper (Heinrich 1956), advocating dedicated safety education in schools and universities. He aimed to raise the status of the industrial safety specialist into a recognised profession.

Heinrich's last paper befitted a man who had dedicated his life to developing the safety domain. That same year, 1956, Heinrich retired at the age of 74 (SHHFI 2008). Heinrich's concepts and his domino metaphor were valued greatly in the

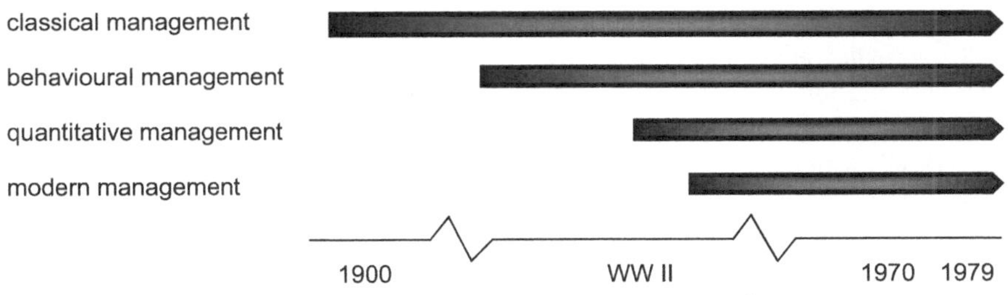

Figure 4.1 Leading management schools in the Western world

Source: Swuste et al. (2016)

Table 4.1 Heinrich's publications, 1951–1980

Books	*Articles*	*Titles*
	1951	The Safety Engineer and Home Safety
	1956	Recognition of Safety as a Profession: A Challenge
	1956	The Accident Cause Ratio – 88:10:2
1959		Industrial Accident Prevention: A Scientific Approach
1980		Industrial Accident Prevention: A Safety Management Approach

literature. For instance, in 1957, Stockdale summed up Heinrich's influence as follows:

> *One of the fundamentals of successful accident investigation is a knowledge of the chain of events that produce all accidents. Undoubtedly, we are all familiar with Mr Heinrich's accident sequence. Briefly stated it is this: 'In every accidental occurrence, there is always a chain of events that occurs in a logical and fixed order. Each link in the chain is dependent upon the preceding link'.*
>
> (Stockdale 1957)

This statement showed that the domino metaphor was not a rock-solid concept, but rather a generally valid process metaphor describing accidents as resulting from a sequence of events. Ruddick also referred to the role of the domino metaphor in incident investigation, an important part of safety management. Ruddick thus stressed the importance of Heinrich in the post-war safety domain:

> *One safety leader assembled the combined thinking of those daring souls, with a considerable amount of his own individual and original thought. H.W. Heinrich of the Travelers Insurance Company will be remembered for his book entitled* Industrial Accident Prevention.
>
> (Ruddick 1957)

The fourth edition of Heinrich's 1959 reference book was mainly a reprint of the 1950 edition (Heinrich and Granniss 1959). The fifth edition, published in 1980, was, however, a completely revised book. Listing Heinrich's basic concepts, this edition also presented a complete review of Heinrich's metaphors. Furthermore, the last edition was subtitled *A Safety Management Approach* (Heinrich et al. 1980), which also stressed the utility of Heinrich's concepts in modern times (post-1980). This 1980 publication was the last bearing the name of Heinrich, who had died 18 years earlier, in 1962.

While the first publications on managing occupational safety appeared just after the Second World War, the term 'safety management' was introduced only some 20 years later. Heinrich's 1950 reference book presented fundamental aspects of accident prevention, which he graphically represented as a ladder, this being a metaphor for a management system (see Chapter 3). Heinrich stressed the exemplary role of the manager in achieving safe and efficient production. This message had already been underscored in the first edition of a publication by Blake and his co-authors five years earlier (Blake et al. 1945).

Damage control

Damage Control by Bird and Germain (1966) also used the domino metaphor, with its focus on unsafe acts as being the primary causes of accidents (Figure 4.2).

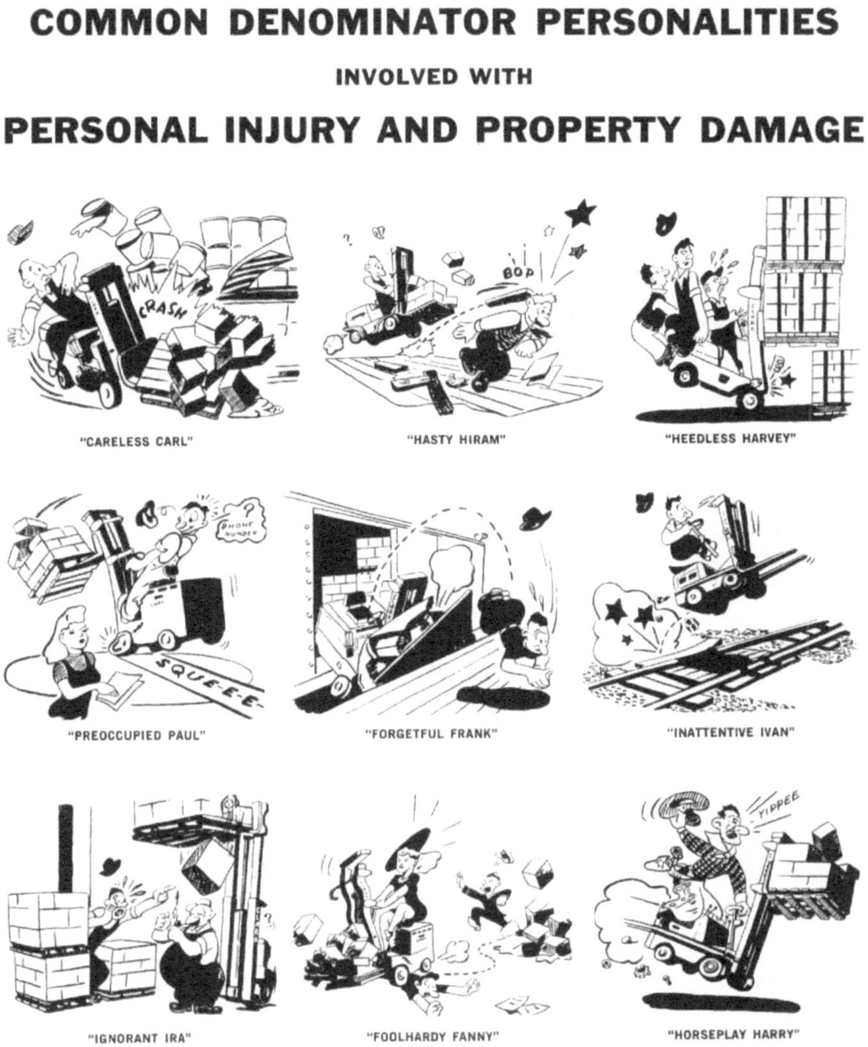

Figure 4.2 Unsafe acts and damage control

Source: Bird and Germain (1966)

The extent of consequences was, however, widened, from the accident process to injury, to near accidents and to damage (Figure 4.3). Both Bird and Germain had worked at the Lukens Steel Co. in Coatesville, Pennsylvania, where they investigated 90,000 accidents with injuries and damage to property (e.g. objects such as equipment) between 1959 and 1965.

Until then, American professional safety literature had not addressed property damage. Accepting Heinrich's ratio on the accident mechanism (300:29:1), Bird and Germain concluded that a large proportion of accidents without injury and near misses could nevertheless cause considerable damage. The costs of property damage caused by accidents were generally higher than the costs of injuries to workers. It was therefore necessary to prevent accidents because the accident mechanisms leading to damage and those leading to injuries were similar. This conclusion led to the creation of the 'damage iceberg', which had ratios differing from those of Heinrich's iceberg, as illustrated in Figure 4.4.

An effective damage control programme thus required a strategy that included accident and damage reporting, work preparation, auditing and cost calculations. The books by Bird and by Bird and Germain gave extensive examples of the kinds of forms used for these reports (Bird and Germain 1966).

Criticism of the psychological explanation of accidents

Many safety publications of the US National Safety Council were based on research funded by insurance companies. In these studies, the costs of safety interventions were always an important topic, as was the influence of psychological factors in the occurrence of accidents. Accidents were primarily seen as an adverse consequence of human behaviour. The justification for these beliefs came from modern psychology, an emerging field that devoted attention to the

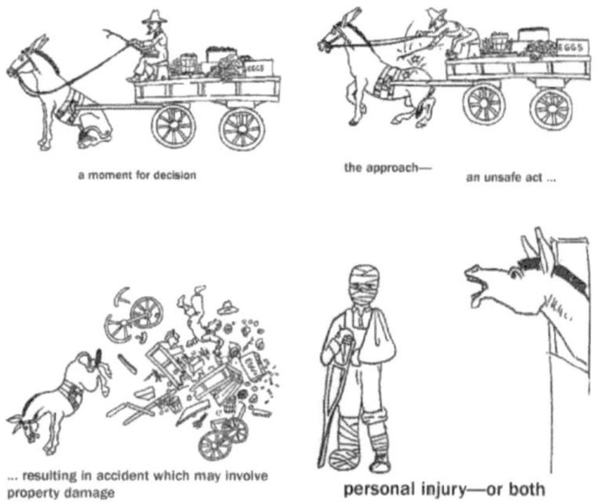

Figure 4.3 Accident process, injury and damage

Source: Bird and Germain (1966)

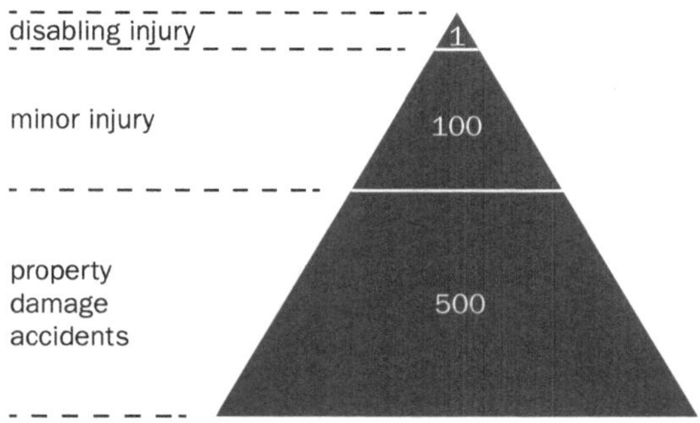

Figure 4.4 The damage iceberg
Source: After Bird and Germain (1966)

non-Freudian reasons for and backgrounds to human behaviour. This approach dominated safety research until the outbreak of the Second World War and was in line with the popular behavioural management school at that time. According to Guarnieri (1992), this justification lasted so long for these reasons:

- *Based on psychological insight, accidents are avoidable. At that time this argument appealed to insurance companies and industry as it fit in with their desire to reduce costs by improving accident prevention;*
- *Psychologists explain that humans and poor training are the causes of accidents and that the selection of workers, education and training are the keys to accident prevention;*
- *The popularity of a 'blaming the victim' attitude in a culture in which risk takers are rewarded and everyone is deemed responsible for his or her own actions and for the consequences, like accidents.*

After World War II, enthusiasm for the psychological approach to accidents began to fade. Long-term effects of interventions based on the psychological approach were difficult to demonstrate. The study designs used were called shotgun approaches, as previously mentioned in Chapter 3. The distribution of accidents in a group of people in a period of time normally followed a J-curve. In a probability distribution, it is to be expected that some people will experience an above-average number of accidents, while others will experience far fewer accidents. Most probably, when the exposed population is large, like a few hundred workers, and, in particular, when the survey period is short, limited to weeks or a few years, the number of accidents will be lower than the number of workers. One would expect that a limited number of workers would experience a large number of the accidents that occurred, when no psychological factors whatsoever were involved. The length of the survey period was one of the main criticisms. Also, the research was retrospective by definition, and based on companies' internal accident reports, whose data reliability was uncertain. Another argument was the vagueness of the definitions used. An accident-prone worker was difficult to define in scientific terms, and this problem became apparent in the low correlations of the results of different tests with accident frequency. The focus was only on the consequences of accidents, bodily harm and the psychological stability of the victim, and not on different accident scenarios or differences in exposure to hazards. These critical comments appeared in the scientific press just after the Second World War, in both American and English sources, and these publications discredited the accident proneness theory. It seemed more effective to make machines and installations fail-safe, so mistakes made by workers would not lead to accidents, rather than to educate workers in safety matters with the help of behavioural programmes (Arbous and Kerrich 1951; Anonymous 1952, 1964; Webb 1955; Froggatt and Smiley 1964; Hale and Hale 1970). In contrast to occupational safety, the individual road safety hypothesis had much support, with evidence from several countries on the existence of accident-prone road users (Shaw and Sichel 1971). But despite the shortcomings of early accident proneness research, the concept of an accident-prone personality appealed to many safety experts in companies and organisations (see, for instance, Fraser 1951; Hale and Hale 1970).

The hazard-barrier-target model

After the Second World War, US medical professionals become interested in accidents. With some disbelief they noted that, in stark contrast to its critical role in improving public health, medical science had failed to reduce accidents substantially. Indeed, improvements in medical science, and other related factors, such as better medical care, the use of antibiotics, clean water supplies, pasteurisation of milk, vaccination programmes, improved surgical procedures and the general improvement in living conditions, had led to huge improvements in public health and to large reductions in mortality in the general population. From a medical standpoint, it therefore seemed perfectly logical to improve safety using a medical approach, which had been so successful in improving human health. Why, then, in contrast to the success of programmes that dramatically improved public health, were the results of programmes to reduce accidents so disappointing? There are, however, a number of reasons, related to the quality of safety research and the complexity of accidents, that provide some answers to this medically oriented question.

Medical research programmes have often been extensive, and diagnoses and therapies are usually tested and evaluated thoroughly. However, compared to many diseases, accidents are obstinate: they are often neither easily controlled nor easily preventable. Only a few laboratories tested safety technique measures. Many of these safety measures were based on a common-sense approach and their effectiveness was not tested scientifically. Accident investigations were straightforward and generally these analyses did not extend beyond immediate causes, typically stopping at human error, unsafe acts or technical reasons (Haddon 1968). Specific hazards were identified, leading to technical adjustments and protection of workers. For instance, rotating parts of machines were enclosed, and protective measures were gradually introduced. In the literature, however, the quality of safety research, in which legal or psychological concepts often prevail, was heavily criticised. Often other causes were not considered, and only the accident itself and the people directly involved were part of the investigation, thereby making accident registrations relatively useless as a source for research. Unsurprisingly, the reduction in accidents virtually came to a standstill. The questions commonly asked were:

Why was research of accidents so obstinate?
Why did results of research have so little impact on accident prevention?

The first answer was that, although safety-related literature was extensive, its quality was generally poor. According to a number of researchers, most of it was not even worth considering (Gordon 1949; Haddon 1963; McFarland 1963).

The second answer was that accidents are complex. The literature presented a number of reasons to explain this lack of progress. For most of the important diseases at the time, there was a clear causative agent, a source, such as bacteria or other pathogens. Without the presence of these agents, there was no disease. This was different for accidents. For a long time, causes of accidents were indistinguishable from their consequences. Amongst researchers this created confusion and despair (Gordon 1949). Furthermore, the recognition of several factors that contributed to accident causality led to a misguided assumption that no common factors were identifiable, and that there was probably no relationship between accidents and injuries (McFarland 1963). Despite the occurrence of many thousands of industrial accidents, causes of accidents were often shrouded in mystery and fatalistically accepted, as if 'it just happened', it was an 'act of God' and other culturally accepted explanations.

Medical professionals introduced the epidemiological triangle in the safety domain (see Chapter 3). This model assumed that by blocking one of the three factors of this triangle, these being victim, agent energy and environment, accident processes could be halted before they caused an accident. It was clear from the literature that the introduction of an epidemiological approach gave a new impulse to research and emphasised interactions among the factors involved. The conceptual and theoretical weaknesses of psychological research, which was descriptive in nature and not focused on aetiology and exposure, could thus be circumvented (Suchman and Scherzer 1960a, 1960b; Haddon 1968; Baker and Haddon 1974). The interaction between the three factors was indicated by vectors or carriers. Haddon's contribution was the translation of the epidemiological triangle into a matrix (Table 4.2). Known as the 'Haddon matrix', it shows the stages of the accident process and its relevant factors. These stages lead to different types of control measures that are similar to the present-day 'occupational hygiene strategy' (Haddon 1968).

A similar classification of preventive measures was published in *De Veiligheid*, referring to 'the Scheme of Zielhuis' (Noort 1952; Malten 1959; Zielhuis 1962;

Table 4.2 The Haddon matrix (after Haddon 1970)

Stages in accident processes	*Victim*	*Energy*	*Environment*
Pre-Event Phase			
Event Phase			
Post-Event Phase			

Treffers 1968). Professor Zielhuis, a Dutchman, was an internationally respected expert on occupational medicine and one of the first scientists to recognise the dangers of industrial lead intoxication.

The epidemiological triangle introduced the concept of barriers, which interfered with vectors, and led to the classical *hazard-barrier-target* model, also known as *hazard-barrier-victim* model (Figure 4.5). This model opened up the possibility of classifying causes of accidents. In medical science, the causes of

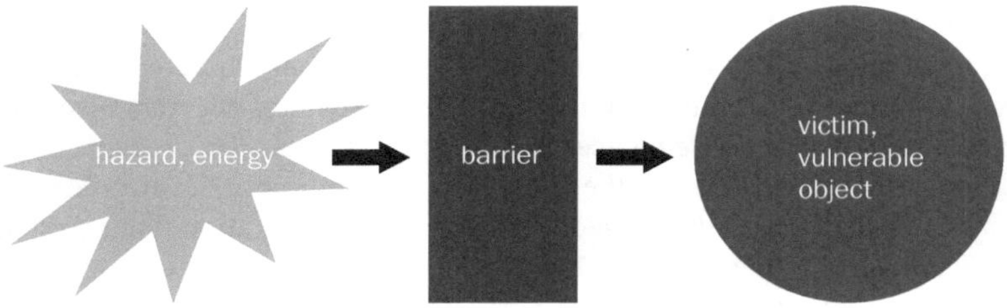

Figure 4.5 The hazard-barrier-target model

Source: Haddon (1963)

events were investigated, as this was necessary for optimal treatment. The key question was thus: *what caused the injury?* It was postulated that the common factor, the cause of an injury, was an abnormal exchange of energy that exceeded the body's natural resistance (Gibson 1961; Haddon et al. 1964; Haddon 1970). This concept of energy exchange had already been proposed, some 40 years earlier, by DeBlois (1926). Using the concept of energy exchange, accidents and accident mechanisms could be classified. In 1958, one classification of industrial and traffic accidents showed the confusion of causes of accidents with the label 'unforeseen event' (Anonymous 1958).

Using this model, Gibson (1961) proposed a detailed classification of types of energy, which acted as external sources of hazards. Gibson distinguished different types of energy, i.e. radiation, potential, kinetic, mechanical, thermal, chemical and electrical energy. These different energy types led to different types of injury (Haddon 1963).

A rather different approach originated from war experiences, when the concept of risk was an element of operations research and reliability engineering. This last topic is addressed in the next section.

The concept of risk

The concept of risk was not commonly used in safety studies in the period 1950–1970. A risk-based approach and the concept of probability were not generally accepted in the safety domain. Yet in this early period, the first ideas for a risk-based approach emerged. Heinrich (1931) described the occurrence of an accident as a separate step in the accident process that led to unwanted effects, namely injuries. This was the basis for his domino metaphor of 1941. Heinrich discussed the causal relations and the severity of effects, but he did not explicitly mention the concept of probability. It was DeBlois who introduced the theory of likelihood, or probability, into safety studies. DeBlois advocated a shift away from a universal belief that *accidents just happen* to a new doctrine in which *accidents were caused, and the relevant causes could be prevented.* He introduced the likelihood concept:

> *Likelihood deals with the maturity of complex events and brings their occurrence under the exact laws as against leaving their happening to random conjecture . . .*
>
> (DeBlois 1926)

DeBlois illustrated the use of probability in the case of a sudden crane failure (probability per year 1 in 10,000), and the probability that an employee who normally works under the load for ten minutes of a ten-hour workday would be hit by the load when the hoisting cable broke. The combined probability of such an accident was the product of both independent probabilities: 1/10,000 x 10/(10x60) = 1/600,000. DeBlois did not mention the term 'risk', but he used some important concepts that would form a central part of the risk-based approach used years later in safety: predicting or estimating combined incident probabilities and the lack of information about the 'number of successful operations'. Unfortunately, DeBlois was ahead of his time, and his ideas on likelihood and probability were not applied in the safety domain at the time.

The concept of risk, however, was already well known in antiquity, and from about the seventeenth century, in the insurance domain. The first references to the concept of risk can be traced back to ship transports in antiquity (Bernstein 1996). Later, the notion of risk was used in the calculation of life insurance premiums in actuarial mathematics. Several historical reviews on risk analysis and safety refer to Pierre de Fermat and to Blaise Pascal, who in the seventeenth century laid the basis for the calculation of probability and thus prepared a path for modern risk assessment (Bernstein 1996; Saleh and Marais 2006). And at the beginning of the twentieth century, the axioms of Kolmogorov provided the mathematical foundation for the calculation of the probability of a particular set of independent events as the product of the separate probabilities for these events (Kolmogorov 1956).

- *For every event A, a subset of F applies a probability $P \geq 0$ (a chance is not negative).*
- *The probability of a set Ω, $P(\Omega) = 1$ (total chance is normalized to one).*
- *For a series of disjunctive events (A_k), $P(U A_k) = \sum P(A_k)$.*

In the period after the First World War, the concept of probability was increasingly applied in domains such as operational research, reliability engineering, human factors, loss prevention and related fields. These domains evolved largely independent of each other.

Reliability engineering

Reliability engineering emerged with the development of mass production and the use of standardised components. It originated from the mass production of weapons and the introduction of interchangeable parts during the American War of Independence (1775–1783) and, much later, in the mass production of automobiles in the factories of Henry Ford and the introduction of fragile vacuum tubes, diodes and triodes, at the beginning of the past century. A second development in reliability engineering started after the Second World War. In 1952 in the United States, the Advisory Group on Reliability of Electronic Equipment (AGREE) was founded, followed by the start of reliability engineering as a discipline in 1957 (Coppola 1984). Reliability engineering made a significant contribution to the collection and analysis of failure probabilities of components in technical systems, like the two technical standards: TR-1100, Reliability Stress Analysis for Electronic Equipment from 956; and H-217, military standards for calculation of reliability predictions from 1961 (Denson 1989). Reliability engineering also formed an important input for risk assessments in safety studies.

First time safe was the reliability engineering credo; it challenged the customary *fly-fix-fly* routine (Roland 1990) (Figure 4.6).

The *Apollo I* fire of 1967, which killed three astronauts, showed that this old motto was unacceptable for complex systems. System safety checked by hazard analysis and fault tree analysis techniques formed the basis of this approach, along with the calculations or estimations of probabilities of component malfunction leading to system errors. It was assumed that 'what can happen will happen when the time is ripe'. Reliability engineering was promoted through a series of seminars, organized by the United States and held in the UK, Germany, the Netherlands, Switzerland and Denmark.

Ergonomics

The focus on ergonomics resulted from developments during the Second World War. The war accelerated the technical development of military equipment, machinery and industrial processes, and had created serious control problems (for a review, see, for example, McIntyre 2000). Weaponry and machines were becoming increasingly complex, thereby creating greater demands on operators' cognition. It is generally believed that human factors, human reliability analysis (Swain 1964) and ergonomics (Singleton 1960) originated during this period, although earlier developments in this domain can be traced to the beginning of the twentieth century. The terms 'human factors' and 'ergonomics' referred to similar

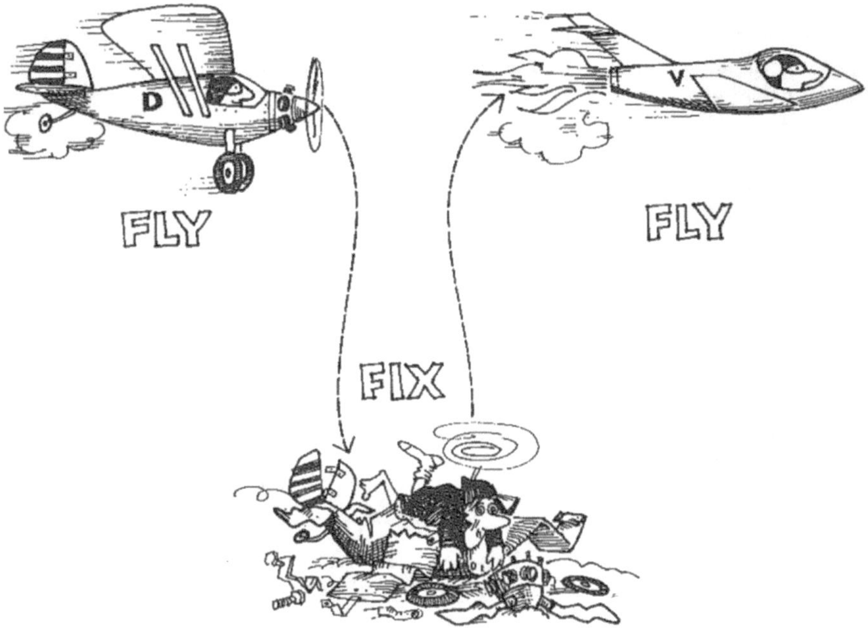

Figure 4.6 Fly-fix-fly

Source: Roland (1990)

domains of applied sciences. The first term is the expression used in North America, and the second in Europe; both referred to the study of man's relationship with machines and with his work. The domain is primarily concerned with the design of equipment, work operations and work environment to optimize worker capacities and to reduce their limitations.

In the United States, human factor engineering strongly emphasised human efficiency by quantifying estimates of human error in man–machine systems as well as quantifying the effects of human error on system reliability. This human reliability assessment in the 1960s was a copy of the approach of reliability assessment, in which failure rates of valves, pumpsand the like were collected. Based on the reliability of hardware failure databases, there was a desire to create comparable (i.e. accurate) databanks on human error probabilities. But the results of initiatives to quantify human error probabilities were less successful than those applied to hardware. The main reason for this was that, in contrast to hardware like valves and pumps, which generally have relatively accurate failure rates, intrinsic variations in, and the complexities of, human behaviour often preclude accurate quantitative prediction of human behaviour, including human failure rates (Kirwan 1994). In complex man–machine systems, problems with data collection and data reliability showed that it was thus quite complicated, and therefore also difficult, to calculate the effects of system failure modes on the probability of successful task completion (Swain 1964). Despite these early limitations, human factor specialists focused on human performance, where safety is regarded as a by-product of efficiency. With a questionable relation between human error and accidents, and a strong engineering orientation in its methodology, there was no demonstrable contribution to theories, models or metaphors in the safety domain. British developments in ergonomics had a different approach and will be discussed in the next section.

Loss prevention and safety tools, FMEA, FTA, energy analysis

In addition to space travel research, the chemical process industries changed significantly after the Second World War. A massive upscaling of the chemical process industries occurred. The control of these processes became more complex, due to higher temperatures, pressures and volumes. *Loss of containment* occurred relatively frequently. This resulted in huge fires, massive explosions and large emissions of toxic substances, often with far-reaching effects beyond the premises of plants, as well as huge financial losses. Widespread public fear of the risks of large-scale accidents and various forms of pollution developed. This development was fuelled by the, at the time, popular publication *Silent Spring*, which dramatically described the detrimental effects on the environment caused by the indiscriminate use of pesticides (Carson 1962). An important development within the safety domain of the process industry to counter this negative development became known as *loss prevention*. Publications on loss prevention first appeared in the United States and the United Kingdom in the 1960s (Association of British Chemical Manufacturers 1964; Fawcett 1965a, 1965b). As with safety tools and human factors, the engineering approach was dominant. The focus was no longer on unsafe actions but rather on preventing, or at least minimising, the effects of 'loss of containment' – keeping dangerous chemicals inside the installation. To better achieve this goal, methods and tools were developed that would improve equipment and process reliability. Some of these tools originated from the military sector, including failure mode and effects analysis (FMEA), and fault tree analysis (FTA); some from traffic safety research (energy analysis); and some from process safety studies in the United Kingdom: hazard and operability studies (Hazop). The aim of these tools was to detect disturbances in the process flow in order to analyse and improve safety in combination with improved operability. At this time, the reliability of complex systems in the military sector and in the space industry was considered to be too low. Also, in the process industry, the earlier mentioned upscaling made many processes unstable.

An FMEA provided an analysis of technical systems, both for the individual components of a system and for the larger functional blocks of a system. In simple terms, the method was designed to answer questions such as *How can a unit fail?* and *What happens then?* (Harms-Ringdahl 1993). Each potential failure mode in a (sub)system is analysed, and the effects on the system are classified by severity. In the late 1940s, FMEA was first described in the United States Armed Forces Military Procedures document MIL-P-1629 (Anonymous 1949). The major breakthrough came in the 1960s, during the development of the manned lunar flights in conjunction with the Apollo project. The FMEA method is currently applied in many industrial sectors.

The energy analysis approach is a general safety analysis technique, based on the hazard-barrier-target model (Figure 4.5). Exposure to the agent energy is the starting point in the case of any damage or accident. With this tool, an installation or plant is first divided into different physical spaces or components, for example a chemical reactor. Energy sources, such as flammable substances, are identified for each space, including the existing barriers. The energy levels are ranked by experts on a four-point scale ranging from 0 (no hazard) to 4 (serious hazard). For the existing and desired barriers, a Haddon matrix classification is used to develop prevention strategies.

FTA (Figure 4.7) is a tool that, like FMEA, originated in the military sector (Ramamoorthy et al. 1977). Several dramatic major accidents, like the detonation in 1958 of four Nike Ajax missiles at a Middletown, New Jersey, missile site, which killed ten soldiers and civilians, made clear the need to improve system safety in the military sector.

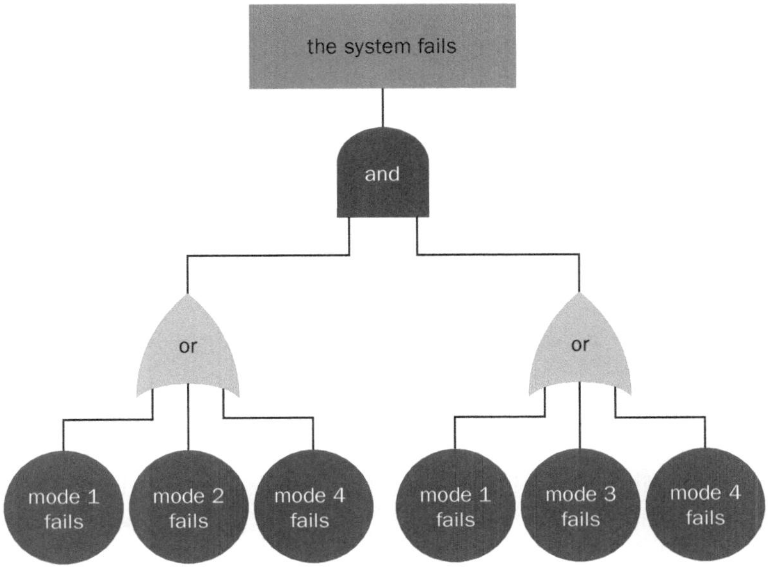

Figure 4.7 Fault tree diagram with specific symbols

Source: After Ramamoorthy et al. (1977)

This major accident was the biggest on American soil in the Cold War period (DeLong 1970; Ericson 1999, 2006). The FTA tool, developed in 1962 by Bell Telephone Laboratories in New York, was commissioned by the US Air Force to prevent accidental launches of Minuteman intercontinental ballistic missiles. Using a tree-like structure, the technique logically organized a series of sub-events that eventually lead to an undesirable event, a top event like an accidental launch of armed missiles, which, obviously, must be avoided. Thereupon, failure scenarios of underlying components of the system are identified. FTA assumes a binary development of events which either can or cannot take place. This is both the strength and the weakness of the method. Errors in complex systems are presented in a simple way with only yes or no options, without any further nuance. FTA uses generally recognized logical symbols to describe events and faults. It is applied mainly to complex technical systems where functional errors could have serious consequences, as in high-risk areas, such as the nuclear and military sectors. FTA is a deductive top-down tool which starts from a top event before going on to analyse the influences of initial errors and events in the complex system. FMEA, on the other hand, is an inductive tool, starting from the effects of the failure of terminal components or functions.

Compared to FMEA, FTA is best at determining the robustness of a system to successfully negate the effects of one or several errors. However, FTA is not suitable for discovering all possible initial errors. Detecting initial errors and their effects on the system is much more the strength of FMEA (Ramamoorthy et al. 1977; Vesely et al. 1981; Ericson 1999). The use of FTA led to the introduction of a system to avoid unauthorized launching of the Minuteman nuclear missiles. There were objections from the US Strategic Air Command, which was afraid that adding an extra layer of protection would delay launching the missiles in the case of a sudden attack. The secret code was set to 00000000, and everybody knew the combination as it was included in a checklist that was widely distributed. Although knowing this code would not immediately allow individuals to start a nuclear Armageddon, it illustrates that application of a seemingly straightforward tool can create discussion about the priority of which events should be avoided (Blair 2004).

United Kingdom

Safety tool Hazop

Hazop originated in the process industry and was developed in 1963 by Imperial Chemical Industries Ltd (ICI). First applied to the design of a phenol plant for the Heavy Organic Chemical Division, Hazop was designed not only to systematically identify design deviations and potential process aberrations but also to determine possible consequences for the whole installation (Lawley 1974, 1976). Hazop became a widely used, formal and systematic review of the design specifications of each process component, this as part of a comprehensive process safety review of new or existing installations. A Hazop session is a group discussion between experts from different disciplines. Guided by a piping and instrumentation diagram (P&ID) of a draft design, the group discusses all possible process safety issues line by line, section by section. The design specifications of each process section are determined as well as how the section is designed to function properly. Then guide words are used in combination with process parameters like temperature, flow, and so forth. These guide words define possible process deviations using terms such as 'no/not', 'more' or 'less'. The guide words are also applied to materials, production functions and layout issues (Harms-Ringdahl 1993; Swuste et al. 1997a; Venkatasubramian et al. 2000; Kjellén 2000).

Human factors and ergonomics

Display configurations of aircraft instruments and controls were too complex for pilots, as early ergonomics research had already pointed out. This complexity had led to many plane crashes. Designers had a natural preference for hardware over human control, not only for military applications but also for factories. Man–machine interfaces could look like clock shops with their many dials and indicators. These interfaces were biased towards the designer's interests and not to the need of machine operators to receive accurate information (Singleton 1969). For ergonomists and human factor specialists, a detailed task description was the starting point for their analysis, and for solutions to the problems they encountered. Their results are used as a stimulus for engineering development (Singleton 1960).

In the 1960s, criticism grew of organisations and companies designed and managed according to the principles of Taylorism. These organisations and companies were not flexible or agile enough and could no longer could meet the demands of time. The quality of work in these organisations and companies, and the short cyclical operations on conveyor belts, which were monotonous and critical, had led to alienation and the psychosocial degeneration of workers. In the United Kingdom, the ergonomic domain developed along these lines, in close relation with human biological sciences: anatomy, physiology and psychology. This British approach differed from the human factor approach predominant in the United States. In the United Kingdom, for example, there was more focus on the well-being and health of workers, which involved topics like the reaction of workers to stress (Singleton 1967a). In the late 1960s, the American systems theory and systems approach entered the British safety science domain and the terms *man–machine systems* and *system ergonomics* became familiar (Singleton 1967b, 1969). Human tasks and human information processing, including their failures, became important, leading to another model explaining accidents: the information accident model (Hale and Hale 1970) (Figure 4.8).

In this period, research on occupational accidents was predominantly carried out by psychologists and ergonomists working at Aston University, Birmingham. Unlike previous accident models and theories, which were based on hazards or on human behaviour, this model focused on the information available to workers just before an accident took place. Hale and Hale started from the assumption that workers involved in accidents were not clumsy or accident-prone. Immediately

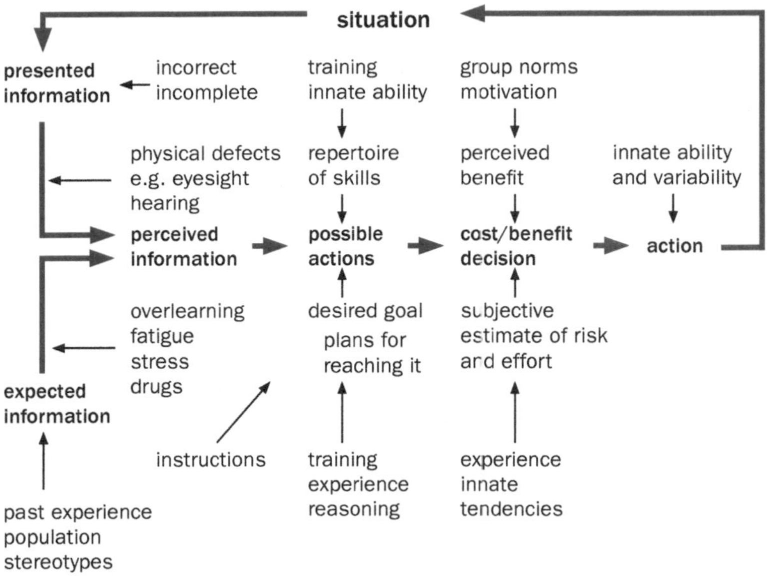

Figure 4.8 The information accident model

Source: Hale and Hale (1970)

before accidents occurred, information flows could be disturbed, or an information overload could occur, or physical limitations preventing adequate responses could arise or a worker could choose an incorrect action strategy (Surry 1969; Hale and Hale 1970). Furthermore, immediate causes of unsafe behaviour were also shaped by systematic causes, such as social norms and pressures, the physical environment and the organisations people worked in. An accident was defined as the failure of a person to cope with the dangers of the actual situation presented to him. The cause of failure was a product of the person and the situation. The information accident model made a distinction between presented information and expected information, the combination of these leading to perceived information and finally to a decision(s) for action. This model classified accidents by specifying the part of the process in which malfunctions occurred. Using this model, ergonomists developed the ergonomics relevant for information processing.

The Netherlands

Task dynamics, a safety theory

Willem Winsemius (1917–1990) presented the main Dutch contribution to the understanding of occupational accidents. Along lines similar to British ergonomists, Winsemius developed the theory of 'task dynamics'. Winsemius had been trained as a medical professional and worked as a scientist at the Department of Mental Health at the Nederlands Instituut voor Preventieve Geneeskunde (Dutch Institute of Preventive Medicine), in Leiden, later known as NIPG-TNO. His 1951 PhD thesis was titled 'The psychology of accident events' (Winsemius 1951). Like other scientists, Winsemius sharply criticised the statistical approach of the accident proneness theory.

According to Winsemius, an accident could be defined as a 'sudden and accidental physical event, causing injury'. Causes of accidents had an active and dynamic character. Winsemius examined 1,300 occupational accidents between 1946 and 1948, registered at the central first-aid post of a Dutch steelworks called Hoogovens. At that time, steel production was already largely mechanised. But unlike the later automation and remote control, machinery in the steel plants was still directly and mechanically operated, which required workers to stand close to their machines. Distances between workers and hazards, therefore, were short. Accidents could be understood only if one knew the conditions and the specific and complex reality in which they occurred. For each accident, its nature, the accident process and outcome were recorded, together with observations at the accident locations. According to Winsemius, one cannot understand accidents by only studying statistics, like the research that had led to the 1919 accident proneness theory.

Task dynamics in Winsemius' view included operations that must be performed prior to achieving a given goal or task. An employee wants to do something and is driven by a force that creates an action. This force, the task dynamics itself, did not explain the accident, but the action taken did. There were three options for this action (Winsemius 1951):

- *To have a faster way of acting, which is only slightly riskier than having a safe path of action, which is a bit longer. The two balance each other out. With normal task dynamics and a normal understanding of hazards and risks, it will not matter which path is chosen;*
- *The quicker route clearly carries a greater, but not an abnormally higher, risk level; the safe path of action is considerably longer. Even with normal task dynamics and an understanding of the hazards, the faster path will be chosen, and risks are taken;*
- *The faster path represents a very high level of risk; in the case of normal task dynamics and understanding of hazards, the safe way will be chosen, even if it is much longer and harder.*

If task dynamics are too great, there is a discharge along the path of least resistance, which is also the fastest path of action. This preference criterion dominates over the least-risk pathway. Even if there is a gap in the knowledge of the relevant hazards, task dynamics choose the fastest path.

The approach of Winsemius was similar to that of ergonomics, with its focus on tasks and actions in complex man–machine systems. Accidents were not caused by simple causal chains but by a combination in time and place of many factors that, taken separately, could not cause accidents. The coincidence of accident-causing factors was an anomaly in, or a disruption of, the normal work schedule, with sudden and unexpected physical events (Winsemius 1958–1960). Process disturbances led to improvisation and reflex behaviour, initiating the shortest and quickest path of action. These reflexes could lead to accidents (Winsemius 1969a). In Belgium, Faverge conducted safety research amongst miners with similar conclusions (Faverge 1967a, 1967b, 1970). Winsemius applied three ergonomic principles to prevent accidents (Winsemius 1965):

- *The safe course of action must always compete with other, less safe paths. The chances of success will be increased when a safe path of action is not a lengthy path of action;*
- *The critical issues of safety are always connected to process disturbances, errors, interruptions and other unanticipated and undesired events during work processes. Minimizing disturbances will thereby increase safety;*
- *Ensure a comfortable position for the worker, an optimal pattern of required actions and, more generally, the best possible ergonomic layout within which the employee has to function during normal and disturbed process steps.*

These three principles were used when designing machinery and installations (Archer 1965; Smallhorn 1967). Winsemius also applied his theory of task dynamics and disturbance to accidents in private homes (Winsemius 1978) and to accidents involving children (Winsemius 1980).

Focus on occupational safety

By the 1960s, a number of articles in the Dutch professional journal *De Veiligheid* (The Safety Journal) had dealt with the accident proneness theory and Heinrich's metaphors. These articles suggested that the Dutch domain of occupational safety was lagging far behind those of the United States and the United Kingdom (Fetter 1947; Gorter 1947; Anonymous 1965). In addition to many articles on classical safety techniques, the modification of dangerous machinery, tools and hazardous sectors such as agriculture, (ship) construction, and the steel and metal industries, there was also extensive coverage of Heinrich's ratios and metaphors, as explained in Chapter 3 (Creyghton 1949, 1952; Kraft 1950; Hart 1950, 1966; Pieters and Hovers 1960; Slob 1961; Harms 1966; Kruithof 1966; Noesen 1966). This was illustrated in one of the more recent drawings by Opland, a well-known Dutch cartoonist (Figure 4.9).

As in the United States, much attention in the Netherlands was paid to the safety education of workers, who needed training in the safety idea and in adequate housekeeping in factories and at workstations (Blaauw 1950; Wallien 1953; Anonymous 1953; Spaan 1956).

Safety posters (Figure 4.10) were widely used to convey the safety message and to hold workers accountable for unsafe behaviour (Sparreboom 1947; Hart 1952;

Figure 4.9 Unsafe acts as a cause of accidents

Source: © Opland (1979)

Slob 1961; Spaan 1961). For example, Slob's safety textbook illustrates accident timescales (Slob 1961):

> *It takes one minute to write a topic for a safety meeting*
> *One hour to handle that topic at the meeting*
> *One week to put together the measures to be taken in the company*
> *One month to implement the measures*
> *It takes a year to see results in the statistics*
> *It takes a lifetime to make somebody a safe worker*
> *All this can be countered by accident in one second.*
>
> (Slob 1961)

Additionally, the Dutch Health Council (Gezondheidsraad 1965) called attention to Haddon's hazard-barrier-target model. The Health Council developed a comprehensive classification system in which causes of injuries – these being abnormal energy exchange – were subdivided in detail into various types of energy and their industrial application. However, a few papers in the professional journal *De Veiligheid* (The Safety Journal) paid attention to the Health Council classification (Boer 1967; Dop 1967).

The Lateiner method

Articles on safety technique in the Netherlands were based largely on German-language publications, while articles on the human factor and ergonomics were based mostly on American and British sources. American ideas, and especially the relatively simple ratios of causes and costs of accidents produced by Heinrich,

Figure 4.10 Dutch safety posters

Source: University museum University of Amsterdam

1941 (Strelitskie): Do not spit

1954: You have it in your own hands

1955a (Mettes): Work safely

1955b: Reduce the number of accidents

1957 (Halsema): Protect yourself during your work

1961a (Slob): 3 causes of accidents: I did not see, I did not think, I did not know

1961b (Slob): By today seize the four O's, ignorance, inattention, carelessness, unwelcomeness

1962: Sources of accidents . . . technique 18%, humans 80%, force majeure 2%

1963 (Halsema): Unsafety, report immediately

obtained an additional dimension in the Netherlands (Lateiner 1958; Lateiner and Heinrich 1969): the number of accidents without victims. Lateiner stressed the need for a correct use of the 300:29:1 (accident mechanism) and 4:1 (accident costs) ratios. He argued that registration of accidents in companies alone was insufficient. Accidents involving non-victims should also be registered, making it possible to decide whether even more stringent safety measures should be taken. In one of these articles Lateiner used the image of an iceberg, most of whose dangerous bulk lies unseen below sea level, as a metaphor for 'the accident problem' (Lateiner 1958):

> *The accident problem is like an iceberg with only one-eleventh of its mass visible. The invisible base – made up of no-injury accidents – is ten times greater. We usually look at an accident critically only when it produces an injury.*

In the late 1950s, the director of the Safety Museum invited Lateiner to the Netherlands. Lateiner brought Heinrich's ideas to the Netherlands by organising a course, called the Lateiner Method. This method was taught until the late 1980s. Lateiner's influence is still noticeable: for example, in the exam requirements of the Contractors Safety, Health and Environmental Checklist (VCA) and the contemporary education of Dutch safety professionals (Bank 2008).

Workers' participation

During this period the organisation of safety, at company level, nationwide and internationally, was emerging. Following in the footsteps of the United States, experiments with safety committees had already started before the Second World War. Workers were also represented in these committees, as the very first forms of worker participation. Immediately after World War II, increasing worker safety became a 'hot topic', and its pros and cons were fiercely debated (Zwaard 2007).

Companies were rapidly setting up safety services. With emerging occupational health services, a collaboration between the 'doctor-hygienist' and the safety inspector was the focus of multiple articles (see, for example, Gerritsen 1957; Bloemen 1967; Kuiper 1969; Peters 1969; Hoeff 1970). These two disciplines had already been cooperating 'on an equal level' from the first years of the Second World War. This cooperation continued until 1966 under the name Consultative Committee. Experts from major companies and representatives of the director-general of Labour, the directorate of the Risk Bank and the executive members of the Safety Museum met a number of times per year, and organized meetings for members. Another development was the involvement of senior executives and managers in operational safety. In a gradual process occurring during the 1950s, it was the direct supervisor who was first trained in occupational safety, and, somewhat later, safety training of senior management followed (Spies 1958; Klunhaar 1964). In the 1960s, the first outlines of what we now call a 'safety management system' became visible (Anonymous 1962, 1968; Serdijn 1962).

Ergonomics and housekeeping

Despite Lateiner's introduction of Heinrich's concepts of safety in the Netherlands, criticism of the accident proneness theory of unsafe acts, similar to that of American and English sources, was also heard in the Netherlands. Winsemius's comments on the statistical approach to accident causation was repeated, as were his comments on the quality of psycho-physiological tests (Winsemius 1951). These tests were largely incapable of selecting accident-prone workers. In line with the ergonomic argument, a fundamental question was raised as to whether humans should adapt to work, or vice versa (Luijt 1948; Frederik 1951; Waart 1951). Winsemius advocated his view strongly, proclaiming that adapting the work process was infinitely more preferable than influencing the behaviour of workers (Winsemius 1946, 1952; Pieters and Hovers 1960; Reij 1962; Leuftink 1964). In a later article, Winsemius argued against what, to his mind, was the exaggerated attention devoted to 'good housekeeping'. Litter in a workplace was not meaningless chaos. Work not only provided a finished product but also caused debris. Debris represented traces of failed task performance and many process disturbances would create much debris (Winsemius 1969b).

The modern management school was on the rise. Total Quality Management in Japan, an application of modern management, was in the 1980s introduced to and applied in Western countries. In the United States, safety included not only preventing personal injuries but also the prevention of property damage. Medical professionals, however, already active in safety research, developed a very different model of accident processes, the 'hazard-barrier-target' model. Initiated by the military, currently known safety techniques, such as fault tree analysis and failure mode and effect analysis, evolved.

This chapter has described the rise of ergonomics in the United States and the United Kingdom and task dynamics theory of the Netherlands. These influenced the development of safety science in the 1950s and 1960s. Human factors, as ergonomics is called in the United States, led to databases of human failures, while in the United Kingdom attention was given to failures in information flows necessary for task fulfilment. The Dutch task dynamics theory developed along British lines. In general, confidence in technology rose to great heights in this period, along with increasing attention to accident prevention.

During this period, reservations regarding the accident proneness theory in general increased and the same applied to Heinrich's safety ratios and metaphors. Only amongst safety practitioners did Heinrich's concepts remain popular. In the Netherlands, Heinrich's concepts were introduced by the 'Lateiner method', which had a long-lasting impact on vocational safety training. The increase in scale of the process industry aided the rise of the high-tech-high-hazard safety methodologies. In the years that followed, it became apparent that these methodologies were undergoing their own development alongside occupational safety. In the following chapters, developments in both domains are discussed separately, in more detail.

5 Risk, safety and organisation – management

1970–1990

DOI: 10.4324/9781003001379-5

The 1970s were a time of turmoil and tension in much of the world.

In the 1970s, a new genre of dance music and a subculture emerged: **disco**. It can be seen as a reaction to both the dominance of rock music and the stigmatisation of dance music at the time. Disco was the last popular music movement driven by the baby boom generation. Disco influenced the development of electronic dance music in general and house music in particular.

The 1973 **oil crisis** began when Arabic countries proclaimed an oil embargo targeted at nations that supported Israel during the Yom Kippur War. The crisis had many short- and long-term effects. Fuel shortages caused long lines at gas stations. Several countries banned flying, driving and boating on Sundays. The crisis changed the nature of policy in the West towards alternative energy research, energy conservation and more restrictive monetary policy to better fight inflation.

Neoliberalism refers to market-oriented reform policies such as eliminating price controls, deregulating capital markets, lowering trade barriers and reducing state influence in the economy, especially through privatisation and austerity. It is often associated with the leadership of prime minister Margaret Thatcher of the UK. Neoliberal thought has been criticized for supposedly having an undeserved faith in the efficiency of markets and in the superiority of markets over centralized economic planning.

In the United States the **Watergate scandal,** in 1972, led to the resignation of president Richard Nixon. The political scandal stemmed from attempts to cover up the involvement of the president in a break-in at Democratic National Committee headquarters in the Watergate building.

By 1970, calculators could be made using just a few chips of low-power consumption, allowing portable models powered by rechargeable batteries. **Pocket calculators** became popular in the mid-1970s as the incorporation of integrated circuits reduced their size and cost. In most countries, students used calculators for schoolwork.

In the 1970s VHS (Video Home System) became a standard for analog **video recording** on tape cassettes. The popularity of VHS led to large-scale use of video tape recorders by consumers at home and changed the economics of the television and movie business.

In the 1970s **abortion rights movements** advocated legal access to induced-abortion services. The issue of induced abortion remains divisive in public life, with recurring arguments to liberalise or to restrict access to legal abortion services. From the 1970s, and the spread of second-wave feminism, abortion and reproductive rights became unifying issues among women's rights groups in various Western countries.

An American epic space opera media franchise, **Star Wars**, began with a film in 1977 and would go on to expand into various other media in later years. The *Star Wars* saga had a significant impact on popular culture. The *Star Wars* film helped launch the science fiction boom of the early 1980s, making science fiction films a mainstream genre.

Disco

Oil crisis

Neoliberalism

Watergate scandal

Pocket calculators

Video recording

Abortion rights movements

Star Wars

Sources:

a. https://commons.wikimedia.org/wiki/ File:ZMF_2015_IMGP_0000.jpg
b. https://en.wikipedia.org/wiki/1970s#/media/File:Line_ at_a_gas_station,_June_15,_1979.jpg
c. https://en.wikipedia.org/wiki/Margaret_Thatcher#/media/File:President_Gerald_Ford_Meeting_ with_ Great_Britain's_Conservative_Party_Leader_Margaret_Thatcher_in_the_Oval_Office.jpg
d. https://en.wikipedia.org/wiki/June_1972#/media/File:Watergate_complex.jpg
e. https://en.wikipedia.org/wiki/1970s#/media/File:HP_35_Calculator.jpg
f. https://en.wikipedia.org/wiki/VHS#/media/ File:VHS_recorder,_camera_and_cassette.jpg
g. www.knkx.org/post/there-will-be-right-abortion-washington-state-even-if-roe-v-wade-overturned
h. https://upload.wikimedia.org/wikipedia/commons/1/18/EmissionNebula_NGC6357.jpg/https:// en.wikipedia.org/wiki/Star_Wars_(film)#/ media/File:Star_wars_1977_us.svg

The 1980s experienced its drawbacks, but also the end of the Cold War.

The **fall of the Berlin Wall,** in 1989, was a pivotal event in world history that marked the disappearance of the Iron Curtain and heralded the fall of communism in eastern and central Europe. The fall of the inner German border took place shortly afterwards. It soon led to an end to the Cold War and the reunification of Germany.

AIDS was first clinically reported in 1981, with five cases in the United States. AIDS had a large impact on society, both as an illness and as a source of discrimination. The disease has become subject to many controversies involving religion, including the Catholic Church's position not to support condom use as prevention.

An American cable TV channel, **MTV**, was launched in 1981 and had a major influence on the growth of music videos during the 1980s. It was instrumental in adding to the booming'80s dance wave. Videos' budgets increased, and artists began to add fully choreographed dance sections.

In the 1980s Nintendo developed and manufactured **Game Boy**, a handheld game console. It soon was one of the most popular game computers, with more than 1,000 games released for it. Later Game Boy became a cultural icon of the 1990s.

In 1987, a **stock market crash** struck the global financial market system. The severity of the crash sparked fears of extended economic instability or even a reprise of the Great Depression. Regulators overhauled trade-clearing protocols to bring uniformity to all prominent market products. They also developed new rules, known as 'trading curbs' or colloquially as 'circuit breakers'.

In 1982, the first **compact disc** (CD) was released. The discs and their players soon became extremely popular. As the price of players gradually came down, and with the introduction of the portable Discman, the CD began to gain popularity in the large popular and rock music markets. CD sales overtook prerecorded cassette tape sales in the early 1990s.

In the 1980s **Voyager 2**, a space probe launched by NASA in 1977, encountered all the giant outer planets of our solar system. The close-up studies of Jupiter, Saturn, Uranus and Neptune provided material that made it necessary to rewrite astronomy textbooks.

In 1980, **Rubik's Cube** conquered the world. On the original Rubik's Cube, each of the six faces was covered by nine stickers, each of one of six solid colours. The cube won several awards for best toy in different countries. Soon it became a craze, and within three years around 200 million cubes had been sold worldwide.

Fall of the Berlin Wall

Aids

MTV

Game Boy

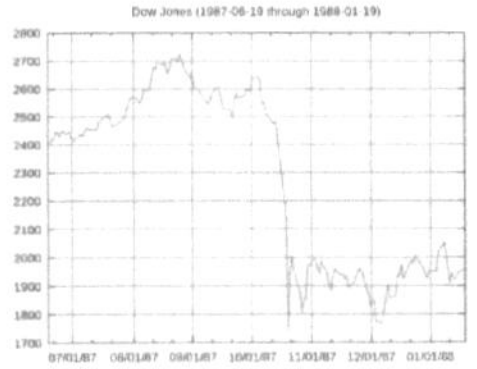

stock market crash

Compact disc

Voyager 2

Rubik's cube

Sources:

a. https://en.wikipedia.org/wiki/Fall_of_the_Berlin_Wall#/media/File:West_and_East_Germans_at_the_ Brandenburg_Gate_in_1989.jpg
b. https://pixabay.com/illustrations/aids-aidsschleife-dis¬ease-health-108235/
c. https://jaegeroslo.no/i-want-my-mtv-and-the-vid¬eos-that-defined-dance-music-for-a-generation/
d. https://en.wikipedia.org/wiki/Game_Boy#/media/File:Game-Boy-FL.jp
e. https://commons.wikimedia.org/wiki/File:Black_ Monday_Dow_Jones.svg
f. https://en.wikipedia.org/wiki/Compact_disc#/ media/File:Compact_Disc_wordmark.svg/ https://en.wikipedia.org/wiki/Compact_disc#/media/File:CD_icon_test.svg
g. https://en.wikipedia.org/wiki/Planet#/media/ File:Jupiter_and_its_shrunken_Great_Red_Spot.jpg/ https://en.wikipedia.org/wiki/Voyager_2#/me¬dia/File:Voyager_spacecraft_model.png
h. https://nl.wikipedia.org/wiki/Rubiks_kubus#/media/ Bestand:Rubik's_cube.svg

Several publications of the 1970s addressed the topic of safety management. During and just after World War II, seven manuals on safety, damage prevention and the management of safety were published in the United States (Heinrich 1941, 1950a; Armstrong et al. 1945, 1953; Heinrich and Granniss 1959; Blake 1963; Bird and Germain 1966). In that period in the United Kingdom, however, only one publication had appeared (Association of British Chemical Manufacturers 1964). All these publications were produced along the lines of two management schools: quantitative management and modern management.

Western Europe and Nordic countries

Quality of legal provisions for occupational management

In the United Kingdom, the government took the initiative of carrying out a survey on the effectiveness of laws and regulations dealing with safety and safety management within companies. The result was the well-known Robens Report (1972). The results were rather shocking. Occupational mortality and morbidity were alarmingly high in the United Kingdom. Furthermore, in addition to the large number of industrial accidents, a high incidence of new occupational diseases, such as bladder cancer and asbestos-related cancers, was also found. The effectiveness of safety legislation was therefore seriously questioned, as would also occur in the United States a few years later (Ellis 1975). The key question was: *What is wrong with the system?* Indeed, that was the title of the first chapter of the Robens Report. And the answer was clear. There were nine groups of laws with as many controlling bodies, spread over five different ministries, and it took an average of 15 years to amend any law. These laws were also impossible to implement: there were simply too many laws that were too detailed and too poorly structured. There were also far too many technical and descriptive regulations, while human and organisational factors remained gravely underexplored.

The remedy was relatively simple. The Robens Committee suggested delegating the technical control of hazards to those who created them – in other words, to industry. Businesses would have to take the initiative, thereby leaving the issue of safety mainly to private parties. The Robens Committee also proposed an organisation with responsibility for research and monitoring. This resulted in the Health and Safety Act of 1974. The act introduced a new system based on less prescriptive and more goal-based regulations, supported by guidance and codes of practice. For the first time, employers and employees were to be consulted and joined forces in the process of designing a modern health and safety system. In 1974, the Health and Safety Commission (HSC) was also created. Its goal was to 'secure the health, safety and well-being of people at work and to protect the public against risks to health and safety arising out of businesses, and to give general direction to the Health and Safety Executive (HSE)' (HSE 2019). The HSE was formed in 1975; it issued requirements, which has to be remitted in practice by the HSE, of the Health and Safety Commission and to enforce health and safety legislation in all workplaces. In a 1976 report, the HSE promoted a humanitarian approach to working conditions, whilst it also concluded that surprisingly few managers had received any form of safety training (HSE 1976). Curiously, the report did not adopt the term 'safety management', in contrast to the American textbooks of that day.

Models of occupational safety

From the 1970s onwards the occupational accident process and scenarios were seen as complex phenomena and therefore not easily predictable, as had been noted earlier by Haddon (Chapter 4), and in the 1980s also by other researchers (e.g. Shannon 1980). Prior to this view, the occurrence of occupational accidents had been seen as a relatively simple phenomenon, the consequence of only one cause. Now more factors were seen to play a significant role. These factors were labelled with colourful names, such as *unsafe acts* or *unsafe conditions*. The utility of relatively straightforward guidance rules for safe and effective production specified by earlier authors (e.g. Heinrich et al. 1980) was questioned. Preventive measures and interventions costing time and money were therefore doomed to fail. This was also true for measures that introduced elaborate working methods and slowed down production speed (Sulzer-Azaroff and Santamaria 1980; Monteau 1983). In this period, models for causes of accidents were often combinations of earlier accident models, such as:

1. *The human error model: this model started from the conviction that sources of human error should be controlled as well as their frequency. The human error model dated back to the Safety First Movement and the accident proneness theory from the 1910s and Heinrich's metaphor of falling dominoes of the early 1940s;*
2. *The sequential accident model: a sequence of events resulted in an accident. This model was also based on the domino metaphor;*
3. *The energy model: the interaction between hazardous energy, the victim or vulnerable object, and the environment shaped the accident process. The interplay of technological, human and organisational factors determined the form of this interaction. The model originated from Haddon's work in 1968, while the concept of hazard had already been published by DeBlois in 1926 and by Blake and colleagues in 1945;*
4. *The information model: an accident was caused by a disruption in communication. This model was produced in the 1970s by Hale and Hale (see Chapter 4) and developed further in Nordic countries. It assumed that workers' access to information and the cognitive processes involved played a central role (see Figure 5.1).*

   Workers at risk combined information they had received with earlier experiences, which influenced how they worked. Various internal factors also played a role, such as physical condition, motivation, intelligence and sensory limitations. This made this model rather complicated (Saari 1984);
5. *The systems model: accidents were outcomes of abnormal system conditions. This model focused on subsystems, components and their interactions to*

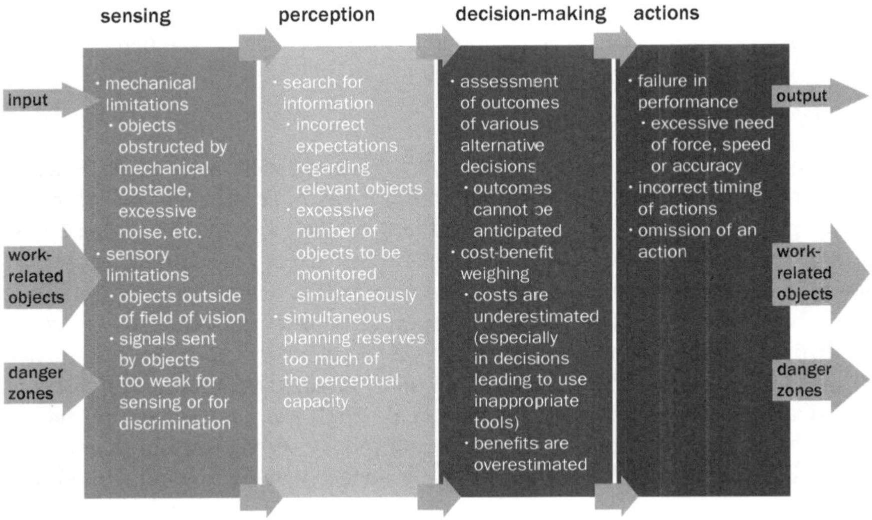

Figure 5.1 Mechanisms leading to errors in human information processing

Source: After Saari (1984)

explain accidents. An accident was an outcome of process disturbances taking place in a dynamic system. Injuries occurred when risk factors coincided with individuals (Tuominen and Saari 1982; Leplat 1984). The International Labour Office (ILO) adopted the systems model approach in its health and safety encyclopaedia and focused on improving occupational systems that were improperly designed or otherwise inherently unsafe (Monteau 1983). The link between design and accidents was, again, a Scandinavian viewpoint (Harms-Ringdahl 1987a). The United Kingdom contributed to this model with their results of fieldwork carried out at the Liverpool Ford manufacturing plants (Shannon 1980; Shannon and Manning 1980). Process disturbances were unforeseen events, like being hit, or ricked, or cut by objects like machines, particles, oil, tools and other moving objects;

6. *The safety climate model: this model was developed in Israel (Zohar 1980a, 1980b). It was assumed that the behaviour of workers was influenced by the organisational climate, which was determined by perceived management engagement, the status of safety experts and the relative priority of safety at work;*
7. *The deviation model: a special type of systems model (Figure 5.2). Human errors were seen as consequences of poor interactions with the production process, in particular inadequate responses to changes in production processes (Kjellén 1984a; Häkkinen 1982). This view was immediately attacked by researchers who emphasised that humans are indeed excellent problem solvers (Hovden and Sten 1984);*

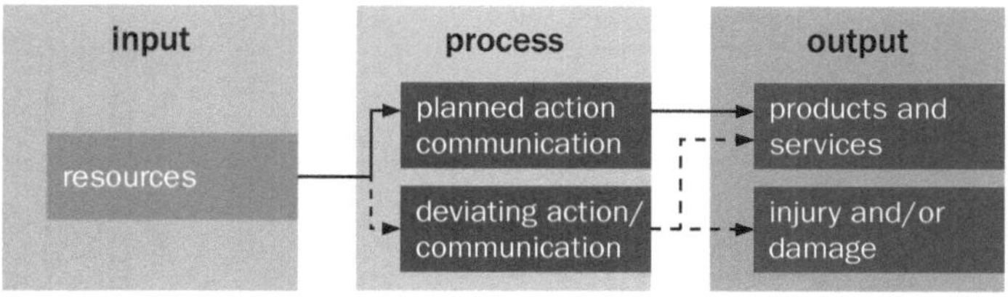

Figure 5.2 The deviation model

Source: Kjellén (1984a)

8. *The accident epidemiology model: this model was developed to study accidents with forklifts in the United States. It viewed unsafety in terms of the metaphor of a contagious disease. In their investigation, researchers studied and classified 88,000 forklift accidents in the period 1983–1985 in terms of gender, age, month of the year, type of injury and job description (Stout 1987). The statistical approach did raise some eyebrows in the scientific domain. In particular, the lack of scenario descriptions was seen as a severe shortcoming, as without these descriptions causal factors could not be established (Purswell and Rumar 1984).*

Models were often combined, leading to new models. The time-sequence approach from the deviation model was combined with other models, such as the energy model (Figure 5.3).

The IJEM factors are examples proposed by the ILO (1988). IJEM factors combined individual, job, environment and materials (Faverge 1983) in order to explain the causes of accidents in a more profound way.

Ergonomics and task dynamics

Ergonomics had become increasingly influential in the safety domain. According to the task dynamics theory developed by the Dutch researcher Winsemius (see Chapter 4), good ergonomic design and redesign of machines and workplaces prevented occupational accidents. Task dynamics was a general model of skilled movement control that was developed originally to explain non-speech tasks such as reaching and standing upright. It was based on general biological and physical principles of coordinated movement but was framed in dynamical rather than anatomical or physiological terms. It involved a relatively radical approach that was more abstract than many more traditional systems, and had proved to be particularly useful, partly because it breaks down complex movements into a set of functionally independent tasks (Hawkins 1992).

The importance of the theory of task dynamics or task momentum was also apparent in a large prospective study titled '2000 accidents', on accidents in the

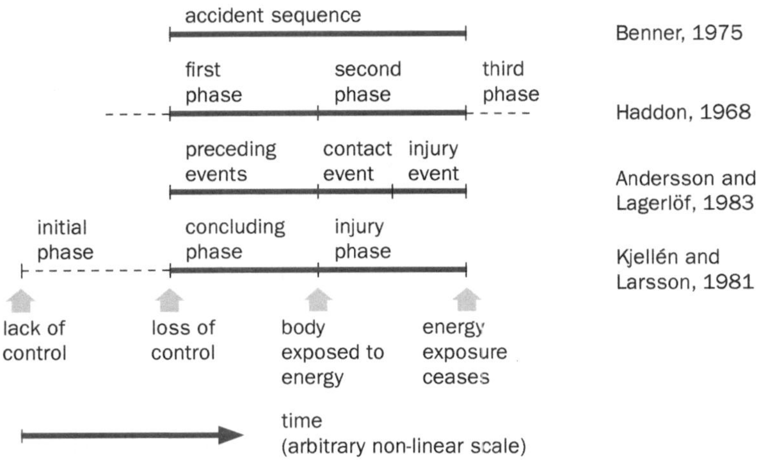

Figure 5.3 The relation between accident process models
Source: After Kjellén (1984a)

metal, assembly and distribution sectors. This study is discussed in the next section. Similar conclusions were also reached in an extensive accident proneness literature review of 80 years of publications on accidents (Hale and Hale 1972). The review also pointed to conflicts at company level between safety and production, to process disturbances being risk factors for accidents and to the relatively low effectiveness of safety training for employees.

The main centre for safety science research in the 1970s was the United Kingdom, later followed by Nordic countries. A number of breakthroughs occurred, such as the multi-causality concept, which stressed that *people, not things, are causing accidents*. In fact, the multi-causality concept still resembled the accident proneness theory (see Chapter 2), but unsafe acts and unsafe conditions were, by then, more clearly explained as symptoms of faulty management rather than causes of accidents (Petersen 1971). The term *accident-prone conditions* emerged in the United States, in sharp contrast to *accident-prone workers* (Pfeifer et al. 1974). Unsafe acts of workers were thereby placed in the context of accident-prone conditions.

Ergonomists developed the *ergonomics of information* (see Chapter 4). They also described the optimal type, presentation form and quality of information to be offered to employees. Ergonomists further classified types of errors that could occur as a result (Singleton 1971, 1972). By breaking down tasks into potential risk areas, this ergonomics-based approach predicted hazardous situations from a simple information flow model of human skilled performance. The model assumed that any breakdown of the flow of information through the system was potentially hazardous (Dunn 1972). The flow model of skilled performance thus distinguished between receptor processes (sources of sensory information), central processes (memory and risk acceptance) and effector processes (preliminary activity related to task performance, ongoing performance of the task and emergency actions in the case of unforeseen events in the system). From this simple model, four possible breakdowns in the information flow, or causes of accidents, were possible:

- *The operator may not, for a variety of reasons, be able to gather all necessary information from the environment, resulting in inappropriate decisions;*
- *There may be an information overload – the operator may be in receipt of too much information at any one time, and be unable to deal with it;*
- *The operator may be using the wrong strategy to deal with the information, and incoming information will not be interpreted correctly, so decisions will not be appropriate;*
- *The operator, due to some physical or physiological limitations, may be unable to perform the requested outputs. The operator knows very well the action required, but due to some inadequacy is unable to achieve it.*

Causes and prevention of 2,000 accidents

A previously mentioned study into the causes and prevention of 2,000 accidents was published in Britain in 1971. What this study made remarkable was its prospective research design. Until then, safety research had relied solely on retrospective study designs or case studies. Winsemius' study showed the general disinterest in safety management in the companies studied. A director could find safety important, but generally it remained simply a message on paper, without further integration into business operations. The study was based on continuous observations for more than a year in four medium-sized manufacturing companies of 100–300 workers in the metal, assembly and distribution sectors (Powell et al. 1971). This research concluded, as Winsemius previously had done, that work-related tasks and corresponding actions were the main determinants of accidents. An increase in production, resulting in an increase in actions, would therefore increase the frequency of accidents. Because of the relatively low degree of mechanisation in these companies, employees worked physically close to the sources of hazards. For example, highly repetitive tasks in a metalworking company, such as manipulating sharp-edged metal plates, inevitably led to many accidents. According to the researchers, these operations should be mechanised, as well as other monotonous high-risk tasks. Researchers did not, however, find a correlation between various machines and different types of accidents: ones with no harm, ones with minor harm and serious accidents. Such a correlation had been suggested by Heinrich with his ratios of the foundation of a major accident (Chapter 2), later known as the famous 'iceberg metaphor' (see Chapter 4) (Heinrich 1929; Lateiner 1958). The ratios varied considerably for each machine. Observations showed that accidents with different consequences resulted from different scenarios, and thus required a different risk-control approach.

The prevention of accidents in these four companies in the 2,000-accident study was a complex issue. In sharp contrast to the claim of management of these companies that safety was an important topic, the reality was that mostly apathy prevailed on the shop floor. 'Accidents are simply part of the work', was the opinion of employees. This indicated a sharp discrepancy between the shop floor and management, the well-known *us-versus-them* dichotomy, and in accident investigations the causes of accidents therefore did not go much further than claiming negligence on the part of the victims. Discussions with insurance companies on the guilt question and negotiations on financial consequences were actually the only forms of 'prevention' that companies undertook. Company safety officers spent so much time on negotiating financial aspects that they barely had time to perform proper accident investigations. Researchers complained that much of the large amount of knowledge available on the hazards of machinery and operations was rarely or never used, due to poor communication between the various parties within the company. An example was the Power Press Regulation of 1965. The hazards and risks attached to presses were discussed in detail in this regulation, including the training requirements for the operators of these machines (Broadhurst 1971). The 2,000-accident study could not demonstrate, however, that these regulations had measurable effects. The conclusions of the 2,000-accident study put forward two recommendations. The first advised a safety-based and ergonomic redesign of installations and workstations. In existing designs, far too little attention was paid to the limitations of human movement. The second recommendation concerned safety training for workers, which appeared to be virtually non-existent. According to the authors, such training, aimed at increasing workers' knowledge and understanding of hazards, would be desirable because it would increase both safety and production efficiency. It was therefore hoped that safety training would lead to a measurable reduction of accidents.

Occupational safety research in the 1980s

The 1980s saw a notable increase in the number of scientific publications on safety. Safety research centres were situated in Nordic countries (Stockholm, Tampere and Trondheim), the United Kingdom (Loughborough, Manchester, Surrey, London Imperial College and Birmingham Aston), the Netherlands (Leiden and Delft), Germany (Wuppertal and Cologne) and France (Paris). In this period more articles appeared on the quality of safety research, as had occurred at the beginning of the 1960s. Some researchers, however, found that progress in accident classification and causality had actually fallen behind (Singleton 1984). For example, on the topic of prevention much goodwill existed, but little theoretical background (Kjellén and Larsson 1981; Saari 1982). This lack of theory was particularly problematic for safety training and education. Safety prevention evaluation studies were often based on statistics on sick leave after accidents, or on sick leave trends before and after implementation of a safety-based intervention. This was not, however, an approach without drawbacks. Firstly, only averages were used and, furthermore, occupational hazards were not the only factor affecting sick leave. These comparisons were prone to regression towards the mean, which meant that statistical differences were wrongly attributed to the effects of interventions or measures, leading to faulty conclusions concerning the effects of interventions (Hauer 1980, 1983). Secondly, sick leave was a poor indicator for accident prevention since it focused on consequences of accidents rather than causes, and it also suffered from registration biases (Menckel and Carter 1985; Purswell and Rumar 1984). Thirdly, little information was available on exposure to hazards and on accident-causing conditions. The concept of exposure was always poorly developed within safety science. The closest organisations came to being aware of the importance of unsafe pre-existing conditions, like exposure, was to stress the implementation of a list of safe practice and safe conditions. These safety practices and conditions could be scored, providing a numerical ranking of the safety level of different factory departments and workplaces. These scores had much in common with the later developed safety indices (Fellner and Sulzer 1984; Frijters and Swuste 2008). But in this time period, the implementation of an index-based approach to safety gained much popularity in organisations, although it would become (much) more popular later on (Saari 1984; Hubbard and Neil 1985).

North America

Structures of organisations

A few publications about decision-making within organisations and the structure of organisations appeared, which changed the prevailing view at the time of organisations and companies. In 1979, the book *The Structuring of Organisations* by Canadian management specialist and engineer Henry Mintzberg (Mintzberg 1979) was published. This book presented a theory describing the division of labour and the coordination of tasks, factors that determined the formal structure of an organisation and its decision-making processes. Mintzberg described five generally applicable coordination mechanisms, ranging from mutual adjustment to direct supervision, and standardisation of work processes, output and skills. According to Mintzberg, every organisation consists of a limited number of components (Figure 5.4).

Every company has a *strategic top*, consisting of top and middle management (line management). Middle management is situated between the strategic top and the *operating core*. This operating core consists of the operators performing the core work that leads directly to production.

An organisation is supported by a techno-structure, namely the various analysts, IT, logistics planning and supporting staff, like a canteen, legal department or public relations department. Different coordination mechanisms, such as a simple structure, a machine bureaucracy, a professional bureaucracy, a divisional structure or an ad-hocracy, will attach a different weight of importance to each of the five main components of the organisation, resulting in different formal organisational structures.

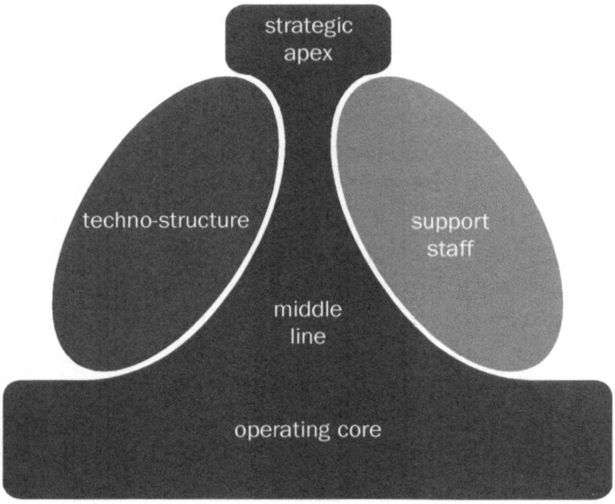

Figure 5.4 Components of organisations
Source: Mintzberg (1979)

A *simple structure* is an organisation in which the strategic top exercises direct supervision and is therefore the central part of the organisation. There is hardly any techno-structure, support staff or middle management. Examples include starting companies with a flat and flexible organisation. A *machine bureaucracy* is a vertically structured but also large organisation in which work processes are highly standardised, hence the term 'bureaucracy'. Executive activities are routine, and the techno-structure forms the central part of the organisation, hence the term machine. In a *professional bureaucracy* skills are highly standardised. The work carried out is more complicated. Therefore, the operating core, consisting of highly trained professionals, forms the most important part of the organisation. Consultancy and accountancy companies and universities are examples of this type of organisation. A *divisional structure* is an organisation with a number of quasi-autonomous divisions that are managed by a central head office. The divisional structure is most common in the private sector of an industrialised economy. In a divisional structure the output is standardised, so that all divisions deliver consistent outputs. Here middle management is the central part of the organisation, leading their divisions toward the common goal. Finally, an *ad-hocracy* has an organic structure with little formalised or standardised behaviour. Work consists of small, market-oriented projects carried out by small groups whose composition can change over time, depending on the expertise required. As with a simple structure, flexibility here is a prime requirement. The difference with the simple structure is the absence of a top, the strategic apex. According to Mintzberg, these different structures of organisations are determinants of an organisation.

These different structures of organisations are crucial for both the design of an organisation and the diagnosis of ineffective organisations. These insights are an indication of the quality of decision-making within organisations, including safety. Within companies, safety departments are generally situated in the techno-structure, sometimes in support services, and are at some distance from the prime production process. Many production companies have either the structure of a machine bureaucracy or a division structure. These different structures combined with the positions of safety organisations in companies lead to different conflicts and benefits. For example, a central organisation responsible for managing safety will be at some geographical distance from producing departments and factories, which is often a major barrier. A dominant position of middle management can also function as a barrier.

In contrast to Mintzberg, Edgar Schein stressed the uniqueness of organisations. The nature of production, a market environment and stakeholders are sector- or even company-specific. Therefore, universal management techniques did not seem to be applicable, and management systems were considered to be unique for individual organisations (Schein 1972).

Risk homeostasis

Only one new theory was developed in this period: the homeostasis theory. The term was derived from the natural sciences and described an organism's capacity to maintain stable internal conditions despite changes in the environment. This dynamic equilibrium in an open system was known as 'steady state', and differed from a physical-chemical equilibrium, which is static and, according to thermodynamics, only applicable in a closed system. The homeostasis theory originated from traffic safety research that explained why drivers seemed to adapt their behaviour to an accepted risk level (Wilde 1982). When drivers felt they drove utterly safely, they started driving more aggressively, and vice versa. Traffic safety research also showed that owners of very safe cars drove more aggressively than those owning less safe cars. The debate about this compensational, homeostasis-based behaviour is extensively discussed in the traffic safety literature.

The homeostasis theory had some practical problems, such as defining a safety level and preventing unsafe adaptations (McKenna 1985a, 1985b). The homeostasis concept was also introduced in occupational safety, where it suggested that very safe working environments could actually elicit **unsafe behaviour** (Wilde 1986; Groeneweg and Ter Mors 2016). However, the homeostasis theory was never widely accepted fully in occupational safety circles.

Occupational safety research in the 1980s

In North America, the safety research centres were in Boston, Chicago, Morgantown (West Virginia), Lubbock, Texas (Texas Tech University) and San Diego; and in Canada, in Hamilton. The United States kept close to its traditional domains, such as the cost of safety (Miller et al. 1987) and technical safety measures, such as the safety of stairs (Templer et al. 1985) and accidents with hydraulic presses (Collins et al. 1986). Comprehensive texts describing safe manufacturing were published: the *Accident Prevention Manual* of the National Safety Council (McElroy 1980) and the final edition of Heinrich's accident prevention book (Heinrich et al. 1980). The increased focus on research showed that safety and accidents were now taken as seriously as occupational diseases (Haddon 1980b), albeit evolving as a research domain with a dominant engineering approach (Fellner and Sulzer 1984).

Some lines of research were of a rather low quality, though. Particularly problematic was research on the effectiveness of safety training and education. Cohen and Jensen (1984) stated that safety training was not very effective since it did not focus on the practicalities of actual job performance. The relationship between the results of accident analyses and accident prevention was also rather obscure.

The focus on workers' responsibility for their own safety was more prevalent in the United States than in Europe. US safety literature focused strongly on human error models, and argued that accident prevention could best be achieved by providing positive feedback and rewarding safe behaviour. The psychological insights of the time dictated that positive sanctions or rewards worked better than negative ones (Sulzer-Azaroff and Santamaria 1980; Heinrich et al. 1980; Fellner and Sulzer 1984; Chhokar and Wallin 1984; Cohen and Jensen 1984). This view was also supported by researchers from Nordic countries (Grondstrom et al. 1980; Vuorio 1982). A key element of this approach was to observe human behaviour. Sulzer-Azaroff (1987) suggested training staff to be observers. These proposals paved the way for behaviour-based safety programmes, which are, to this day, quite popular in industry.

Prevention of accidents

The Haddon matrix (Table 4.2) led to preventive measures that could be divided into five large groups, all of which were aimed either at preventing energy exchanges or reducing their effects. First, energy accumulation had to be prevented (Strategy 1). If unsuccessful, then a second type of measure was aimed at preventing energy emission or changing the energy emission to a less dangerous form (Strategies 2–4). Another option was to separate potential victims from the energy release source (Strategy 5), and another was to create a barrier between victims and the energy emission source (Strategy 6). Finally, if Strategies 1–5 all failed, care of the victims had to be organized to occur as quickly as possible (Strategies 6–10) (Haddon 1963). In later publications, Haddon reformulated his preventive measures by listing ten different strategies designed to prevent, or at least reduce, injury (Haddon 1973, 1974, 1980a, 1980b):

1. *Prevent the build-up of energy: mechanical, thermal, kinetic, radiation or electric;*
2. *Reduce the amount of energy that can build up;*
3. *Prevent the release of energy. This refers to the Old Testament book of Deuteronomy 22:8. 'When you build a house, put a parapet along the roof, or you will bring the guilt of bloodshed on your house if anyone should fall from it';*
4. *Reduce the rate and spatial distribution of the energy release;*

5. *Separate the energy output of the host in time or space;*
6. *Place a physical barrier between the energy output and the host;*
7. *Limit the contact surface for the host;*
8. *Strengthen the resistance of the host;*
9. *Detect and evaluate the damage as soon as possible and take action;*
10. *Stabilize the host.*

Safety training was often based on Haddon's ten strategies for injury control. These strategies were more like rules of thumb and were not based on scientific research (Compes 1982). The ten strategies were furthermore only applicable to relatively simple occupational accidents. The strategies were therefore not effective in reducing accidents of a more complex nature (Barnett and Brickman 1986).

Occupational safety management systems and auditing

In the United States, several safety reference books on safety management were published in the 1970s (Petersen 1971, 1975, 1978; Bird 1974; Bird and Loftus 1976). Petersen coined the term 'safety management' in the titles of his books. He was the first to use this term. Models of accidents and their prevention in manuals by both Bird and Petersen looked very similar; both authors were indebted to Heinrich's domino metaphor. Bird used a modified version of the dominoes, changing the first three dominoes and introducing root and direct causes (see Figure 5.5 and Table 5.1).

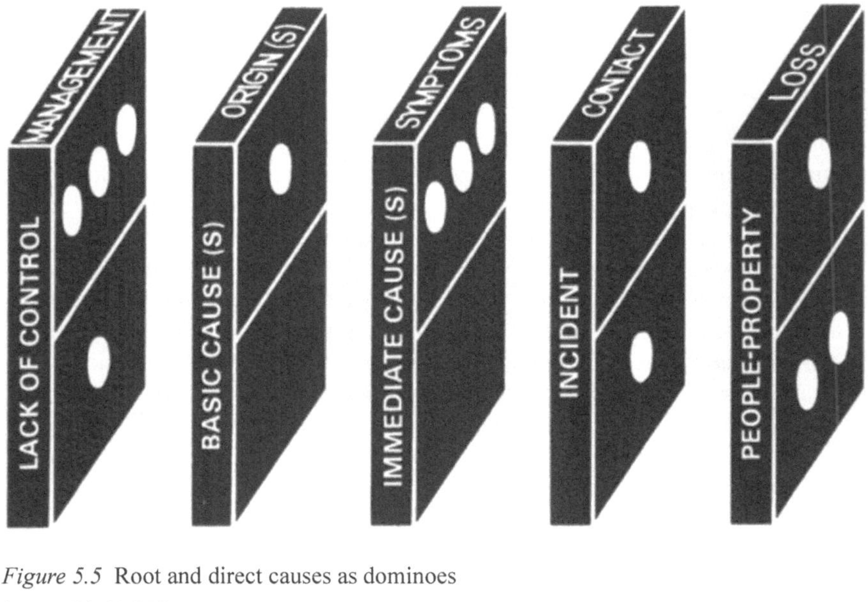

Figure 5.5 Root and direct causes as dominoes

Source: Bird (1974)

Table 5.1 Immediate causes of accidents (after Bird 1974)

Unsafe conditions	*Unsafe practice*
inadequate guards or protection	operating without authority
defective tools, equipment, substances	failure to warn or secure
congestion	operating at improper speed
inadequate warning system	making safety devices inoperable
fire and explosion hazards	using defective equipment
substandard housekeeping	using equipment improperly
hazardous atmospheric conditions: equipment gases, dusts, fumes, vapours	failure to use protective equipment
excessive noise	improper leading equipment
radiation exposures	improper lifting
inadequate illumination or ventilation	taking improper position
	servicing equipment in motion
	horseplay
	drinking or drugs

Root causes were a combination of personal factors and work-related causes. Personal factors were skills, motivation and mental or physical problems. Work factors were the different standards for tasks, design, maintenance, insufficient purchasing or simply wear and tear. The direct causes resembled the factors published by Heinrich in 1941 (Table 5.1).

Instead of social environment and fault of person, the first two dominoes of Heinrich, Bird's first domino symbolized a lack of management control of the company in question. This change in the domino metaphor resulted in a safety programme consisting of regular inspections, task analyses, safety procedures and training, as well as personal contacts with employees who functioned badly. The manual by Bird, and later by Bird and Loftus, took the reader by the hand and was written in a very practical and hands-on way, including extensive chapters on psychological insight into behaviour and motivation of employees.

The employer had to make safe working possible, but even more important was creating and maintaining a correct attitude on the part of the employee with regard to his work and safety. 'Motivation' was the term increasingly used (Zwaard 2007). Especially when combined with insights of American psychologists, this motivation-based approach to work situations became highly practical.

Maslow's pyramid of needs (Figure 5.6) was also a source of much inspiration (Maslow 1973; Mulder 2012). The pyramid illustrates that safety is an essential human need.

This topic was also discussed in Petersen's reference books. Both Bird and Germain were convinced that safety had to be integral to business and a central part of work procedures. This conviction is articulated in the third point of Table 5.2 (Petersen 1971). Again, both Bird and Petersen provided many examples of forms

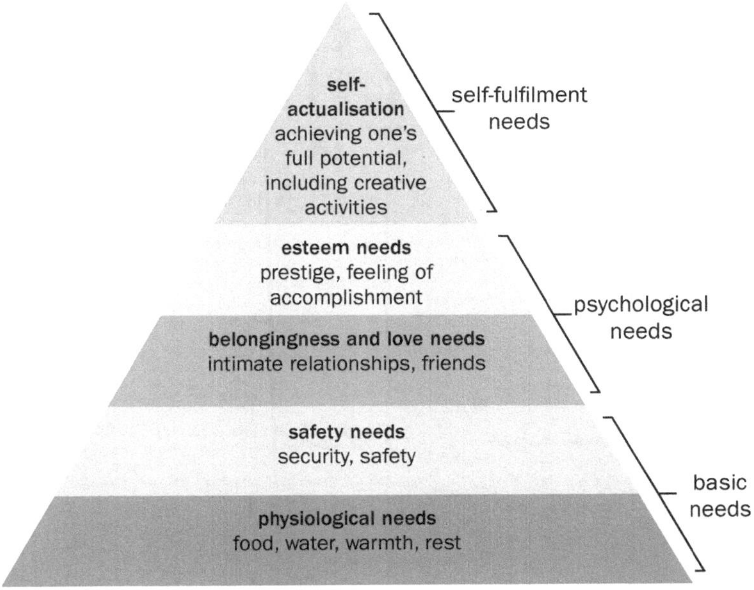

Figure 5.6 Maslow's hierarchy of needs

Source: After Maslow (1973)

Table 5.2 Safety management (after Petersen 1971)

New principles of safety managemen
1. An unsafe act, an unsafe condition, an accident: all these are symptoms of something wrong in the management system.
2. Certain sets of circumstances can be predicted to produce severe injuries. These circumstances can be identified and controlled: Usual, non-routine High-energy sources Non-productive activities Certain construction situations
3. Safety should be managed like any other company function. Management should direct the safety effect by setting achievable goals, planning, organising, and controlling to achieve them.
4. The key to effective line performance is management procedures that fix accountability.
5. The function of safety is to locate and define the operation errors that allow accidents to occur. This function can be carried out in two ways: 1) by searching for root causes of accidents, and 2) by asking whether certain known, effective controls are being utilised.

that could be used for accident investigation reports, safety inspections and other initiatives.

Combinations of these topics turned into a first draft of what was later called a 'safety audit'. Petersen introduced SCRAPE, a systematic model for measuring

TOR TECHNIC OF OPERATIONS REVIEW

1 COACHING

10	Unusual situation, failure to coach (new man, tool, equipment, process, material, etc.)	44, 24, 62
11	No instruction. No instruction available for particular situation	44, 22, 24, 80
12	Training not formulated or need not foreseen	24, 34, 86
13	Correction. Failure to correct or failure to see need to correct	42, 20, 30
14	Instruction inadequate. Instruction was attempted but result shows it didn't take	15, 16, 42
15	Supervisor failed to tell why	44, 24, 83
16	Supervisor failed to listen	11, 81
17		
18		
19		

2 RESPONSIBILITY

20	Duties and tasks not clear	44, 34, 14, 53
21	Conflicting goals	80,
22	Responsibility, not clear or failure to accept	26, 14, 54, 82
23	Dual responsibility	47, 34, 13
24	Pressure of immediate tasks obscures full scope of responsibilities	36, 12, 51
25	Buck passing, responsibility not tied down	44, 26, 55, 60
26	Job descriptions inadequate	80, 86
27		
28		
29		

3 AUTHORITY (Power to decide)

30	Bypassing, conflicting orders, too many bosses	44, 13
31	Decision too far above the problem	36, 83, 85
32	Authority inadequate to cope with the situation	81, 83
33	Decision exceeded authority	20, 26, 14
34	Decision evaded, problem dumped on the boss	36, 14, 85
35	Orders failed to produce desired result. Not clear, not understood, or not followed	40, 46, 13, 15
36	Subordinates fail to exercise their power to decide	26, 12, 83, 85
37		
38		
39		

4 SUPERVISION

40	Morale. Tension, insecurity, lack of faith in the supervisor and the future of the job	15, 56, 64, 80
41	Conduct. Supervisor sets poor example	13, 84
42	Unsafe Acts. Failure to observe and correct	24, 11, 52
43	Rules. Failure to make necessary rules, or to publicize them. Inadequate follow-up and enforcement. Unfair enforcement or weak discipline	25, 34, 12, 52
44	Initiative. Failure to see problems and exert an influence on them	22, 34, 30
45	Honest error. Failure to act, or action turned out to be wrong	10, 12, 15, 81
46	Team spirit. Men are not pulling with the supervisor	40, 21, 56
47	Co-operation. Poor co-operation. Failure to plan for co-ordination	23, 25, 15, 66
48		
49		

Figure 5.7 Example of a TOR management audit

Source: Petersen (1971)

5 DISORDER

51 Work Flow. Inefficient or hazardous layout, scheduling, arrangement, stacking, piling, routing, storing, etc. 41, 24, 31, 80

52 Conditions. Inefficient or unsafe due to faulty inspection, supervisory action, or maintenance 21, 32, 14, 86

53 Property loss. Accidental breakage or damage due to faulty procedure, inspection, supervision, or maintenance . 43, 20, 80

54 Clutter. Anything unnecessary in the work area. (Excess materials, defective tools and equipment, excess due to faulty work flow, etc.) 44, 36, 80

55 Lack. Absence of anything needed. (Proper tools, protective equipment, guards, fire equipment, bins, scrap barrels, janitorial service, etc.) 44, 36, 80

56 Voluntary compliance. Work group sees no advantage to themselves 40, 15, 41

57

58

59

6 OPERATIONAL

60 Job procedure. Awkward, unsafe, inefficient, poorly planned 44, 32

61 Work load. Pace too fast, too slow, or erratic 44, 51, 63

62 New procedure. New or unusual tasks or hazards not yet understood 43, 44

63 Short handed. High turnover or absenteeism 80, 40, 61

64 Unattractive jobs. Job conditions or rewards are not competitive 81, 46

65 Job placement. Hasty or improper job selection and placement 80, 86

66 Co-ordination. Departments inadvertently create problems for each other (production, maintenance, purchasing, personnel, sales, etc.) 45, 35, 13

67

68

69

7 PERSONAL TRAITS
(When accident occurs)

70 Physical condition — strength, agility, poor reaction, clumsy, etc. 44, 26, 65

71 Health — sick, tired, taking medicine 44, 24, 65

72 Impairment — amputee, vision, hearing, heart, diabetic, epileptic, hernia, etc. 44, 24, 65

73 Alcohol — (If definite facts are known) 80

74 Personality — excitable, lazy, goof-off, unhappy, easily distracted, impulsive, anxious, irritable, complacent, etc. 44, 13

75 Adjustment — aggressive, show off, stubborn, insolent, scorns advice and instruction, defies authority, antisocial, argues, timid, etc. 44, 13

76 Work habits — sloppy. Confusion and disorder in work area. Careless of tools, equipment and procedure 44, 13

77 Work assignment — unsuited for this particular individual 42, 65

78

79

8 MANAGEMENT

80 Policy. Failure to assert a management will prior to the situation at hand 24, 81, 83

81 Goals. Not clear, or not projected as an "action image" 83, 86

82 Accountability. Failure to measure or appraise results . 36

83 Span of attention. Too many irons in the fire. Inadequate delegation. Inadequate development of subordinates 12, 86

84 Performance appraisals. Inadequate or dwell excessively on short range performance 20, 65

85 Mistakes. Failure to support and encourage subordinates to exercise their power to decide 36

86 Staffing. Assign full or part-time responsibility for related functions 66

87

88

89

Figure 5.7 (Continued)

the safety efforts of foremen, such as through safety inspection rounds, safety training and meetings, and accident investigations (Petersen 1971). These activities were recorded and given numerical scoresthrough which management could award a weekly score to each leader. A second system was the Technic of Operations Review (TOR), which, after an accident, a near accident or damage, enabled the detection of safety deficiencies in an organisation. An example of this is given in Figure 5.7.

The reference book by Petersen (1975) provided an extensive report of the psychological and management models that existed on motivation, behaviour and different ways of managing a safety programme. Another audit system, the International Safety Rating System (ISRS), was developed by Bird and Germain (1985). The ISRS was based on the International Mining Safety Rating System, used by the Chamber of Mines of South Africa to measure safety improvements. The audit consisted of 20 elements, each with a dozen questions (Figure 5.8). (The questions are not shown in the figure.)

Answers to these questions produced a score allowing benchmarking, and also guiding management actions in the case of questions answered negatively. Unfortunately, a clear relationship between accident occurrence and answers to the audit questions was never found (Eisner and Leger 1988; Guastello 1991). Following these ISRS elements, the role of management was not fundamentally different from that of the 1950s. Cost control and production efficiency remained central. A combination of a safety analysis of each task, task-specific training and ongoing safety observations, otherwise known as the JSA JIT SO system (job safety analysis, job instruction training, safety observations), gave companies a high degree of freedom of choice. From then on, the safety record of a plant was seen as a measure of the quality of its management, its safety system and its safety audits. As Heinrich had stressed, the foreman was central. If a foreman could not properly organize safety, then his control of costs and general product quality and production level was also doubtful. The responsibility for safety was

Figure 5.8 Elements of loss control management – ISRS

Source: Bird and Germain (1985)

placed fairly and squarely on the shoulders of middle and lower management. At the same time, the system approach was directly related to training and design. If an industrial system failed, it was because those operating the system were insufficiently trained to deal with the design of the system (Johnson 1970; Cleveland et al. 1979). This view was reflected in the definition of an accident, in which the 'energy transfer' and 'barrier' concepts referred directly to the models of Gibson (1961) and Haddon and his co-authors (1964):

> *An accident is the result of a complex series of events, related to energy transfer, failing barriers and control systems, causing faults, errors, unsafe acts, and unsafe conditions and changes in process and organisational conditions.*
> (Johnson 1970)

Workers' well-being

In the United States, attention was paid to a topic that later in the Netherlands became known as the *humanisation of labour*. Short-cycle work on conveyor belts was described as being monotonous and demotivating. In the literature, a comparison was made between existing factory situations and those tragicomically portrayed in Charlie Chaplin's 1936 classic satiric film *Modern Times* (Swain 1973). Higher wages, strict employee selection, training and motivational programmes, and punishment of unsafe behaviour – none of these measures had any demonstrable effects, or, if they did, effects were only short-lived. The suggested solutions were to seek job enrichment, to match tasks to humans and to give workers greater autonomy over the organisation of their work (Pfeifer et al. 1974; Cohen et al. 1975). The need for active involvement of top management of companies was also emphasised. Such involvement not only saved time and money but additionally showed demonstrable management care for the well-being of employees. The concept of *workers' well-being*, which originated in the United Kingdom, was introduced to the United States (Ellis 1975; Cohen 1977; Cleveland et al. 1979; Nye 2013). With this new concept, the previously mentioned safety initiatives as the generally accepted determinants of safety were criticized for their poor scientific basis. Cohen also commented on the extensive amount of literature on safety training that lacked evaluative research (Cohen et al. 1979). Due to the low scientific quality of the research on which they were based, Ellis (1975) doubted the effects of safety legislation, inspections, statistics and government standards on safety in companies.

Safety and changing technology

Safety had changed beyond recognition during the second half of the twentieth century. Purswell and Rumar (1984) tried to capture this changing relationship between employees and their work (Figure 5.9). They showed that the evolution of industry led to significant changes in safety. From a direct coupling between employees and their tasks, as in manual labour before 1800, when employees

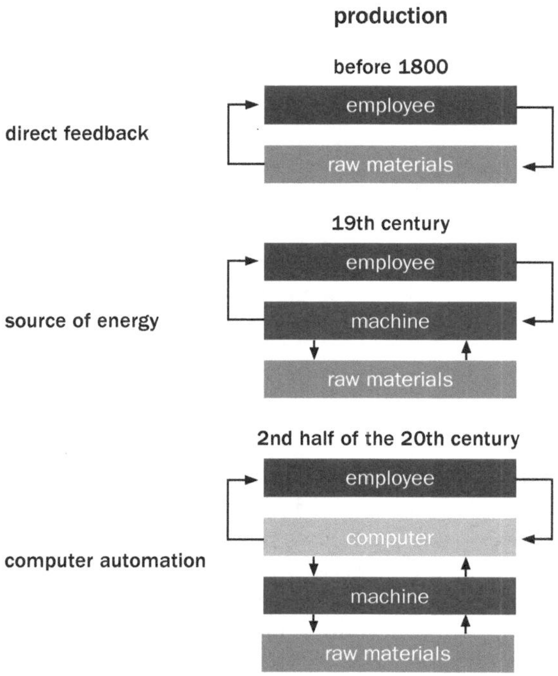

Figure 5.9 Evolution of industry

Source: After Purswell and Rumar 1984)

determined their own safety and production speed, the early 1800s saw the introduction of machine-powered manufacturing. This development increased production speed substantially and also changed the relation between the employee and his work fundamentally.

The employee was now the controller of machinery and was therefore expected to solve problems himself, which increased the need for job concentration and mental processing. The physical distance between employee and power machine was also short, which led to many accidents in factories. Furthermore, managers started to depend on employees' knowledge and experience, rather than their own, a factor that distanced managers from production processes. While mechanisation of processes made tremendous progress, employees' tasks remained fundamentally unchanged. The responsibility for safety was now shared between designers and managers, who determined the interaction between machines and employees.

This relation changed again during the second half of the twentieth century, when automation and remote control were introduced. Production processes were now directly controlled by computers, which increased both productivity and complexity significantly. Employees' tasks were now reduced to monitoring processes and intervening during process disturbances, the latter occurring only rarely. Both literally and figuratively, the distance between workers and hazards increased. The need for the human controller to have greater cognitive capabilities increased as well, even if this cognitive power was required only infrequently. Thus, while employees experienced extensive periods of boredom, they also experienced infrequent, short periods of frantic situations, during which their mental models of the complexity of the process they were monitoring often proved to be inadequate. The responsibility for safety was still the responsibility of designers and managers, but the responsibilities that lay between employees' actions and the occurrence of accidents were still not very clear. Situational factors and various human shortcomings played a role in safety, but how these interacted was not well understood. At the same time, both the number and seriousness of major accidents increased, often extending into public spaces, making safety a political issue.

The Netherlands

In the Dutch professional journal *De Veiligheid* (The Safety Journal), from the 1930s onwards the focus on accident proneness theory and Heinrich's notions of lack of safety had become dominant, but by the 1970s this focus had somewhat diminished. Other developments originating from the safety science domain and mainly from the United States and the United Kingdom were increasingly discussed. There was extensive coverage of the theories and models expounded by Winsemius (1951), task dynamics (Anonymous 1974; Andriessen 1974; Wijk 1977), prevention strategies and Haddon's hazard-barrier-target model (Bergsma 1974), loss control management proposed by Bird and the reference books of Petersen (Pope 1976; Bird 1978; Fletcher 1978; Wright 1978; Leij 1978, 1979a). The first book in Dutch on safety management was also published in this period (Zwam 1978). The review of Petersen's safety manual marked the introduction of the term 'safety management' to the Netherlands. The regular control of process deviations and organisational change was viewed as a method of prevention (Dop 1977, 1979; Radandt 1979). The topic of the humanisation of labour, which addressed the negative effects of an extensive division of labour, and the separation of management and operations, or *head and hand*, were the consequences of scientific management (Strien 1978). Conflicts over responsibility in safety management were also evident to Working Group 13 of the Dutch Society for Safety Science (NVVK). This working group addressed the dilemma of company interests versus the interests of employees. The role of a safety officer was difficult to combine with the promotion of the well-being of workers; at least that was the argument (Meertens and Zwam 1976; Kraan and Schenke 1976; Oosterom 1979). The debate about the ethics of the safety profession was not unique to the NVVK. At universities, the societal role of scientists and engineers was being discussed in newly established chemistry and society courses and applied in practice in science shops (Lemkowitz 1992; Zwaard 2007).

For safety professionals in the Netherlands, the mid-1980s was a tumultuous period. The term 'occupational safety' seemed old-fashioned with the introduction of the 1983 Working Conditions Act and was being replaced by the more modern term 'working conditions'. In 1986, for example, the monthly journal *De Veiligheid* made way for the new *Monthly Journal on Working Conditions*, and two years later a scientific journal, the *Tijdschrift voor toegepaste Arbowetenschap* (TtA; the Journal of Applied Occupational Sciences), was born. At first this magazine appeared as a separate section of the *Monthly Journal*, at a frequency of five to six times a year; ten years later it became an independent journal. Its editors signalled a growing gap between researchers reporting largely theoretical scientific development of work, safety and health and the practicalities of working in the professional field itself. Researchers at universities and research institutes usually published in English-language scientific journals, which were rarely read by Dutch safety professionals. The *Journal of Applied Occupational Sciences* sought to close this gap, firstly by challenging researchers to make their publications accessible to a more practically oriented audience, and secondly by encouraging safety professionals to publish their research in a scientific format (Korstjens 1988; Vernooy 1988). In 1987, the Safety Institute merged with the *Coördinatie van Communicatie met betrekking tot gegevens inzake Onderzoek Ziekteverzuim* (CCOZ; Coordination of Communication on Research on Sick Leave) to become the *Nederlands Instituut voor Arbeidsomstandigheden* (NIA; Dutch Institute for Working Conditions). Within a few years, the Dutch safety domain lost three important fundamentals: a law, a journal and an institution.

Human error

In the Netherlands, the protection of workers was often discussed in this period. Ergonomics and inherently safe workstations and equipment were also frequently reported (Stassen 1981; Poll 1983, 1984; Jong and Poll 1984). A popular slogan was *'Expect trouble if you allow safe people to work in an unhealthy environment'* (Boudri 1979). Also, attention shifted from the victim to the context in which the victim worked (Keyser 1979; Redactie 1986). Task dynamics increased in importance (Kraan 1981; Zuuren 1983), as did ergonomics and safe design (Comeche 1979; Zwam 1979b).

Other issues concerned safety departments and safety management systems, and the focus on the so-called soft side of safety: organisational factors, particularly management, perception and responsibility. This notion of a soft side started in the 1970s with studies on risk perception and safety climate (e.g. Zohar 1980a). Zohar's work was translated into Dutch in *De Veiligheid* (Zohar 1980b). The importance of safety meetings, training, audits and regulation was also stressed (Zwam 1979a; Putman 1986). An early type of safety audit was presented, which showed quantitatively the different maturity levels of safety in organisations. This audit proposed safety indicators, such as the number of safety audits, the frequency of safety meetings, the number of training courses and the number of task analyses (Leij 1979a, 1979b).

In regard to the question of responsibility for safety, two schools of thought existed in the Netherlands. The first school adhered to the American-style human factor approach: workers were themselves responsible for their own safety and the safety of their less senior colleagues. Training and exchange of experiences in management teams could facilitate this approach. The Swain project report disseminated this approach in the Netherlands (Anonymous 1982; Blijswijk and Mutgeert 1987). The second school of thought adhered more to the British approach, in which the responsibility for worker safety was primarily that of organisations (e.g. companies) and their managers. This approach followed the recommendations of the Robens Report of 1972, which had blamed failing management and poor management systems as the primary causes of inadequate safety (Oirbons 1981).

Humanisation of labour

Related to the management school of thought, discussions of the humanisation of labour were centred on the introduction of a new law on working conditions. This Dutch *Arbowet* (Working Conditions Law) dealt with occupational hazards, safety organisation and safety responsibilities. Its philosophy was based on creating a harmonious collaboration and maintaining active discussion between labour and management about safety issues. *Humanisering van de arbeid* (humanisation of labour) meant, specifically, the endeavour to change labour and labour conditions in order to minimise the chances of injuries to body and mind (Roos 1979). The *Arbowet* also introduced the term 'well-being' (Dutch: *welzijn*) as a central concept within Dutch labour law. Additionally, the Minister of Labour at the time specifically stated that women and immigrants should not carry the uneven burden of substandard working conditions. Discussion and personal development were to be promoted, monotonous work to be avoided and craftsmanship to be stimulated (Anonymous 1979, 1980a; Sluis 1984). Perhaps not surprisingly, in practice these idealistic goals were hard to achieve. Publications in Dutch occupational safety journals, such as the *Risicobulletin* (Risk Bulletin), showed that many corporations often failed to reach these idealistic goals. In many publications, labour unions, like the Industriebond Federatie Nederlandse Vakbeweging (IBFNV; Industry Union Federation Dutch Trade Union Movement), criticised companies that failed to meet the requirements of the new law. These publications were often based on cooperation between unions, universities and non-governmental bodies, like the Advisory Group on Work and Health (Adviesgroep Stichting Arbeid en Gezondheid). An example was research conducted by the University of Leiden Science Shop, scrutinising the American corporation Cyanamid for exposing workers to unacceptably high levels of toxic chemicals (Beek et al. 1982).

Unions played a central role in the discussion on the humanisation of work and in the critique on practices of many corporations. The latter consisted of many reports describing poor working conditions in corporations and factories, such as chemical process industries, rubber factories, transport companies, food processing plants, construction, electrical engineering and printing (Vreeman 1982; Buitelaar and Vreeman 1985). These reports described unsafe installations and unacceptable levels of exposure to toxic chemicals, vibration, noise, electrical shocks and toxic dusts. Additionally, many problems with salaries, leave, product quality and low frequency of work meetings were identified (Hattem 1980). For the Dutch unions at that time, occupational safety was an important issue.

An unusual example was the DuPont company, which was actually criticised for stressing safety too much. DuPont's safety management extended even into the private lives of its workers (Duyvis 1979), which created resistance amongst those workers. DuPont's justification for its policy was the high frequency of domestic accidents of its workers, which reduced labour performance. DuPont therefore encouraged its employees to follow corporate safety rules even during their holidays (Sluis 1983). Safety procedures at DuPont existed for more or less everything, ranging from generic ones, department-specific ones, section-specific ones and procedures, right down to individual tasks. Weekly inspections created numerical 'unsafe act index scores'. With so many procedures it was evident that at least one was broken or violated when an accident occurred. This pressure to follow DuPont's extensive rules created an overly competitive atmosphere, with negative consequences for pay rises and advancement. While DuPont paid a lot of attention to personal protective equipment, implementing a more basic prevention-based safety approach, such as redesigning machinery to reduce risk, was out of the question (Onderzoeksgroep Veiligheid en DuPont 1982; Boonstra 1983, 1983/4). On the whole, this type of behavioural focus on safety, popular in the US, was not embraced in the Netherlands (Anonymous 1980b, 1983).

Risk and occupational safety

In the 1970s and 1980s, *De Veiligheid* published its first articles on the concept of risk and the benefits of a risk-based approach. Both risk and probability were unknown concepts in the occupational safety domain in the Netherlands at that time. Only the term 'hazard' (Dutch: *gevaar*) was used in laws and publications. Smit (1971) argued that accident statistics were inadequate for policy purposes and were even misleading. He introduced the concept of risk very carefully, stating, for example, 'We could call risk the product of severity multiplied by probability'. A major impact of the risk-based approach was the necessity of accepting a particular risk, as zero risk was not possible. When discussing acceptable risk criteria, Smit referred to flood risks in the Netherlands, which for decades had been used as one of the criteria for determining the heights of dykes that protected the country against flooding (see Chapter 4). From this point onward, acceptable risk criteria and the impossibility of zero risk comprised important elements of the safety debate.

In the Netherlands, this debate resulted in a risk-ranking method being introduced in occupational safety (Henstra 1992; Zwaard and Goossens 1997) (see Table 5.3). Occupational safety within most companies usually involved risks with relatively high probabilities of occurrence but relatively minor effects. The risk-ranking method used in the Netherlands was based on publications by Fine and by Kinney and Wiruth. In 1971, Fine presented a method for preparing a simple and semi-quantitative risk assessment suitable for these risk categories (Table 5.3).

Table 5.3 Risk ranking in occupational safety (after Fine 1971)

Likelihood	*Value*
Might well be expected	10
Quite possible	6
Unusual but possible	3
Only remotely possible	1
Conceivable but very unlikely	0.5
Practically impossible	0.2
Virtually impossible	0.1
Exposure	*Value*
Continuous	10
Frequent (daily)	6
Occasional (weekly)	3
Unusual (monthly)	2
Rare (a few per year)	1
Very rare (yearly)	0.5
Possible consequence	*Value*
Catastrophe (many fatalities, or > 10^7 US$ damage)	100
Disaster (few fatalities, or > 10^6 US$ damage)	40
Very serious (fatality, or > 10^5 US$ damage)	15
Serious (serious injury, or >10^4 US$ damage)	7
Important (disability, or > 10^3 US$ damage)	3
Noticeable (minor first aid accident, or > 100 US$ damage)	1
Risk situation	*Risk score*
Very high risk; consider discontinuing operation	>400
High risk: immediate correction required	200 to 400
Substantial risk: correction needed	70 to 200
Possible risk: attention indicated	20 to 70
Risk: perhaps acceptable	<20

The method was further developed by Kinney and Wiruth in 1976 (Fine 1971; Kinney and Wiruth 1976). In two brief reports, these researchers presented a method for determining the relative seriousness of hazards to support the setting of priorities for control measures. In particular, Fine discussed the evidence supporting the method. The method was tested, comparing the risk ranking performed independently by several experts, thereby calibrating the ranking method. Based on these risk rankings, the different levels of risk were classified by giving them a certain weight and valuation. Fine believed that this classification would require adjustment in the course of time due to changes of insights. He also stressed that the method was suitable for ranking risks within a given organisation. The risk score was calculated using the following formula:

$$Risk\ score = probability \times exposure \times consequences$$

The severity of the risk was assessed based on the potential consequences of an incident, the exposure to or frequency of occurrence of a dangerous situation that can lead to an incident, and the probability that such a dangerous situation could lead to an incident with these potential consequences. The outcome is a risk score. The terms ‘scenario’, or ‘accident scenario’, were not used. But the method did not allow the assessment of several different consequences to be combined, because the frequency of exposure and the probability of consequences would differ for different scenarios. The method was semi-quantitative. Thus, the probability was not calculated as a product of the probability of successive events but was estimated as the probability of a sequence of the respective events, each occurring one after another. The method presented a rating scale of six to seven categories for each of the three factors in the formula above. For each factor the two extremes were first determined, such as the most serious and least serious consequences. The intermediate categories were then used for the other possible consequences, with the distance between the seriousness categories plotted on a logarithmic scale. The assessment of the possible consequences included both personal injury and property damage. Fine and Kinney, as well as Wiruth, presented the method to be used as a tool for decision support to prioritise interventions. Additionally, Fine (1971) argued that the cost of control measures should be justified on the basis of risk instead of the persuasiveness of a safety expert.

Acceptability of risks, standards for occupational exposure to carcinogens

Acceptability of risks remained a difficult subject, including in occupational hygiene and medicine. Zielhuis (1984) addressed a number of difficult questions in determining the acceptable risk from occupational exposure to carcinogens. What was needed were quantitative standards for acceptable exposure limits, based on a reliable standard-setting procedure that could be reviewed and validated by third parties. It was insufficient to state that the exposure *should be as low as possible.* Standards for exposure limits should be based on transparent criteria, such as preventing unacceptable health effects in general or, more specifically, preventing specific health effects for a given calculated fraction of employees at an exposure of eight hours per day, five days a week, for an entire working life. Earlier the Dutch Health Council had published a report on the carcinogenicity of chemicals (Gezondheidsraad 1979, 1980), and one of the first recommendations of the Dutch Expert Working Group (Werkgroep van Deskundigen, or WGD) was also the assessment of the risks of carcinogens. Zielhuis, who was involved in both publications, discussed in detail the many questions involved in setting standards, such as variation between individuals, variability in daily exposure and extrapolation of the results of animal experiments to humans. Zielhuis (1984) ended his publication dramatically, as follows:

> *From both the Dutch Expert Commission of the Ministry of Social Affairs (WGD) and the occupational physicians in practice, a response is expected in areas that even angels are reluctant to enter.*

The Health Council of the Netherlands and the Dutch Expert Working Group (WGD) distinguished two categories of substances (Health Council of the Netherlands 1979):

1. *Substances causing an irreversible self-replicating effect, i.e. causing a permanent change of structure in the DNA: initiators;*
2. *Substances that are active through other mechanisms: promoters or co-carcinogens.*

This distinction was crucial for setting acceptable limits, as it explicitly stated that for the first category of substances no safe threshold could be determined, while for the second category a safe threshold could be established, based on a no-effect level. For exposure to the first category of carcinogenic substances the WGD presented a basis for calculating acceptable risk: exposure to x micrograms/m^3, for eight hours per day, five days per week, for 35 years will result in a probability of developing a specific form of cancer in 1 in 10,000 employees. By linear extrapolation from this basis and after balancing various interests, the Dutch National MAC committee (the national committee for occupational standard setting) could establish an exposure limit that corresponded to a certain accepted risk. Zielhuis (1984) referred to the Scientific Committee of the Food Safety Council of the United States. This council bases its standards on a 'near zero' risk of 10^{-6} to a lifetime exposure of the general population. The Food Safety Council proposed this value as a measure of a negligible risk. Later this same value (10^{-6}) appeared in many different policy areas as a criterion for acceptable risk (Kelly and Cardon 1991). On the origin of the value of 10^{-6} per year as a negligible risk level, many different views can be found in the literature. Wildavsky and Wildavsky (2008) argued that the 10^{-6} criterion had largely a symbolic value. Others stated that the limit was based on the belief that the acceptable risk of an industrial activity should not exceed the risks of natural hazards that the public accepts (Kletz 1981), such as floods, earthquakes and lightning, which are around (or less than) 10^{-6} (Kletz 1981).

This chapter has shown the limited development of theory in the occupational safety domain in the 1970s and 1980s. Only one theory, risk homeostasis, was published. This theory gave insights into the occurrence of road accidents, but these insights could not be applied to occupational safety. A major development came from the United Kingdom. The Robens report showed the ineffectiveness of safety laws and regulations and handed back

the responsibility for providing sufficient worker safety to industry, as the obvious party creating hazardous conditions. The report also initiated a move from descriptive legislation to goal-based legislation. Additionally, a humanitarian approach to creating safe working conditions was proposed. In the Netherlands a few years later, this development led to the incorporation of a new concept into the new Labour Law, the 'well-being' (*welzijn*) of workers. In the United Kingdom, the first prospective safety study was carried out. The study showed a severe disinterest in the management of safety, and, furthermore, in existing safety legislation having only a *paper reality* – having no connection to safety in practice. This study also questioned Heinrich's ratios proposed with his iceberg metaphor.

During this period many safety models were developed, whose origins lay in western Europe and Nordic countries and North America. British ergonomists conducted a lot of research on information flow breakdowns that led to accidents. In North America, a management theory on the structure of organisations and its influence on decision-making, including safety, was published, in which the focus on human error remained prominent. North America had had a strong focus on management for a long time. During this time period, audits were developed using scores to gain insight into improvements made and to benchmark in order to compare results with those of other companies. In the Netherlands, initial discussions on risks and risk acceptance, and on safety climate, also started in this period. Risk, risk acceptance and safety climate all received a lot of attention, these topics being an essential part of discussions on high-tech-high-hazard safety, as will become apparent in the next chapter.

6 Risk and management, safety by organisation

1960–1990

The 1970s and 1980s were the heyday of safety science for the high-tech-high-hazard industry. The trend was amplified by frequent major accidents, which attracted a lot of media attention and gave rise to various models and theories concerning safety. The concept of 'risk' gained acceptance, but there was a clear countermovement. Slowly the optimism regarding technology bringing salvation to humankind from the pre-war period and the 1950s changed into the general view that technology can fail with terrible consequences. This notion led to various safety techniques in the 1960s and became prominent starting in the 1970s. In this chapter we will discuss only a few major accidents and their impact on knowledge development. To make the information manageable, the presentation of these major accidents in western Europe and Nordic countries will be divided into two parts, one until 1980, and the other 1980 to 1990. As in earlier chapters, the focus is on western Europe and Nordic countries, North America and the Netherlands, but major accidents happening in the former Soviet Union, India and Mexico are also discussed.

DOI: 10.4324/9781003001379-6

Western Europe and Nordic countries

Some major industrial accidents in the 1960s and 1970s

Major industrial accidents occurred in western Europe in this period and before the time frame of this chapter. The accidents had a profound impact on European legislation and knowledge development. In retrospect, these developments looked like a roller-coaster ride and meant the start of the 'golden years' of safety science. This chapter deals with 14 accidents that caused a turning point in the safety sciences. Many more took place. An overview of post-1990 accidents can be found in the appendix of this book. For the pre-1988 period the handbook of Frank Lees provides an overview (Lees 1980, 1996; Mannan 2005, 2012). Two industrial sectors had the most accidents. The first was the nuclear sector, which started just after the Second World War. After some nuclear incidents in the late 1950s and in the 1960s, it was the nuclear sector that was the first to publish visual presentations of an accident process. The second industrial sector was the chemical process industry. After some less well-known major accidents in the early 1970s, the process industries launched loss prevention, a new movement in the process industry (Rademaeker et al. 2014). The chemical engineer Frank Lees (1931–1999) published his monumental *Loss Prevention in the Process Industries* (Lees 1980), and the chemist Trevor Kletz (1922–2013) published a huge number of books and articles on practical safety issues in the process industries, including inherent safe design (Kletz 1976, 1978, 1981, 1982, 1984a, 1984b, 1985a, 1985b, 1985c, 1986, 1988a, 1988b, 1988c). In effect, they founded this movement. Lees and Kletz limited their field of interest to chemical and petrochemical processing, and their approach provided useful guidance in an industry where production volumes, instrument dimensions and the technical complexity of chemical processes increased constantly (Kletz 1999).

The boiling liquid expanding vapour explosions (BLEVE) at Feyzin, France, in 1966, the explosion at Flixborough, United Kingdom, in 1974, and the BLEVE at the Los Alfaques camping site, in Spain, in 1978 emphasised the necessity of this movement. In both the nuclear sector and the process industry, the concept of risk became a major issue, based on operations research and reliability engineering (see Chapter 4). This led to a quantitative risk assessment research project at Canvey Island in 1978, an industrial cluster close to a residential area at the mouth of the River Thames, to calculate the risk of a major industrial accident for residents of this island. The dioxin emission at Seveso, Italy, in 1976 led to European Council Directive 82/501/EEC of 24 June 1982 on the major-accident hazards of certain industrial activities, commonly known as the 'Seveso I directive'. Both of the major accidents at Feyzin and Flixborough were examples of domino effects, a topic that was taken on board much later in academic safety research. The so-called forgotten mine accident at Aberfan, Wales, in 1966 was one of the examples of the 1976 'disaster incubation theory' of the sociologist Barry Turner. According to Turner, large-scale accidents such as the one at Aberfan are often years in the making before they finally materialise into disaster. Below, we review a few of the major industrial accidents that happened in Europe between 1960 and 1980.

Feyzin, 1966

The major accident at the Feyzin refinery, which took place about 10 kilometres south of Lyon, did not attract much publicity outside France at the time. A large fire and several explosions started in the storage tank of this refinery. A liquid petroleum gas (LPG) emission, caused by a gas leak during sampling, created a gas cloud, which was ignited by a car from the adjacent road. After an hour, the first BLEVE from an LPG tank was a huge fireball spanning 250 metres. Eighteen people died, 81 were injured and there was extensive damage to the site and the neighbouring village of Feyzin (Figure 6.1).

Today this major accident is recognised as one of the first recorded accidents where a domino effect took place (Reniers et al. 2005a; Swuste et al. 2020c). Domino effects were defined as 'a cascade of events in which the consequences of a previous accident increased through successive events, both spatially and sequentially, and led to a major accident' (Reniers et al. 2005b). The blast wave of the explosion travelled through the Rhone Valley, breaking windowpanes up to a distance of 8 kilometres. The fragments of the explosion caused severe damage to neighbouring LPG storage spheres and equipment. After 45 minutes a second BLEVE occurred at an adjacent LPG tank. This domino effect created additional fires and fragments that weakened the legs of surrounding spherical storage tanks and tilted them. Fortunately, these tanks did not explode, but a number of petroleum and crude oil tanks caught fire. The water spray system was activated but did not function adequately (Lees 1980; IChemE 1987; HSE 2010; Török et al. 2011).

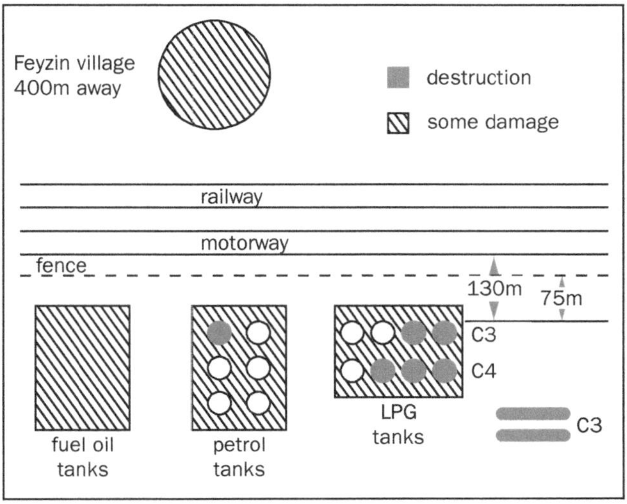

Figure 6.1 LPG facility, Feyzin, France

Source: IChemE (1987)

Aberfan, 1966

The second major accident occurred in the village of Aberfan, Wales, 25 kilometres north of Cardiff. On 21 October 1966, after three weeks of heavy rain, a colliery spoil tip slid down the mountain slope over the Welsh village, near Merthyr Tydfil, killing 116 children and 28 adults. The death toll for the children was high because the mudflow engulfed Pantglas Junior School, in the middle of the village (Duell 2016) (Figure 6.2). It was the first major industrial accident for which quite a lot of footage was shown on British television.

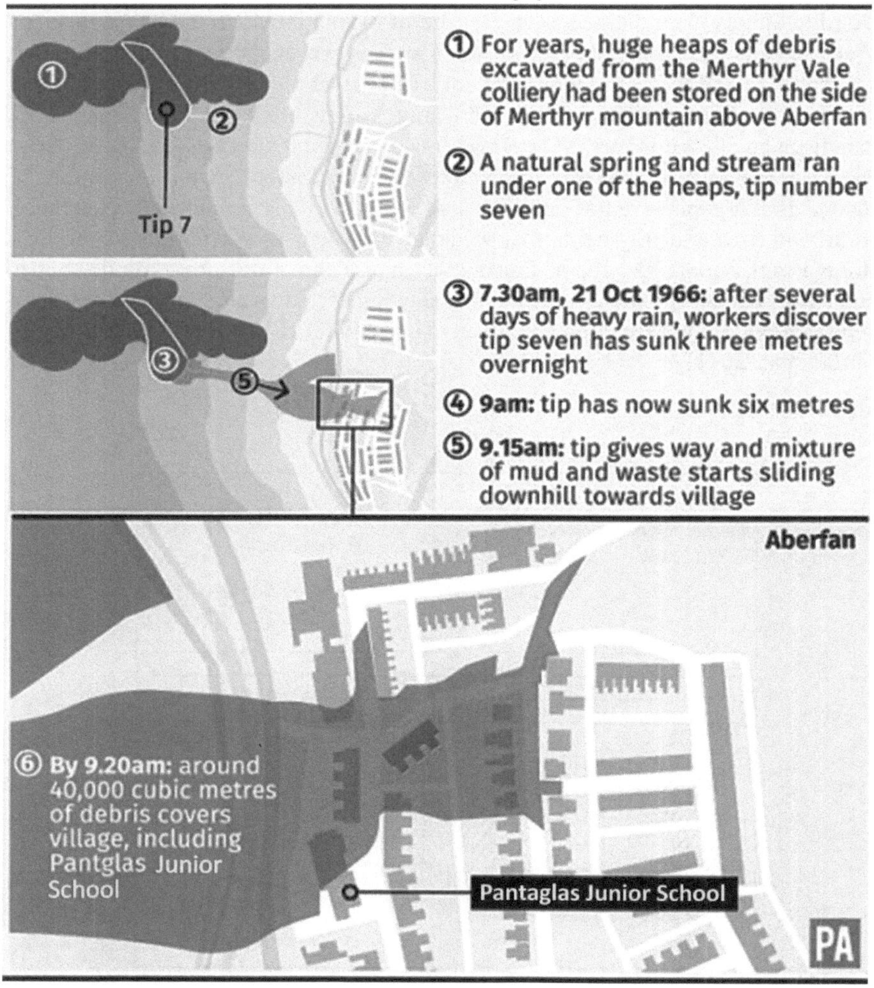

Figure 6.2 The Aberfan disaster: how it happened

Source: Duell (2016)

The National Coal Board (NCB) operated the colliery and the seven spoil tips on the slopes above Aberfan. Tip 7, the one that slipped onto the village, was created in 1958 and was 34 metres high by the time the accident happened. The hazard of coal slurry being tipped at the rear of Pantglas Junior School in Aberfan had already been mentioned in a letter from the Borough and Waterworks Engineer (Jones 1963). In contravention of the NCB's own official procedures, the tip was partly laid over underground water springs, which made the tip unstable to begin with. The tip was soaked with water after the long rainy period. Approximately 110,000 cubic metres of spoil slid down the side of the hill and covered around 40% of the Pantglas area of the village. The main building hit was Pantglas Junior School, where lessons had just started. Five teachers and 109 children were killed almost instantly. The nickname 'the forgotten accident' came from the lack of response from both the Queen and the director of the NCB in the immediate aftermath of the accident. In hindsight, the Queen saw this as her biggest regret (Lees 2014; IJsendoorn 2016). Also a special phenomenon was observed:

> *One of the local GPs, Dr Arthur Jones, noticed two phenomena that at the time were poorly understood. There was a tension between the families where children had died and the families where everybody survived. Playing was frowned upon by some of the families who lost children. The other families experienced 'survivor guilt'. The extreme stress the families were put under also led to a poorly understood syndrome: post-traumatic stress disorder (PTSD). Dr Jones made sure that the lessons of Aberfan were not forgotten and he contributed with the lessons learned to the guidelines on PTSD in the UK.*
>
> (Jones, 1963)

Flixborough, 1974

The third major accident attracted a lot of publicity. A massive blast hit the Flixborough works of Nypro Limited, a joint venture between Dutch State Mines (DSM) and the British National Coal Board. The accident took place in North Lincolnshire, 70 kilometres east of Leeds (Figure 6.3). A newly installed bypass between two reactors burst open during start-up and a large amount of cyclohexane escaped and exploded. Subsequent fires and subsequent explosions blew up a large part of the factory, killing 28 people, many of them in the plant control room, which collapsed. Fires started on-site and were still burning ten days later.

About 1,000 buildings were damaged within a 5-kilometre radius of the site, including in the village of Flixborough and neighbouring villages. Images of the major accident appeared on television soon after the initial explosion, and the devastating effects of the blast shook society. The scientific community was equally surprised and started developing models and theories with a view to understanding gas and vapour explosions (Roberts and Pritchard 1982; Baker 1982). This research took a turn toward a more generic interest in the safety of complex technical installations. It also showed that the so-called internal domino effect was a major contributing factor (Parker 1975; Lees 1980; Khan and Abassi 1998; Høiset et al. 2000; Venart 2004).

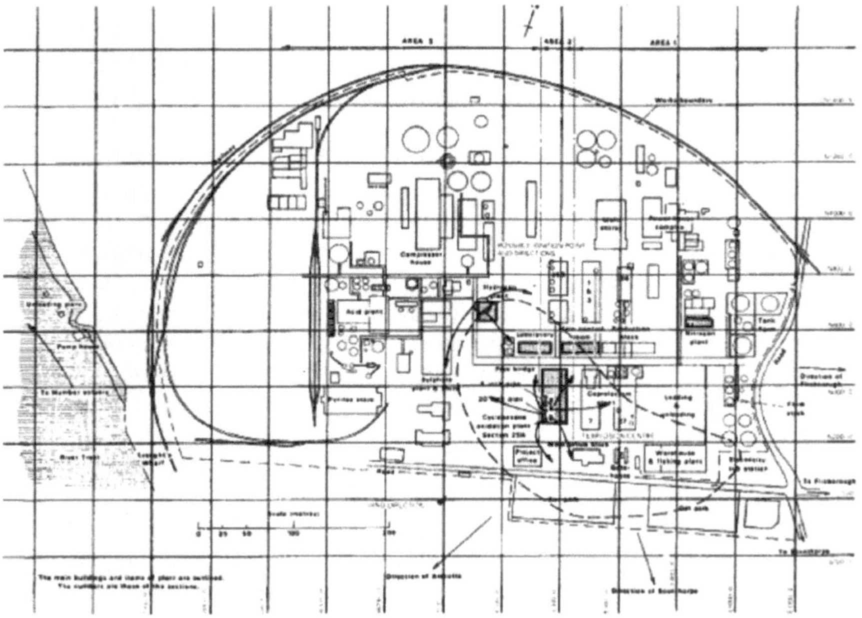

Figure 6.3 Simplified site plan of the works of Nypro Ltd at Flixborough. The dotted lines show the estimated dimensions of the vapour cloud, and the arrows the initial blast

Source: Lees (1980)

Seveso, 1976

Two years later another major industrial accident occurred near the town of Seveso, 24 kilometres north of Milan, Italy. On 9 July a chemical reactor that produced the herbicide 2,4,5 tri-chlorophenoxyacetic acid exploded at the Icmesa chemical company in Meda (Pesatori et al. 2003) (Figure 6.4).

The excessive heat produced the highly toxic TCDD, 2,3,7,8-tetrachlorodibenzodioxin, of which something between 1 and 30 kilogrammes of dioxin was vented through a safety valve. A toxic white cloud drifted over a part of the town and the rain brought the cloud down to earth. Even though there seemed no direct consequences, all farm animals in the region had to be destroyed because they ate contaminated grass. Despite there being no direct human fatalities, many cases of skin infections were reported (chloracne) along with a high number of spontaneous abortions in pregnant women (Lees 1980; Pesatori et al. 2003, 2009).

Los Alfaques, 1978

On 11 July a road tanker carrying liquefied flammable gas ran into – and exploded in – a campsite in Los Alfaques, south of Tarragona on the east coast of Spain. The

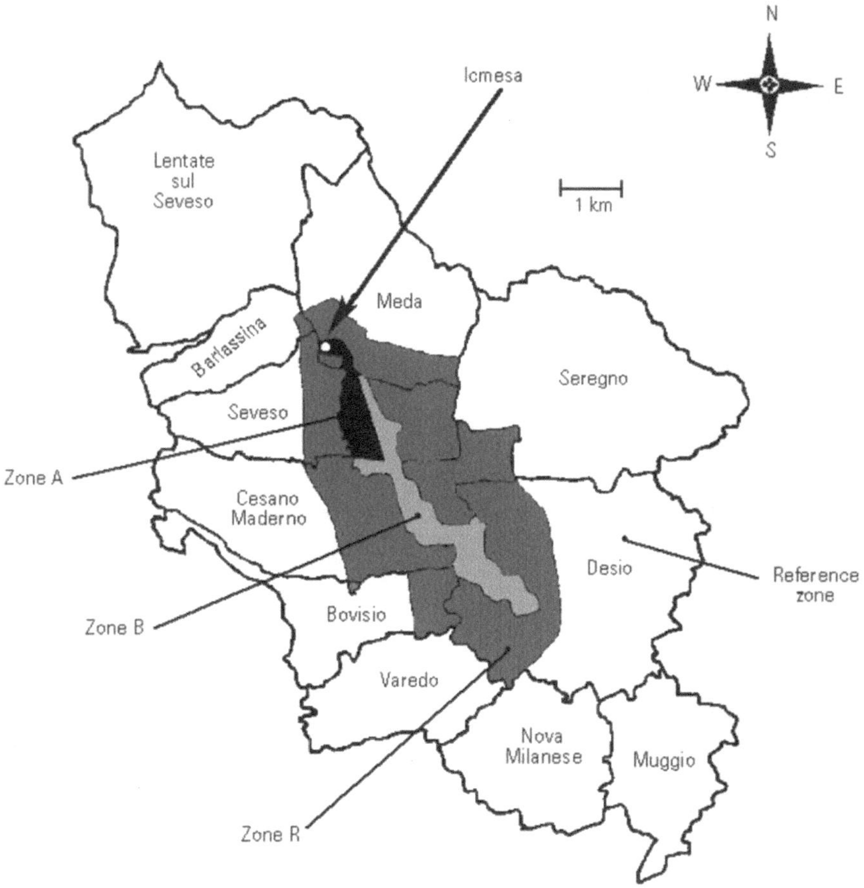

Figure 6.4 Map of Seveso and surrounding towns

Source: Pesatori et al. (2003)

Zone A: heavily contaminated 15.5 μg/m²–580.4 μg/m² TCDD

Zone B: further contaminated zone, levels not exceeding 50 μg/m² TCDD

Zone R: low-level contamination, generally below 5 μg/m² TCDD

road tanker was carrying about 45 cubic metres, or 23 tons, of liquefied propene (propylene). The maximum load had been exceeded by almost 4 tons and the truck had no emergency pressure relief valves. For unknown reasons, the truck swerved off the road (Arturson 1981) (Figure 6.5). The tank itself had actually twisted itself loose and continued over the wall. The tank split and was burst by the overpressure into three parts. The cloud of gas, mixed with air, was ignited, causing a violent explosion: a BLEVE.

The rear part of the tank was blasted away, propelled by the rocket effect across the main road, and came to rest against the wall of the Cerromar restaurant, about

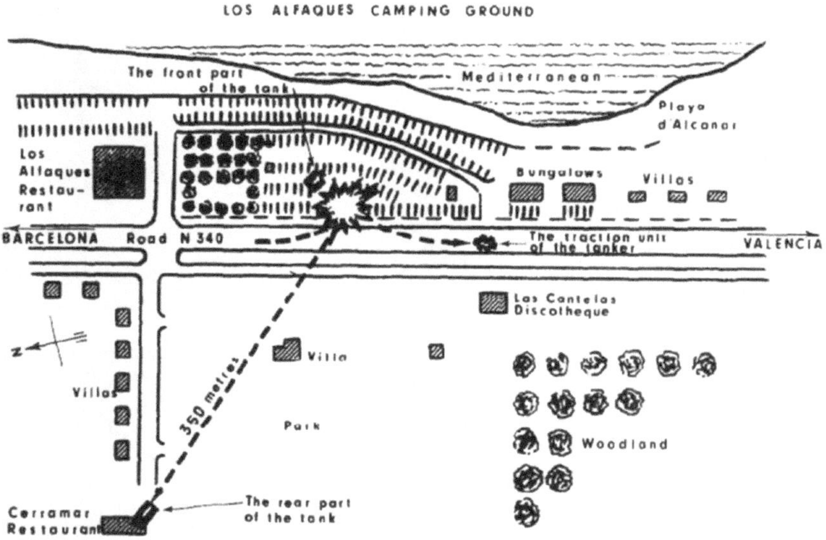

Figure 6.5 Los Alfaques site. The dotted line shows the route of the tanker and its parts

Source: Arturson (1981)

350 metres from the campsite. A total of 217 people were killed and 200 more severely burned; almost all of them were Dutch, Belgian, French and German tourists (Arturson 1981). After the disaster the transport of dangerous goods through densely populated areas was prohibited in Spain as a measure intended to reduce the risk of terrible consequences after an explosion.

Impact of these major industrial accidents

The nuclear sector

The nuclear industry, which started taking off after the Second World War, experienced some major accidents in western Europe from the 1950s onwards. The Windscale accident is an early example. This facility, in the United Kingdom, suffered a reactor fire in 1957 that resulted in a radioactive cloud, which spread over the UK and Europe. This accident, more than anything, made Windscale a hated symbol for environmentalists and opponents of nuclear energy, something that barely changed even when British Nuclear Fuels (BNFL) decided to try and banish the bad memories by changing the plant's name to Sellafield in 1981. The Nuclear Energy Agency, a specialized agency within the Organisation for Economic Co-operation and Development (OECD), classified the accident at 5 on a seven-point severity scale, signifying an accident with wider consequences.

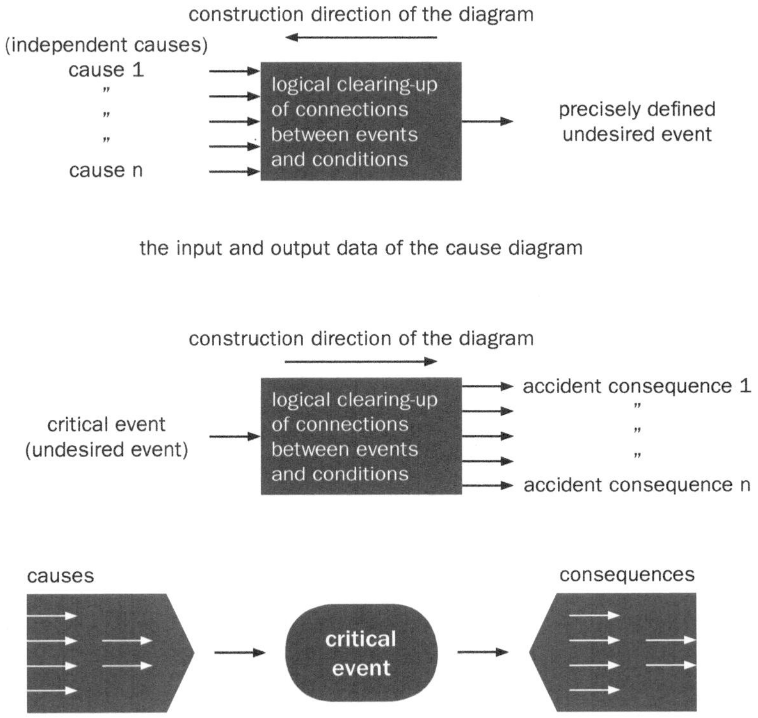

Figure 6.6 The input and output data of the consequence diagram

Sources: Nielsen (1971); after Johnson (1973a)

A level 4 accident (local consequences) occurred in 1969 at the Saint-Laurent Nuclear Power Plant in France. The nuclear core of the power plant partly melted down but there was no cloud this time (NEA-OECD 2010). Maybe due to the general technology optimism in the 1950s and 1960s, these accidents did not receive much international media attention, but nuclear safety became a serious issue for the nuclear community. It adopted a systems approach, previously advocated by ergonomists in the safety domain (see Chapter 4).

Nuclear engineers preferred technological solutions for solving safety problems, partly because they were easier to define. The Electronics Department of the Danish Atomic Energy Commission, at Risø, Roskilde, Denmark, started working on a technical visualisation of the accident process in the nuclear industry. The department was headed by the electrical engineer Jens Rasmussen. In 1971, the consequence diagram was applied to complex nuclear installations and disseminated to a wider audience. The model proposed an expedient presentation of the logical connections between a spectrum of accident causes and one of consequences. It showed scenarios of how fission products released and what controls were in place to prevent the release. Moreover, the model was the basis for probability calculations and consequences for individuals. The main feature was the *critical event*, which was often depicted as the transgression of a key safety limit for a nuclear reactor. The critical event was preceded by a cascade of failures in a fault tree, as was the development of consequences after the critical event. The logical connection between the causes of the critical event and the highly unlikely, but still possible, consequences of it was a new concept in safety at that time (Nielsen 1971; Nielsen et al. 1975) (Figure 6.6). The concept of a critical event in this model will return in Chapters 7 and 8 when we discuss the later presentation of accident processes, the bowtie metaphor.

Loss prevention

With larger chemical plants built in the 1950s and 1960s, and higher temperatures, pressures and consequently more energy, the response of the process was a bit faster. The scale of possible fires, explosions and toxic releases had grown significantly, including for people outside the company's gates. This led to a new movement that approached safety in a more systematic and technical way, known as 'loss prevention'. Several events are associated with the start of loss prevention: the first was the Hazards Conference, in Manchester, UK, in 1960; the second was the first annual Loss Prevention Symposium, in 1967, of the American Institute of Chemical Engineers; and the third was the first European Conference on Loss Prevention, held in Newcastle, UK, in 1971. In fact, loss prevention developed during the 1960s as a response to an increased number of fires and explosions and a worsening fatal accident rate (Kletz 1999). Loss prevention was a new movement with a systems model at its core. It mapped risks through system analysis and quantified risks with various probabilistic techniques and physical-chemical reliability models. The identification of hazards and major accident scenarios with tools like Hazop, FMEA and FTA (see Chapter 4) was promoted together with safety audits and the introduction of inherently safer design. A more thorough investigation of incidents was encouraged as better incident reports and more in-depth studies on explosions, runaway reactions and the dispersion of leaks emerged (Lees 1980; Kletz 1989). Similarly, to the nuclear industry, risk and quantitative methodologies were pivotal. Lees defined risk as:

> *(A) measure of economic loss or human injury in terms of both the incident likelihood and the magnitude of the loss or injury.*
>
> (Lees 1980)

Lees' inclination towards a quantitative definition of risk became, and still is, the prevailing definition of risk in the process industry. According to Lees, the extension of process safety required a reorientation of the approach to 'safety and loss'. Lees systematically and comprehensively described the relevant body of knowledge, methods and techniques required in order to prevent damage to property and adverse health effects. He described the methods for quantification of risks and applications of reliability engineering. He also criticised the traditional standards and codes for the design of installations because safety considerations were not given explicit attention. The systematic and comprehensive approach in the first edition of his handbook illustrates the revolutionary change in approach that the introduction of risk brought to safety engineering.

Kletz, being more inclined toward disseminating knowledge and educating chemical engineers, argued that process plants could be made safer by considering three *E*s: equipment, that is, creating safer process Equipment; Equations, to calculate risk; and Experiments, to investigate whether controls worked adequately and to explore unknown risks (Kletz 1988c). He also stressed that loss of containment, or LOC, was the most important process disturbance of chemical processes. In order to carry out comprehensive and reliable quantitative risk assessment (QRA), extensive intercorporate and international databases were established, listing risk data in the chemical processing industries (Lees 1980, 1996; Mannan 2005, 2012). An example is Appendix 1, which lists major incidents in this era. For this appendix, publicly available information was consulted. This table gives only an impression, as there is an unknown level of under-reporting, especially because under-reporting varies by country, time and sector. The United States, Japan and Europe seemed to have many more accidents than those reported in the rest of the world. The authors assume this is a perception introduced by our focus on English-language articles and reports. In fact, except for Africa, all continents of the world are represented in the list, which shows that chemical safety is a global problem. The large proportion of accidents on the list involving storage and transport is striking (Lees 1983; Kletz 1984b, 1985a, 1986).

Another innovation from loss prevention researchers was the focus on process deviation parameters. Table 6.1 shows some of Lees' process deviation parameters.

Table 6.1 Some deviations of operating parameters from design conditions (after Lees 1980)

Process variables	Pressure, temperature, flow, level, concentration
Pressure system	Mechanical stress, loading, expansion, contraction, cycling effects, vibration, cavitation, resonance, hammer; corrosion, erosion, fouling
Chemical reactions	Reactions in reactors; nature and rate of main reactions and side reactions; catalyst, behaviour: reaction, regeneration, poisoning, fouling, disintegration; unintended reactions, elsewhere: explosion, heating, polymerisation, corrosion
Material characteristics	Vapour density; liquid density, viscosity; melting point, boiling point; latent heat, phase change; critical point effects; solids physical state, particle size, water content
Impurities	Contaminants; corrosion products; air; water
Localised effects	Mixing effects, maldistribution, adhesion, separation, vapour lock, surging, siphoning, vortex generation, sedimentation, fouling, blockage, hot spots
Time aspects	Contact time, control lags, sequential order
Process disturbances	Operating point changes, changes in linked plants, start-up, shutdown, utilities failure, equipment failure, control disturbance, operator disturbance, blockage, leakage, climatic effect, fire
Constructional defects	Plant not complete, not aligned, not level, not supported, not clean, not leak-tight; materials of construction incorrect or defective
Loss of containment	Leakage, spillage

Four factors were deemed relevant for deviation analysis with LOC events: the quantity of materials released, the energy they carried, the speed in which energy was released and dispersion of harmful materials. *Quantity* was simply the mass and/or volume of the release. *Energy* pointed to the energy that released hazardous substances needed to transform into a toxic, flammable or explosive mixture (e.g. energy for evaporation) and the energy content of the released materials. For compressed flammable gases, the energy that could potentially be released was immense because an LOC could create an explosion, but for cryogenic liquids the energy was lower because some of it was lost by extracting heat from the environment.

Speed was related to the time it took for a material to be released and the total duration of the spill. *Dispersion* concerned the type of release (e.g. liquid versus gas) and how far the release could travel from the point of release. Risk analyses used these four factors for their estimations of risk. Scandinavian researchers wrote extensive reviews about risk-based safety analyses and methods. They tended to focus on the reliability of data, for instance, for failure frequencies of individual components (Suokas 1985; Harms-Ringdahl 1987b). An engineering approach that was prevalent in this school of thought emphasised that good design and redesign of technical installations would be the best solution for a safer industry.

The explosions at Flixborough and the Los Alfaques campsite proved to be game changers for loss prevention. They changed the relatively academic approach that was set in the first Newcastle Loss Prevention Symposium (1971) into a frenzied scramble for knowledge. It was blatantly obvious that the devastating effects of vapour cloud explosions were poorly understood (Parker 1975; Ficq 1976; Sadee et al. 1976). This shortcoming initiated a lot of research aimed at understanding the dangers of gas clouds (see, for example, Nettleton 1976, 1976/1977; Helderslot 2009). As a result, loss prevention saw an increase in the number of publications in both professional and scientific journals. Also, the daily operations of process plants were scrutinised. On industrial sites, offices and control rooms were either situated in the vicinity of process installations or were dotted around the plants, which invariably put workers in the blast radius of any serious flammable substance LOC.

The official Flixborough report, published by the Department of Employment (1975), was quite mild about the company's management of safety: it said that management had been safety conscious and there were no indications that production had taken precedence over safety. A minor point mentioned was the understaffing of the technical support facilities. But other reports and articles came to very different conclusions. They noted the appallingly low standards of safety management, and referred to the way in which production prevailed over safety and to the ineptitude of the local authorities' licensing officers (Lees 1980; Carson and Mumford 1979; Harvey 1979). Trevor Kletz (1976), from Imperial Chemical Industries (ICI), was the most vocal critic. He denounced the way in which management was fascinated by occupational accident rates as a gauge of safety but completely overlooked process safety. For him, loss of containment and analyses of near accidents were much more important than recording slips, trips and falls.

For Kletz, the reliability of plant components presented a more realistic picture of the safety of any process.

But the impact of Flixborough resonated beyond the loss prevention movement. In the immediate aftermath of Flixborough, the Health and Safety Commission (HSC), which was installed after the 1972 Robens Report (Chapter 5), produced two reports on 'major hazards' (HSC 1976, 1979). The reports featured an inventory of all British companies using toxic, flammable, unstable and reactive materials that were subject to legislation when they were stored in certain quantities. They also featured an overview of explosions in all corners of the world. As the Robens Report did before, these reports reiterated the importance of top company management commitment for safety, and particularly process safety. Going forward, companies had to demonstrate that their management systems had a tangible impact on safety, and hazard and risk analyses became mandatory. This echoed the Health and Safety Executive's focus on adequate safety management for occupational risks.

Canvey Island study

Two oil companies, Occidental Refineries Ltd and United Refineries Ltd, had been granted planning permission for the construction of two oil refineries. The construction of the Occidental Refinery started in 1972 but was halted in 1973 pending a major design study review. United Refineries had valid planning consent but had not started construction. Several companies had installations at the petrochemical cluster of Canvey Island (Figure 6.7). When the permits were

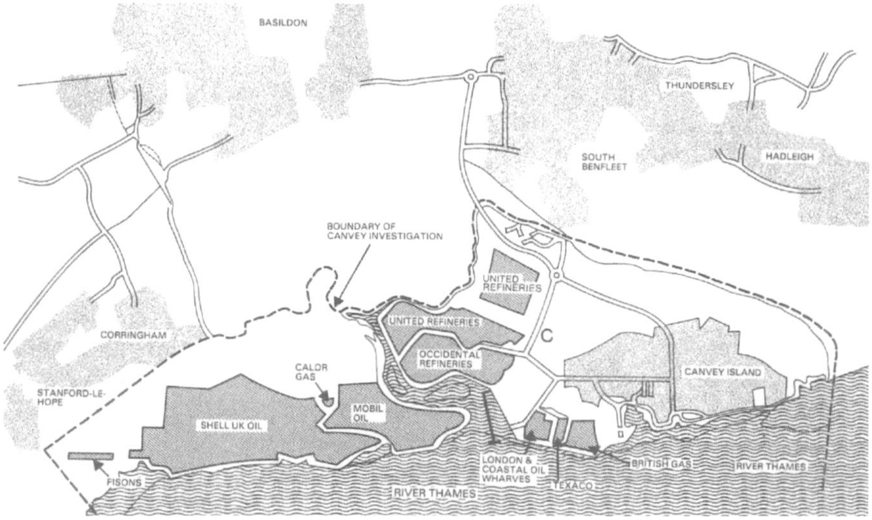

Figure 6.7 Canvey Island and surrounding residential areas

Source: HSE (1978)

retracted, a thorough investigation was launched: the Canvey Island study. The report, entitled 'An Investigation of Potential Hazards from Operations in the Canvey Island/Thurrock Area', was the most comprehensive hazard assessment of non-nuclear installations in the United Kingdom (HSE 1978) at the time.

The proximity of towns to the chemical complexes posed a geographical problem. Figure 6.7 illustrates the complexity of the situation (Cremer and Warner 1980). British Gas operated a methane terminal for importing and storing liquefied natural gas (LNG) from Algeria; Texaco had a petroleum storage, as did London & Coastal Oil Wharves (which also stored flammable and toxic chemicals); Mobil Oil was a refinery with a pressurized LPG and hydrogen fluoride (HF) storage; the Shell refinery stored pressurized LPG and liquid anhydrous ammonia; and Fisons was an ammonium nitrate plant that stored ammonia. Transport of chemicals was by road, ship, rail and pipeline.

The study was a public inquiry on the risks for residents of Canvey Island. In the end, two reports were issued. The first Canvey Report centred mainly on two aspects: the methodology for risk assessment and the magnitude of the risks. Historical failure data were gathered and fault trees were made for installations such as vessels, tanks, pipelines and so on, and for activities like transport of chemicals and staff transport with helicopters. The report estimated that the average probability of a resident being killed by a major accident at Canvey Island was 5.3 10^{-4}/year. Several measures were proposed, such as water spray systems, design adaptations to plants and tanks, that were expected to reduce the risk. According to the HSE, none of the installations would be required to shut down and they concluded that the new installations would not add significantly to the risks that were already there. Yet, even though the measures were implemented, some hazards persisted. Spontaneous failure of storage vessels, jetty incidents and ship collisions could not be made less risky with the proposed measures and would still contribute significantly to the risks. Paradoxically, the report itself expressed some reservations on the risk estimates by stating: 'Practical people dealing with industrial hazards tend to feel in their bones that something is wrong with risk estimates as developed in the body of the report.' This reflects some of the fear of uncertainties that fuelled the fierce opposition surrounding the Flixborough explosion (HSE 1978; Mannan 2005).

But local people came together in action groups, with the support of national action groups, and protested loudly in the months following the publication of the report. It was not so much that they disagreed with the risk numbers, but they simply did not want any more factories on their island. They were supported by the Oyez Intelligence Report, produced by Cremer and Warner (Cremer and Warner 1980). The report stated that the basic philosophy and the risk assessment of the Canvey Report were sound but lacked analysis of the acceptability of the risks. It was found that the risks were implicitly deemed acceptable after the implementation of measures, but this was not explicitly verified. Another point that Cremer and Warner made was that the probability estimates were very low but the uncertainties were large (Cremer and Warner 1980; Warner 1981). They stressed that QRA-based reports should state the uncertainties and include the influence of human factors. Following this advice, Griffiths summarized this as follows (Griffiths 1981):

- *Application of event and fault tree analysis can never be shown to represent every possible outcome; there can always be some failure sequence that has not been accounted for. This is like throwing a pair of dice with an unknown number of faces;*
- *The magnitude of the consequence cannot be exactly calculated; various sources of uncertainty enter the problem. This is like throwing dice where the numbers on the faces are not exact but are thought to be within certain ranges. The number actually on one particular face is not revealed until the throw is made;*
- *Only certain elements of the system can be experimented with to test the model. Reliability data for some components may be available, but other necessary information is inaccessible. This is like throwing dice where only some of the numbers on the faces are known in advance.*

Expert judgement had a great influence on the estimation of the failure probabilities. This was a shortcoming that had already been noted, by DeBlois in 1926 (see Chapter 2), but somehow remained a problem, and it remains a weak point to this day. Databases on the reliability of system components were and remain essential for carrying out a QRA. In England, process safety companies, the Ministry of Defence (UK) and the power industry made their data available through the Systems Reliability Service (SRS) of the UKAEA (United Kingdom Atomic Energy Authority). In the first-generation models for quantitative risk assessment, component failure probability data were limited, and, out of necessity, were extrapolated for wider application (e.g. failure probabilities for pressure vessels and process vessels) by expert judgement (Goossens and Cooke 1997, 2001; Goossens et al. 2008; Pasman 2011). Depending on expert judgement remained a weak point because it is the quality of the numerical values of failure probabilities that largely determines the correctness of the QRA analysis.

The second Canvey Report was issued by the HSE three years after he first one. The agency recalculated risk levels after the implementation of safety measures. This included changes to plants, equipment and methods of operation, physical improvements to LNG ships, cessation of ammonia storage by Shell and reduced risk of ammonia spills at Fisons. The average individual risk of death or serious injury from an industrial accident to a resident on Canvey Island was reduced from 5.3 10^{-4}/ year (reported in the first report) to 0.4 10^{-4}/year: a 14-fold improvement. According to the HSE, the review of operations on Canvey Island had a positive effect on the safety practices of the firms concerned. The exercise led to the removal of redundant plant and pipework, and to the identification and resolving of troublesome operations. The measures improved safety and led to a reduction of the overall risks but remained difficult to quantify for individual measures. The industry as a whole benefitted from the Canvey Reports: they were a stimulus for the development of a systematic risk assessment methodology (HSE 1981).

Inherent safe design

'What you don't have can't leak' was one of Kletz's mantras (Kletz 1978, 1988a). The mantra was repeatedly used to explain inherently safe design: keep the design simple, ensure the designer familiarises himself or herself with the tasks of the operators and verify the design with post hoc hazard and risk analysis such as Hazop (Kletz 1982, 1984a; Clarke 2008) (Figure 6.8). Scandinavians adopted a similar approach for nuclear power plants and process industries (Rasmussen 1980, 1985; Hollnagel 1983; Suokas 1985, 1988).

Kletz's inherent safe design illustration tended to be most useful for individual processes and subprocesses rather than for entire sites (Kletz 1985c). For reasons of confidentiality, litigation and the fear of negative publicity, the findings of safety research in the process industries were generally not publicly available. Kletz proposed that transparency would be preferable for several reasons: firstly, because morally it was preferable to inform society ('If we know, we must tell'); secondly, and more pragmatically, organisations would be better off learning from each other's mistakes; thirdly, more sharing would be economically beneficial as

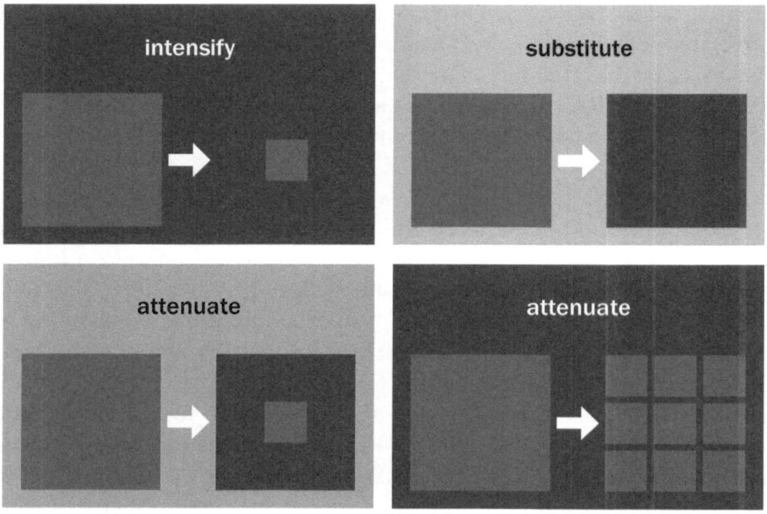

intensify	use smaller amounts of dangerous goods, which limit the effects of unintended release.
substitute/replace	use safer alternatives, e.g. less toxic or less flammable.
attenuate	if a dangerous substance is essential, dilute it, try to operate with less severe conditions or use it in a more stable form.
attenuate	reduce effects, change the design to limit the effects of individual releases.

Figure 6.8 Principles of inherently safe design

Source: After Kletz (1985c)

good practice could be shared; and finally, if any one organisation had an accident that could have been prevented by knowledge in another organisation, the entire sector would suffer unnecessarily (Kletz 1988c).

The disaster incubation theory

In the 1970s, certain sociological studies were published that dealt with the complexity of production processes (Reeves and Turner 1972) and their organisation but also their internal codes, rituals and socialisation processes (Turner 1971). In three medium-sized to large companies, a relationship was established between the organisation of the work, the technology of the production process and the control that management had over production. In the early 1970s, the automation of production in the manufacturing sector was limited and production was mainly organized as batch-wise processes. These processes possessed a high degree of complexity due to the numerous subproducts and many process steps, which made any production planning virtually impossible. Foremen and middle managers had to resolve production, planning and safety problems in an ad hoc fashion.

The description of socialisation processes in companies came as a direct result of extensive sociological studies and was in line with a focus on organisational causes of occupational accidents. The British proactive research on 2,000 accidents by Powell and co-authors and the new principles of safety management by the American Petersen were examples (see Chapter 5). Through informal interviews and observations, researchers had been able to put themselves in the shoes of company employees, describing organisational characteristics using the grounded theory approach (Glaser and Strauss 1967). These studies were relevant to occupational safety but especially to safety in high-tech-high-hazard sectors. A sociological approach did not focus on the individual behaviour of employees, or on technical malfunctioning, but on decision-making in the organisation. The investigators found that batch production, due to its complex nature, was the main problem in high-tech-high-hazard organisations. However, batch-wise production was 'loosely coupled', a technical term referring to the presence of a buffer, or actionable space, between production steps, as well as a degree of variability between these steps. Similar investigations in the US came to a different conclusion. They found that batch production yielded a loosely coupled production system that introduced flexibility, a capacity to respond to local needs and the opportunity to restore faults in production, and had a lower vulnerability than did tightly coupled continuous production processes (Weick 1976).

As sociological studies gained traction, a new research question appeared in the industry: 'What is going wrong in organisations?' In 1976, Turner focused his attention on organisations outside the process industries (Turner 1976). He analysed the slag of a colliery tip on a mountainside in Aberfan, Wales, that slid down into the village in 1966; the collision of an express train carrying exceptional freight in Hixon, Staffordshire (1968); and the fire at a resort in Douglas, Isle of Man (1974). As a socialist, he assumed that no single human error could be accountable for such accidents. The cause had to originate from the complex and diverging chains of events and decisions made within the organisation (Table 6.2). Despite the wide differences in the types of accidents, they all suffered from the same shortcoming: they had great difficulty in recognising deviant signals. For instance, the NCB focused on underground hazards and not on the hazards of their tips. The importance of signs from outside the organisation on the sliding risks of tips were not accepted (see stage I in Table 6.2).

But big accidents remained virtually impossible to predict. The analysts had the benefit of hindsight where, in retrospect, the disturbances confronting workers seemed fairly clear and well defined, but they were evidently not so clear before the accident. For example, deviations during production could be quite diverse

Table 6.2 The stages of major accidents (after Turner 1976)

Some deviations of operating parameters from design conditions	
process variables	pressure, temperature, flow, level, concentration
pressure systen	mechanical stress, loading, expanding, contraction, cycling effects, vibrations, cavitation, resonance, hammer, corrosion, erosion, fouling
chemical reactions	Reactions in reactors: nature and rate of main reactions and side reactions; catalyst, behaviour: reaction,
material characteristics	
impurities	
localised effects	
time aspects	
the sequence of events associated with a failure of foresight	
Stage I	**Notionally normal starting point:** (a) initially culturally accepted beliefs about the world and its hazards (b) associated precautionary norms set out in laws, codes of practice, mores and folkways
Stage II	**Incubation period:** the accumulation of an unnoticed set of events which are odd with respect to beliefs about hazards and the norms for their avoidance
Stage III	**Precipitating event:** forces itself to attention and transforms general perception of stage II
Stage IV	**Onset:** the immediate consequences of the collapse of cultural precautions become apparent
Stage V	**Rescue and salvage – first stage adjustment:** the immediate post-collapse situation is recognised in ad hoc adjustments which permit the work of rescue and salvage to be started
Stage IV	**Future cultural readjustment:** an inquiry or assessment is carried out and beliefs and precautionary norms are adjusted to fit the newly gained understanding of the world

and poorly understood or ignored, which made the early signs of major accidents very unapparent for workers. Turner tried to capture that concept in stage II, the 'incubation period' (Turner 1976) (see Table 6.2). The incubation period resembled incubation in the medical sense, but in later theories this would be termed 'latent organisational factors', a decisive problem in the prevention of accidents. Turner extended his research to include 84 accidents in aviation, water heaters, trains, ships and mines and was the first to coin the term 'man-made disasters' in his 1978 publication (Turner 1978).

Man–machine interactions

Major accidents in this period led to increased attention to human performance. The United States and western Europe had different views on human performance (see Chapter 4). On the one hand, human factors (HFs) dominated in the United States. This branch of research dealt with the quantification of human errors. The United Kingdom and Nordic countries, on the other hand, focused on ergonomics to understand human performance in complex technological systems. These topics were discussed extensively during NetWork (New Technologies and Work) sessions, held by a multidisciplinary working group of experts and scientists with different backgrounds and supported by the Werner Reimers Foundation (Bad Homburg) and the Maison des Sciences de l'Homme (Paris). This group published two books during the period covered by this chapter: *New Technologies and Human Error* (Rasmussen et al. 1987) and *The Meaning of Work and Technological Options* (Keyser et al. 1988). The engineering approach demanded a separation between workers and hazards as much as possible. Automation and remote-controlled operations were key to this strategy. In some industries, such as car manufacturing, energy production and hazardous processes, this trend was already prevalent. In the 1980s, even more processes were automated, and computers controlled machines in the production process. The distance between the worker and the production process increased and detailed knowledge of the production process and craftsmanship diminished (Singleton 1984; Rasmussen 1980; Hollnagel et al. 1981). Workers' activities were reduced by automation to troubleshooting during process disturbances (Figure 6.9).

Factors influencing human performance in technologically complex environments were not well understood in the 1980s (Eberts and Salvendy 1986). Human errors were still viewed as the results of incompetent operators, and the psychology of fault generation was underdeveloped. Rasmussen proposed that errors were actually normal variations of human behaviour: variability was a mechanism explaining how operators learned to run the system and deal with unexpected circumstances (Rasmussen 1982; Rasmussen and Lind 1982). Human errors were viewed as unsuccessful experimentations that produced unacceptable consequences. According to Rasmussen, an unsupportive workplace caused errors because workers had difficulty in understanding and analysing the process. The traditional single-indicator-single-response, where an action was required from an operator and he or she had to discover the state of the system based upon indicators and their training, was considered to be ineffective. Too little attention was given to ergonomic principles of process control systems. An operator needed

Figure 6.9 Control room of potato mill
Source: © Michel Pellanders (1980)

sufficient time and information to react correctly to process deviations. Information had to be presented in terms of human mental functions that operators could easily comprehend and that enabled them to perform control functions. It was found that humans were capable of integrating technically incompatible sources of information and recognising patterns but were at a loss to adequately respond to a one-sensor-one-indicator disruption (Rasmussen 1983; Rochlin 1986).

Rasmussen developed a taxonomy for operator behaviour, which he based on empirical evidence gathered in the 1970s while analysing problem solution strategies of maintenance operators. He published his well-known skill-rule-knowledge model for human behaviour in his 1982 and 1983 papers (see Figure 6.10). The individual levels were not alternatives, but they interacted (Rasmussen 1983). Some years later, this theory would be referred to as the Rasmussen-Reason model (Rasmussen and Reason 1987). An operator's behaviour in a high-risk, complex technological work environment was captured in three different categories:

Skill-based: automated and subconscious processes that are prevalent in skills internalised by operators' experiences;

Rule-based: behaviour that is prevalent when operators use explicit rules to execute a specific task;

Knowledge-based: operators use their intelligence to judge situations and find solutions for problems that are new or not within the explicit remit of their task; this behaviour requires knowledge, attention and concentration.

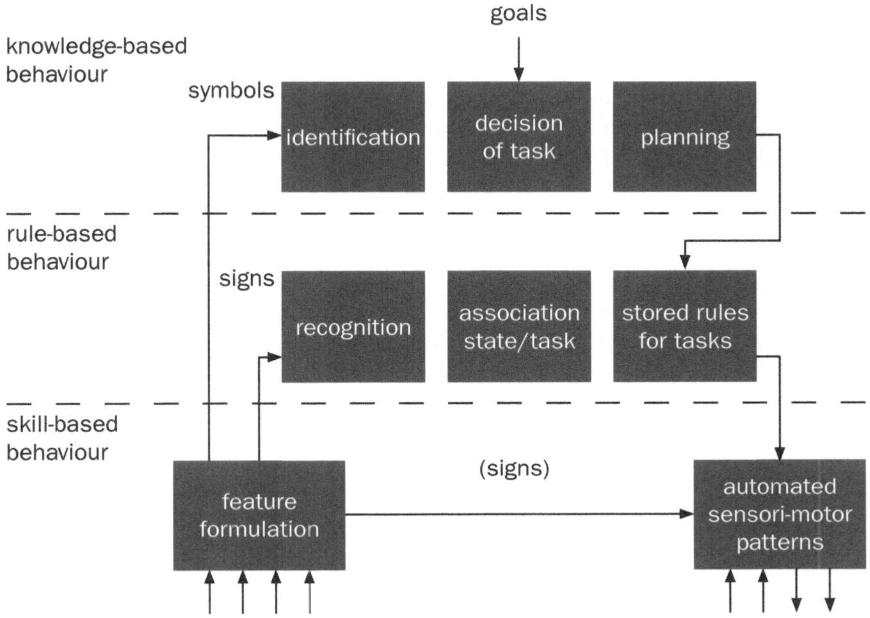

Figure 6.10 Simplified representation of Rasmussen's mental model for operator performance

Source: After Rasmussen (1983)

Le Coze et al.'s (2014) paper discussed the origin of this model and pointed to the concept of cognitive engineering and its relationship with process design. Cognitive engineering is the design of the man–machine interface to support operators in their complex task of dealing with unexpected process disruptions and deviations. This taxonomy of behaviour was the first initiative to make human performance part of the design of controls and so initiated cognitive engineering.

Seveso I

The Seveso I directive, issued in 1982, was an attempt to obtain a minimum standard of uniformity amongst European member states on major hazard regulations (Duffield 2003). The handling of dangerous substances was a central point of the regulations, and a list of hazardous substances was included. Seveso I stipulated that each European member state should appoint a competent authority (CA). This CA would be a focal point in the collection of major hazards of installations, of installations with large volumes of listed chemicals and of major accidents. A company would have to produce a safety report that provided information on dangerous substances that were in use at the site and the emergency plans. People who could be affected by an accident needed to be informed 'actively' about what safety measures were in place and what was expected from them in the case of an accident (Kirchsteiger et al. 1998).

The safety report was a relatively successful tool that gave a description of the company's production processes, safety systems to control these processes and measures to avoid emissions, exposures and major accident-generating conditions. Partly due to its success, the directive was expanded to include additional industries after major industrial accidents of the late 1980s. Initially, LPG stations and storage sites had to produce safety reports, followed by Seveso companies. After the 1986 Sandoz accident in Switzerland, which released tons of agrochemicals into the air and the Rhine, killing almost all river wildlife, the storage of pesticides and agrochemicals was incorporated into the Seveso legislation. Subsequently, ammonia refrigerating units, (rail) transport of hazardous chemicals and airport safety also came under the attention of the regulators (Swuste and Reniers 2017).

Some major industrial accidents in the 1980s

By the end of the 1970s it seemed that much had improved and our understanding of safety and major accident processes had greatly increased. That may have been true, but it did not prevent several major accidents that drew the world's attention. Three are discussed below. The first major accident was the nuclear meltdown of a power plant at Chernobyl in 1986, followed by the explosion of the oil platform Piper Alpha in 1988 and the train collision at Clapham Junction. In line with the 'man-made disaster' approach of Turner, the investigations focused on the impact of management and organisational flaws, but surprisingly, they abstained from even mentioning Turner's work. In his article on Chernobyl, Reason came up with the metaphor of a 'resident pathogen', destroying technical systems. Lord Cullen in his Piper Alpha report noticed the severe shortcomings of safety systems. And Hidden, in his Clapham Junction report, showed the difference between a paper on safety and the reality. The reports edged toward the 'defence in depth' approach that became popular. The nuclear industry was the first to adopt the strategy, which steadily propagated into the chemical process industries. The concept consists of safety redundancy by installing multiple layers of safety barriers, each capable of stopping the accident on its own. This would create a more successful accident prevention strategy, even if some defences were undermined by an overlooked incubation process. Rasmussen (1988a, 1988b) pointed out that the approach was not without its shortcomings. Whilst referring to Reason (1987), he argued that multiple barriers might not necessarily be visible to operators, meaning that failure scenarios, or incubation processes, were partly activated and could lead to accidents without operators ever knowing what was going on. In literature, this became known as the 'fallacy of defence in depth'.

Chernobyl, 1986

The major nuclear accident in Chernobyl occurred on 26 April in the No. 4 nuclear reactor at the Chernobyl Nuclear Power Plant, near the city of Pripyat, in the north of the Ukrainian Socialist Soviet Republic (SSR). It was considered the worst nuclear accident in history and was one of only two nuclear energy accidents rated at 7, the maximum, on the International Nuclear Event Scale, the other being the 2011 Fukushima Daiichi nuclear accident, in Japan. This accident started during a safety test on the nuclear reactor. The test was a simulation of an electrical power outage to aid the development of a safety procedure for maintaining cooling water circulation until the backup generators could provide power. This operating gap was about one minute and had been identified as a potential safety problem that could cause the nuclear reactor core to overheat. During the test preparation, the reactor power unexpectedly dropped to a near zero level. The operators were able to restore the power level, but in doing so they put the reactor in an unstable condition. They proceeded with the test even though the power level was lower than prescribed in the procedure. Upon test completion, they triggered the reactor shutdown, but a combination of the reactor design and construction flaws caused an uncontrolled nuclear chain reaction instead. A large amount of energy was suddenly released, vaporising superheated cooling water and rupturing the reactor core in a highly destructive steam explosion. This was immediately followed by an open-air reactor core fire that released considerable amounts of airborne radioactive contamination that precipitated into parts of the Union of Soviet Socialist Republics (USSR), western Europe and Nordic countries, before being contained after nine days. The fire gradually released almost the same amount of contamination as the initial explosion. About 49,000 people were eventually evacuated from the area, primarily from the neighbouring town of Pripyat, and a further 68,000 people were evacuated from the wider area. It proved to be hard to estimate the number of fatalities. Model predictions vary, from 4,000 fatalities when solely assessing the three most contaminated former Soviet states to about 9,000 to 16,000 fatalities when assessing Europe in its entirely (Wikipedia 2019).

After Chernobyl, Reason became interested in the human and organisational factors contributing to accidents. He created a new metaphor to understand accident processes. He analysed the nuclear incident at Chernobyl and introduced the medical metaphor of a *resident pathogen* (Reason 1987). Similar to the human body, a technological system could carry pathogens with the potential to destroy it. Unfortunate events could trigger such pathogens, and several of these pathways, or scenarios, could combine and become major accidents, even if the individual events did not pose a direct threat. This also explained why accidents rarely had a single cause. Reason also introduced the 'swamp' metaphor. Trying to eliminate human errors was equivalent to 'swatting a mosquito': not very useful in the long term but very attractive to managers with a short-term focus. Organisations had to search for the reasons why mosquitoes were so prevalent: the swamp. The most effective way to improve safety was therefore to drain the swamp in which they breed (Reason and Hobbs 2003; Reason 2000). Reason paved the way for a new understanding of accidents and catastrophes.

Piper Alpha, 1988

An explosion in the gas compression section initiated the catastrophe on the North Sea oil-drilling rig Piper Alpha on 6 July. The rig was connected to two other rigs in the North Sea (Figure 6.11).

The explosion occurred when a pump that had been taken out of operation because of a leak was inadvertently switched on (Paté-Cornell 1993). It was a consequence of a faltering 'permit to work' system. The first explosion destroyed fire-resistant walls, the control room, the communication system and the energy supply system. The fire extinguisher was out of operation because divers were working near the water inlet, and a series of explosions aggravated the situation until it was completely razed. It also took a long time to stop the flow of oil and gas from the Tartan and Claymore platforms, which fed a fire that created a lethal black smoke in the living areas where most workers had fled.

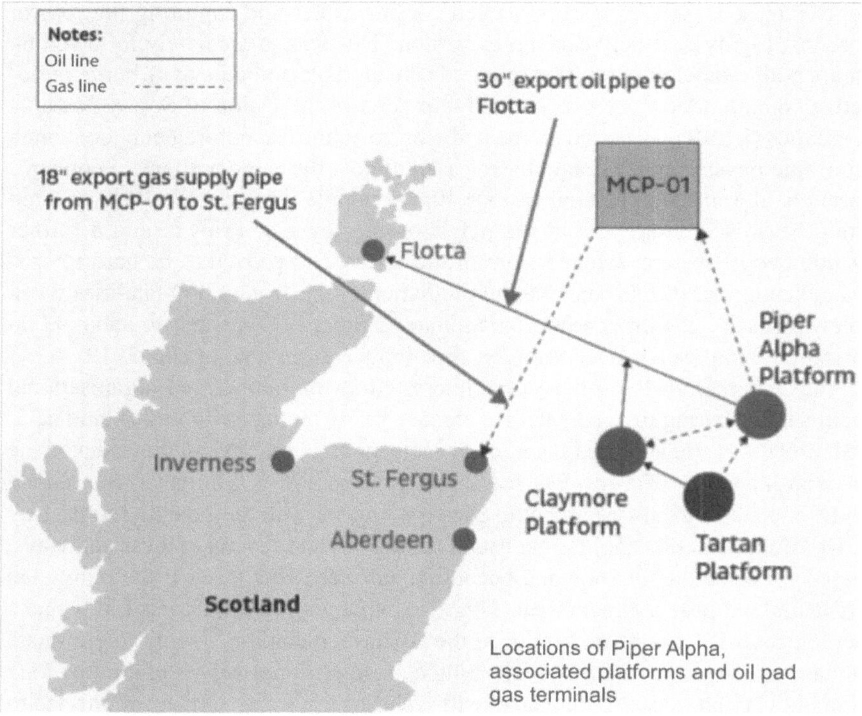

Figure 6.11 Connections between Occidental rigs

Source: BHGE (2018)

The catastrophe was the most lethal in offshore history after the capsizing of the Alexander L. Kielland platform in 1980. On Piper Alpha, 167 people died, and more than 60 survived by jumping off the platform, some from a height of 30 metres. Lord Cullen (1990) was asked to lead the investigation. His report contained the most comprehensive accident investigation that had ever been performed in the UK. Cullen found that the oil- and gas processing techniques had become exceedingly complex, and safety management systems were too fragmented and too inconsistently administered to cope with the complexity of the process installations. The report contained extensive technical information about the accident and how it had occurred, about the safety management system operated by Occidental, about the operator of the rig and about the role of the regulator, the Department of Energy. Occidental managers lacked adequate safety training and knowledge, such as basic knowledge of fire control. The permit-to-work system was fatally flawed, and, additionally, there was only rudimentary feedback from audits and findings from earlier safety issues. The regulator had failed to detect such shortcomings in its inspections, which was attributed to a lack of knowledge on the part of the regulator. The catastrophe laid bare the fact that, despite years of research, safety systems were still too underdeveloped to prevent such catastrophes.

Clapham Junction, 1988

On 12 December a high-speed train that was traveling from Poole to Waterloo collided at the railway intersection at Clapham Junction in Wandsworth, near London, with an empty train going from Waterloo to Basingstoke. The collision pushed some carriages onto the adjacent track where at that time a train from Waterloo to Halsemere was passing. One-third of the Poole train was destroyed (Figure 6.12).

Six hundred people were injured and 35 died. The unthinkable had happened: a technical malfunction of the signalling system had occurred. The British Railways Act of 1971 required a formal investigation with its conclusions addressed to the Secretary of State for Transport. On the day after the accident, this task fell to barrister Anthony Hidden (Hidden 1989). Hidden worked on the case for nine months. In his description of the legal process, he was clearly shocked by two other train accidents that actually occurred whilst his investigation was still underway. The first was at Purley Station, in Croydon, London, on 4 March 1989. This crash killed six passengers and injured 94. The train driver was convicted of manslaughter. The second accident occurred at Bellgrove, Glasgow, on 6 March of the same year. One passenger and one train driver lost their lives (HMSO 1990). Hidden's report was remarkably uncritical of the mechanic who had left loose wires that had caused a false connection and deviant signals, but he pointed the finger at British Rail, at that time a national institution. There was insufficient independent oversight of the actions of staff, and no corrective interventions for clearly erroneous and/or obsolete working methods (Rasmussen 1994). The report also extensively reported on the role played by working excessive overtime and resulted in

Figure 6.12 Clapham Junction
Source: Press Association (1988)

stricter regulation of the working hours of staff in critical safety positions. There was no effective project control of the Waterloo Area Resignalling Scheme at all (Hidden 1989). Hidden criticised the organisation for its poor internal safety processes. The image of British Rail's safety, an important aspect for the organisation, was unacceptably damaged.

Much was going wrong at British Rail at the time of the rail accident (see also Maidment 1997). The department responsible for safety was reorganized four times in the years prior to the accident: in 1982, 1984, 1986 and 1988. During the last reorganisation, key safety positions were filled by people with relatively little experience in the operational testing of safety systems. Because of the reorganisations, the organisation struggled to attract and retain qualified personnel. This was especially problematic because at that time British Rail was under considerable time pressure to quickly automate rail systems. Communication between management and the shop floor was inadequate and prioritisation and political motives hindered efficient execution of the modernisation process. The *coup de grâce* was the fact that safety was deemed important on paper but of little importance in daily operation. Hidden called it 'the devaluation of the organisational culture of British Rail'. Traditional rail culture was based upon competence, professional pride and safety. But it had changed to a financially driven business culture moving toward automation. This pushed personnel away from doing their work well, and Hidden blamed British Rail's management. The majority of the recommendations were aimed at British Rail to adjust its internal organisation. Quite a few recommendations were directed at the government, to play a better supervisory role, and a fair number of the recommendations were meant for the emergency services, who had shown insufficient crisis control. The main findings of the report addressed key aspects of modern safety thinking: dealing with immediate technical and organisational shortcomings, managing the consequences of collisions with better equipment and promoting a better safety culture. Unfortunately, Hidden never made it clear why he was so intent on the organisational aspects of the accidents, which makes it hard to see whether he was aware of Turner's or Rasmussen's work.

North America, India, and the former USSR

Management, including management of safety, already had a long tradition in the United States, as previous chapters have shown. The quantification of human reliability was another area of research, although this line of research provided only limited insight into the processes of major accidents. This changed when properties of complex technologies were included, as advocated by the sociologist Perrow with his normal accident theory. An alternative approach came from a research group located at the University of California, Berkeley. This group developed an organisational theory and specific interventions in potential accident processes, leading to the high reliability theory. Although no reference was made to the structure classification of organisations by Mintzberg (see Chapter 5), this theory was a further elaboration of his classification and of the consequences for safety organisations of complex tasks or production processes. This section will first discuss a managerial approach to major accidents in the nuclear sector before presenting three controversial major accidents in North America and India: Three Mile Island (1979), Mexico City (1984) and Bhopal (1984). The section ends with some remarks on the consequences of these major accidents.

Management oversight and risk tree

Prompted by nuclear plant accidents, for instance the explosion at the National Reactor Testing Station in Idaho Falls, Idaho, in 1957 and the meltdown of some fuel elements at the Enrico Fermi Nuclear Generating Station at Frenchtown Charter Township, Michigan, Johnson (1973a) presented an integrated management model for occupational and process-related accidents in the nuclear sector. Accidents were multi-causal and resulted from a relatively long sequence of changes and errors: the accident scenario. Any components of the industrial process could be part of this sequence: management, design, the environment, machinery, equipment, supervision and employees. Before accidents occurred, scenarios were often partly developed because of interaction and changes in both process and organisational conditions, as mentioned by Petersen (see Table 5.2 in Chapter 5). These conditions simply added to the complexity of the accident, which only in retrospect could be seen as having led to the major accident scenario (Figure 6.13).

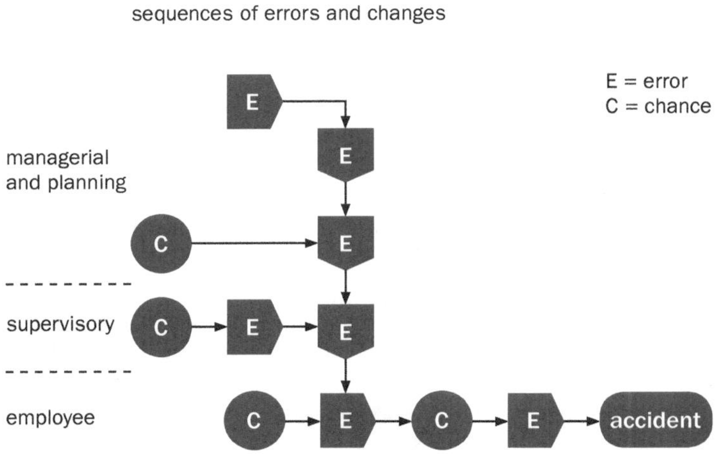

Figure 6.13 The errors and changes leading to accident scenarios

Source: After Johnson (1973a)

This concept of accident causation led to the MORT technique: the Management Oversight and Risk Tree. MORT was developed by the United States Atomic Energy Commission to establish a better safety management system by combining accident models with quality systems. MORT was also useful as an in-depth analysis technique for system failure. MORT comprised a number of fault trees, starting from the functional failure of a system (Johnson 1973b). The use of these kinds of error trees had been previously represented by Nielsen with his logical connection between causes and consequences (Figure 6.6).

The top event in MORT is an accident or damage. Directly below the top event there are four different fault trees that are used to analyse the event: (1) assumed risks; (2) the Haddon (1963) energy model; (3) a feedback and control system; and (4) the system life cycle (Figure 6.14). Like a fault tree, MORT was based not on systematic research but rather on a logical description of the functions required for an organisation to manage its risks effectively.

Johnson's report concluded that, overall, the safety programs seen in companies were far from optimal, similar to the literature on safety tools for major accidents: safety programmes were poorly defined, information was flawed and resources were incomplete, leaving middle managers and workers with the blame for accidents. In this respect, they found that little progress had been made since the beginning of the century. What was needed was a systematic understanding of the predominant probable major accident scenarios, including their likely consequences. The next issue was the controllability of risks and, if residual risks were present, what the arguments were for rejecting the measures to reduce risk.

Finally, there was the question of the quality of the company's safety programme: was it as effective as it was designed to be? The report also gave a

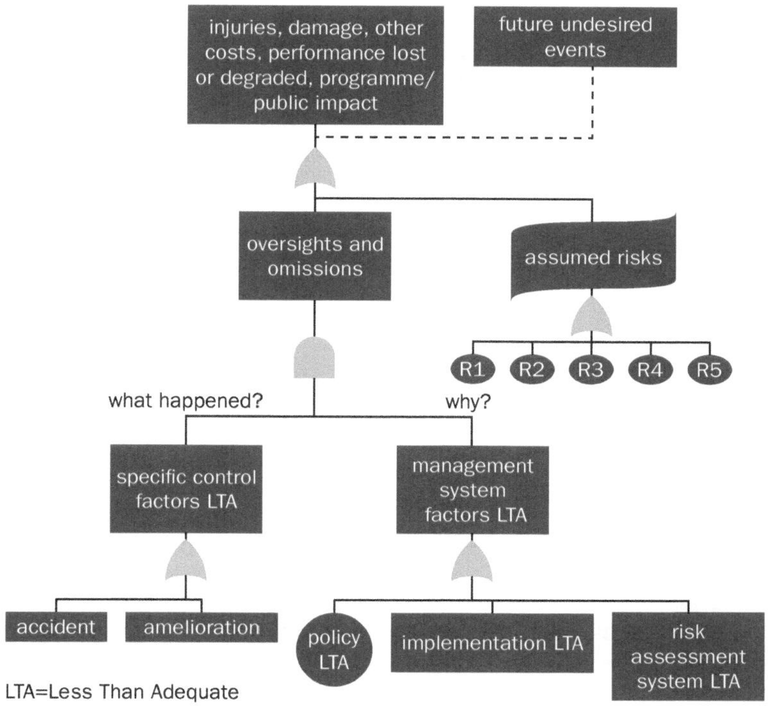

Figure 6.14 MORT

Source: After Johnson (1973b)

characterisation of five safety programme quality levels and linked that to the risk of a major accident (see Table 6.3). The origin of the probabilities presented was not given. The only thing mentioned was that companies had sufficient data to justify an order of magnitude difference between the successive safety levels. The nuclear energy sector reactors fell into the fifth level.

Some major industrial accidents in the 1970s and 1980s

As in Europe, the quality of safety management and safety programmes in many companies was not up to scratch. Major accidents in this period still occurred in the Americas and India as well as outside these geographical regions. America was not devoid of problems in the nuclear sector either. The Three Mile Island accident, in Harrisburg, Pennsylvania, frightened the country and forced people to review safety practices. This accident was classified as a category 5, an accident with wider consequences, by the Nuclear Energy Agency. The two other major industrial accidents discussed here occurred in the process industry, in Mexico

Table 6.3 The quality of safety programs and disaster probability (after Johnson 1973b)

Safety Program Level	*Disaster Probability*
sub-minimal, less than minimal compliance with regulations	1×10^{-3}
minimal, minimal compliance with enforced regulations	5×10^{-4}
manuals, applications of manuals and standards	1×10^{-4}
advanced, improved programming, exemplified by leading industries	1×10^{-5}
systems, system safety	1×10^{-6}

City and Bhopal. These accidents had an extreme impact on residents around the facilities.

Three Mile Island, 1979

On 28 March, a defect in the secondary cooling system came close to causing a meltdown of a nuclear power reactor at Three Mile Island, near Harrisburg, Pennsylvania. A failure in a water pump started the process. It led to an increased pressure in the primary cooling circuit, and radioactive gases were able to escape through a pressure safety valve that had been inadvertently left open. This particular combination of conditions had not been identified in a safety analysis. Also, the control room (Figure 6.15) was not adequately designed to react to a crisis. Radioactive gases were vented into the atmosphere, but a nuclear meltdown was prevented. The incident was classified by the media as a 'disaster' despite there not being any injuries or deaths or any other demonstrable adverse effects on public health in the Harrisburg area. The report of the president's commission on this accident stated that the accident was due to technical failures and human error. Also, the management procedures and emergency response were found to be deficient, and the organisation's safety management system was inadequate (Kemeny 1979; Lees 1980). Operators were not trained well, procedures were contradictory and safety management focused on only a few major risk scenarios. A comprehensive overview of possible major accident scenarios was lacking. One peculiar effect of the absence of a comprehensive system was that more than 100 audio alarms were continuously ringing in the control room. This left operators struggling to identify which alarms were the most important ones.

Managers were several hundreds of kilometres distant from the actual site, which made communication laborious and mostly ineffective. Managers had perceived safety as just another bureaucratic task; as such, lacking priority, it had not received sufficient attention. A tragic side note is that Lord Cullen, 11 years after Three Mile Island (in 1990), in his reporton the Piper Alpha disaster, described similar shortcomings. Cullen reported that the oil- and gas-processing techniques had become exceedingly complex, and safety management systems were too fragmented and too inconsistently administered to cope with the complexity

Figure 6.15 Control room TMI-2 during the accident
Source: Kemeny (1979)

of the process installations. So, nothing pointed to any kind of safety learning whatsoever.

Even though public resistance to the dangers of industrial accidents had taken off as early as the 1960s, these accidents led to high media coverage: people were fed up with disasters and wanted improvement. The chemical and nuclear industries had a serous image problem and action groups published articles in *Nature*, some of them as early as 1977 (Anonymous 1977b). In the case of Three Mile Island, there was a peculiar coincidence. A major Hollywood movie called *The China Syndrome* depicted what could happen at a nuclear power plant and was released, ironically, in 1979, just 12 days before the accident occurred at Three Mile Island.

Starring Jack Lemmon, Jane Fonda and Michael Douglas, the movie told the story of a near meltdown of a nuclear reactor. The title was a metaphor for a meltdown at such a plant. If this were to happen, it was predicted that the repercussions would ultimately reach China. The impact of films on the discussion of risks was also recognized in the chemical industry. In 1986, the film *Acceptable Risks* showed how unsafe practices and a lack of regard for blue-collar workers in combination with a lack of concern from the management and local authorities could lead to disasters. It was loosely based on the Bhopal major accident. The movie was repressed in the US because of pressure from the very powerful chemical industry. It only ran once and was never broadcast again.

Mexico City, 1984

Just two years before Chernobyl one of the most lethal major industrial accidents on record occurred in San Juan Ixhuatepec, a northern district of Mexico City. The staggering loss of life ensured worldwide media coverage, and process safety, once again, took centre stage. On 19 November the LPG terminal of Pemex LPG leaked large amounts of flammable vapour into a densely populated area through a flange leak (Figure 6.16).

The terminal was used for the distribution of LPG, which came by pipeline from three different refineries. Adjoining the Pemex plant were distribution depots owned by other companies. The gas cloud ignited, damaging supply lines to the storage

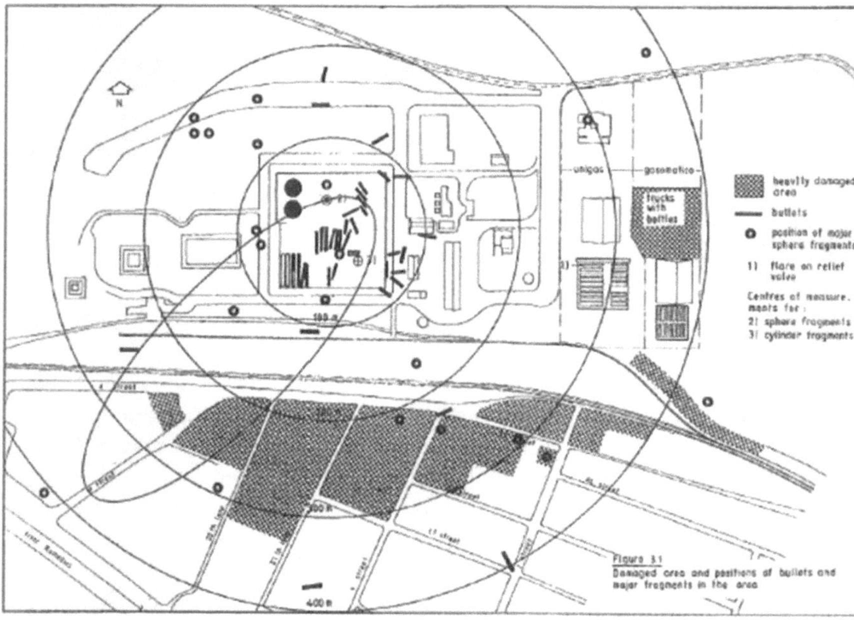

Figure 6.16 Pemex location, including damaged area and the initial gas cloud

Source: Lees (1996)

tanks, and the subsequent fires caused multiple boiling liquid expanding vapour explosions (BLEVEs), followed by an explosion. A rain of LPG droplets fell on the area and were set alight by the heat of fireballs. There followed another series of explosions as vessels suffered BLEVEs. In all, there were some 15 explosions. BLEVEs occurred in the four smaller storages and many of the cylindrical vessels. Numerous steel vessels were launched like missiles by bursting. Many of these were large and travelled far, 100–1200 metres. The flying vessels caused damage with their impact and their temperature, which was high enough to set houses alight. This was the worst ever recorded domino effect, with some 650 people losing their lives and an unknown number of people being injured (Lees 1996; Pietersen 2009).

Bhopal, 1984

On 3 December a relief valve lifted on a tank containing highly toxic methyl isocyanate (MIC) at the Union Carbide India Ltd works at Bhopal, Madhya Pradesh, India. A cloud of MIC gas was released onto housing, including shantytowns, adjoining the site (Figure 6.17).

The accident at Bhopal was by far the worst major industrial accident that had ever occurred in the chemical industry. Its impact had been felt worldwide, but it affected India and the United States the most (Mannan 2005).

The works was in a heavily populated area. Much of the housing development closest to the works had occurred since the site began operations in 1969. In the process, monomethylamine reacted with excess phosgene in the vapour phase and produced methylcarbamoyl chloride plus hydrogen chloride, and the reaction products were quenched with chloroform. The unreacted phosgene was separated by distillation from the quench liquid and recycled to the reactor. The scenario that the investigators used to explain the events was as follows. The content of the MIC tank was initially at 15–20^{0} C. Some 5–900 litres of water entered the tank for reasons unknown. The exothermic reaction between the MIC and the water led to an increase in temperature and also in pressure due to the evolution of carbon dioxide. The higher temperature and presence of chloroform caused accelerated corrosion. The iron in the vessels catalysed the exothermic trimerisation of the MIC, and 45 tons of MIC leaked from the storage tank through the stack of the factory. It took hours before the leak was detected. As MIC spread over the city, more than 2,000 people died within hours and tens of thousands of people were injured. A decade after the catastrophe, the number of direct fatalities had tripled, and an official 30,000 were listed as permanently injured. Union Carbide consistently claimed that sabotage was the cause and persistently denied that engineering designs could be at fault.

Impact of these major industrial accidents

Risk approach and risk perception

The accidents drew people's attention to process safety and two trends emerged: reliability engineering and loss prevention. Reliability engineering originated in

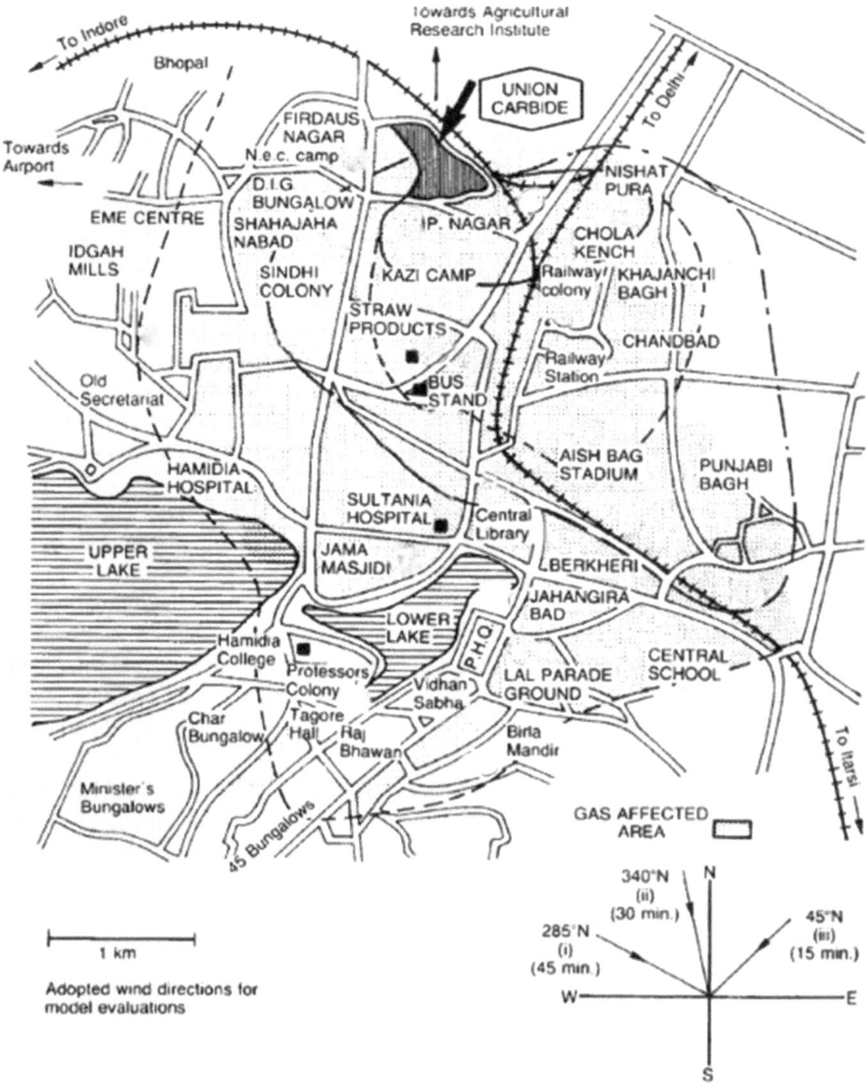

Figure 6.17 Bhopal, the plant and the estimated gas cloud dimensions (dotted line)

Source: Lees (1996)

the military, nuclear and aerospace domains (Barlow and Proschan 1975; Rasmussen 1975) and loss prevention came from the processing industry (Chapter 4) (Lees 1980). These two movements, together with Rowe's 1977 first systematic, book-length presentation of risk assessment, *An Anatomy of Risk*, introduced the concept of risk in the safety domain. In the nuclear industry, probabilistic risk assessment became an accepted safety technique (INSAG 1988). Also, the process industries adopted the approach, which entailed functional analyses, fault trees, event trees and risk profiles. In the United States, a 'risk triplet' was introduced that included information about the scenario in the risk calculation. This approach clarified what could go wrong and how bad the consequences could be (Kaplan and Garrick 1981):

$$R = \{\langle s_i, p_i, x_i \rangle\}$$

s_i: scenario i – which accident scenarios i are relevant?
p_i: what is the probability of occurrence of scenario i?
x_i: what is the damage that can be expected from scenario i?

But these methods suffered from already known weaknesses in the sense of concerns about uncertainties, model uncertainties in the form of unknown risks and uncertainties concerning the quality of data, such as component failure rates. But there was also a lack of data on the consequences of exposure to toxic substances. Thus, the quantitative approach to risk, suggesting an absolutely correct numerical answer and an acceptable level of safety, slowly changed to a best-practice method in cost-benefit analyses (Oostendorp et al. 2016; Paté-Cornell and Boykin 1987).

A bit ahead of the curve, the United States pioneered the public debate about the acceptability of risks of commercial nuclear power plants. The discussion reached its apex in the 1975 Reactor Safety Study of the United States Nuclear Regulatory Commission, known as the WASH-1400 report (Rasmussen 1975). The report used probabilistic risk assessment (PRA) to demonstrate that the risks associated with nuclear power plants were relatively small. The WASH report and its various subreports were of great importance for the development of quantitative risk assessment in Western countries. But they did not solve the acceptability of risk. For one thing, comparing alternatives was not always straightforward with QRA models. Consider alternatives A and B in Figure 6.18. Both have the same risk, but alternative A has a higher probability of a low consequence and B vice versa – which is better?

Pioneered by Rowe, the discussion about acceptability had to take the context into account: who benefits, who suffers, who pays for it, what if a completely different industry could provide similar societal benefits? One thing was clear: few people were willing to take risks if the benefits were unclear to them (Rowe 1977; Conrad 1980; Fischhoff et al. 1981; Short 1984; Covello et al. 1987). A major point of criticism from psychology concerned the comparison of different types of risks in, for example, the WASH-1400 report (Figure 6.19). This report used a graph comparing common risks, accurately known because of widely available data, with the calculated individual risk of an incident with a nuclear power plant, based on QRA results.

The graph clearly showed that the probability of death among the public caused by an incident occurring at a nuclear power plant was acceptable. This risk was

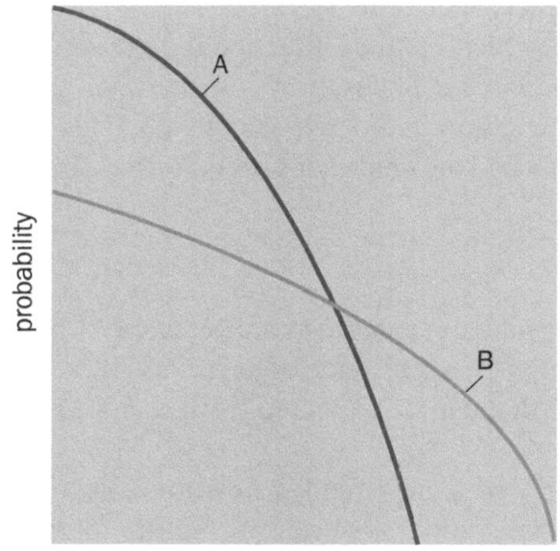

Figure 6.18 Comparing alternative risks
Source: Kaplan and Garrick (1981)

much smaller than other generally accepted risks associated with, for instance, dam breaks or air travel (Rasmussen 1975). This graph was used by many proponents of nuclear energy to prove its safety, and thus the acceptability of nuclear energy. Indeed, the argument was that those who opposed nuclear energy on grounds of safety issues were not rational. This argument drew criticism because the risks were not comparable (Griffiths 1981). Another point of criticism was the incompleteness of the scenario analysis, but this objection was also applicable to the risk analysis of industrial risks, with which the nuclear risks were compared by Rasmussen (Kaplan and Garrick 1981). The fundamental flaw of this use of QRA results, when comparing calculated results whose accuracy could not be checked, was its use to justify a chosen policy.

Baruch Fischhoff, who was trained in both mathematics and psychology, argued that there was no definition of risk that fitted all applications. The choice of a particular definition was actually political (Fischhoff et al. 1984). Risks could not be distinguished from the perceived benefits for citizens of a particular activity (Stallen and Tomas 1985). The debate on nuclear power was thus part of a battle royal over social choices for future energy supply.

Normal accidents

Extensive research in the United Stated led to the development of new safety theories. For Charles Perrow, the nuclear incident at Three Mile Island was the starting point for developing indicators that predicted major accidents. Like Barry Turner, Perrow was a sociologist, and he was interested in indicators of

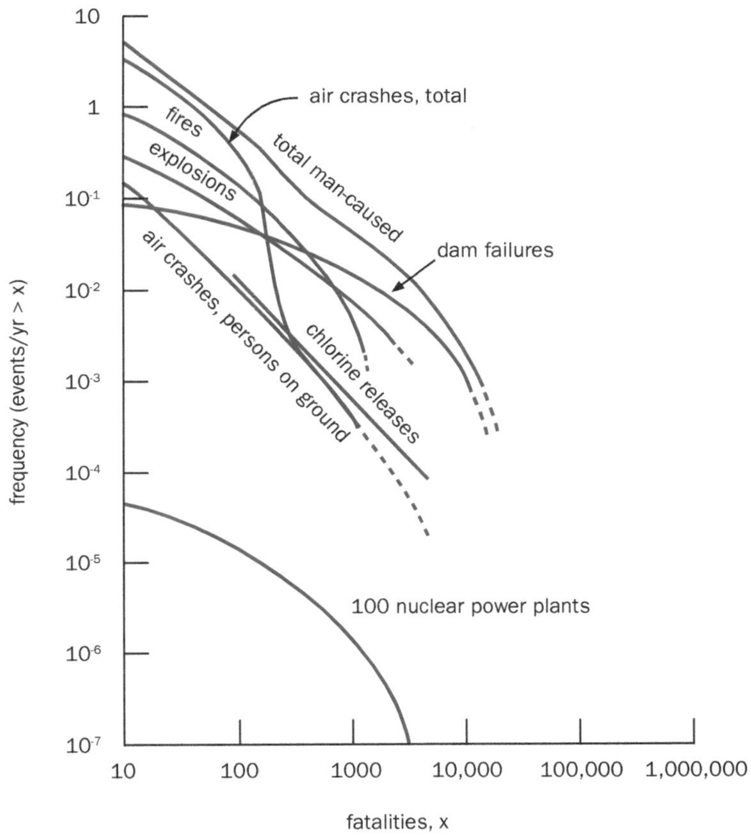

Figure 6.19 Risk comparison of human-induced events

Source: NUREG 1975

major accidents. He systematically investigated process industries, maritime and air transport, space travel, dams, mining, the use of DNA techniques and nuclear power plants. Similarly to Kletz, he concluded that major accident events could start from small beginnings. Additionally, high-risk industries shared special characteristics that explained these major accidents: coupling and interactions. With a tight coupling and absent buffers, a fault in one subsystem propagated into other subsystems, and system collapse was more likely. If the system was complex, if there were strong relationships between the subsystems, system collapse was highly probable. Perrow concluded that major accidents were inevitable because of tightly coupled processes with complex interactions. This was his seminal concept of *normal accidents* (Perrow 1984).

Table 6.4 shows some of the coupling parameters that Perrow considered relevant. Loose systems had slack and allowed for delays. They offered flexibility and time for recuperation and restoration.

Manual processes were generally loosely coupled, and delays could be dealt with. Production pressure or any other form of pressure that infringed on time increased the level of coupling. Table 6.5 shows some interaction parameters.

In Perrow's view there were two extremes: linear and complex interactions. Complex interactions were common mode functions. When a process controlled several functions, failure effects multiplied to affect multiple subsystems. When many such control functions existed, the system became complex. Also, transformation processes were complex if these processes were not fully understood, such as certain chemical or nuclear reactions or flying in the stratosphere. Complex reactions, multiple product pathways, feedback loops and unexpected changes – this complexity became visible in control rooms, in the many parameters needed for process control, as at Three Mile Island (Figure 6.15). Perrow assessed several industries along the complex versus coupling axes to yield Figure 6.20, which became famous amongst safety researchers.

Major accidents originated from non-observable and usually unforeseeable deviations that cause complex chains of events. With the installation of the Chernobyl nuclear plant, society opted for cheap, tightly coupled production systems of nuclear power, which yielded complex nuclear plants that, by their nature, were prone to failures. The human tunnel vision that prevented understanding the reactor was just the last straw in the accident process. But also, management was monolithic, distant and slow, and operators had only a limited understanding of the process they had to control. It was recognized that the same conditions were equally, abundantly present outside Russia.

A peculiar contrast appeared with Turner's British studies in the 1970s: he thought batch processes were too complex to control, but Perrow concluded that

Table 6.4 Tight and loose coupling (after Perrow 1984)

Tight and Loose Coupling Tendencies	
Tight Coupling	**Loose Coupling**
delay in processing not possible	processing delays possible
invariant sequences	order of sequences can be changed
only one method to achieve goal	alternative methods available
little slack possible in supplies, equipment, personnel	slack in resources possible
buffers and redundancies are designed in and deliberate	buffers and redundancies fortuitously available
substitutions of supplies, equipment, personnel limited and designed in	substitutions fortuitously available

Table 6.5 Complex and linear interactions in systems (after Perrow 1984)

Summary Terms	
Complex Systems	**Linear Systems**
proximity	spatial segregation
common-mode connections	dedicated connections
interconnected subsystems	segregated subsystems
limited substitution	easy substitution
feedback loops	few feedback loops
multiple and interacting controls	single-purpose, segregated controls
indirect information	direct information
limited understanding	extensive understanding

they should be safer because of decoupling. Strangely, these blatant opposing views were not widely discussed.

Man–machine interactions, THERP and high reliability theory

The focus on human error was still dominant in the United States and differed significantly from European concepts like the Rasmussen-Reason skill-rule-knowledge model. The 'technique for human error rate prediction', or THERP, originated in the 1960s and was based on a probabilistic approach to human error called human reliability assessment (HRA) calculations. HRA estimation of probabilities of human error were combined with a fault tree approach, similar to a fault tree for technical systems. A large database was developed for quantification of the fault trees by Sandia Laboratories for the US Nuclear Regulatory Commission (Swain and Guttmann 1983). An HRA method was developed there using an expert system (Raafat and Abdouni 1987).

Quantification of errors was not restricted to the US alone. In the UK it was ascertained that many human errors occurred in so-called human-factor-poor working environments (Hawkins 1987). Many authors expressed concerns about quantification. The reliability of information was questioned, and therefore its predictive power. A reliable database containing observable and reversible errors was all but impossible, which was also the case for fault predictions. A more fundamental concern was the view that human behaviour was similar to technical system components. It just did not represent the way in which people in technologically complex environments were dealing with process deviations (Singleton 1984; Rasmussen 1982).

Organisational psychologists, such as the American Karl Weick (1974, 1979), also doubted the quantitative approach. Weick stated that managing a company, or any organisation, was a fundamentally social process and not a technical one. In his view, people in an organisation had to deal with practical organisational

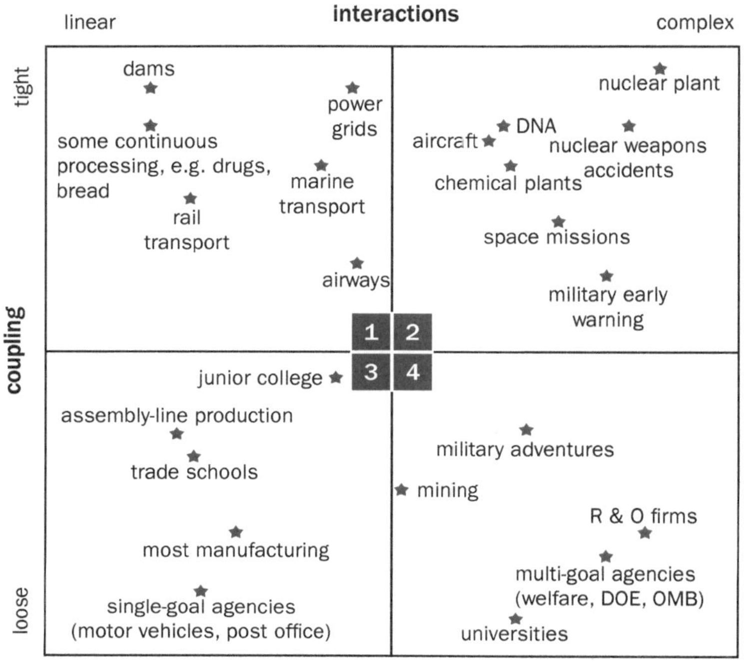

Figure 6.20 Industry sectors scaled along coupling and interaction
Source: Perrow (1984)

puzzles and try to make sense of their predicament. 'Control', 'system' and 'feedback' were terms that carried no value in their decision-making processes; their actions and decisions were rational only in retrospect. It would be more sensible to focus on organisational processes. However, knowledge about how these organisations functioned was relatively low (Daft and Weick 1984; Rochlin 1986; Weick 1987). But there was agreement on the observation that 'balance', 'control' and 'feedback' – terms from a system approach – would not explain organisational dynamics.

By the end of the 1980s a new research direction had taken off, namely 'high-reliability organisation' (HRO). HROs were organisations that defied Perrow's broadly accepted safety beliefs because they dealt safely with very complex processes that were tightly coupled and experienced very few major accidents. Flight operations on aircraft carriers and air traffic control were prime examples. They performed very reliably, despite operating dangerous processes. Reliability, in this context, was characterized by learning, understanding and correcting complex processes to prevent accidents. The approach was diametrically opposed to the traditional, mechanistic, efficiency-focused management approach. Weick found that the supposedly efficient traditional management techniques yielded situations in which operators would simply not have enough time to respond to complex deviations (Weick 1987; Roberts 1988). It was actually claimed that '[w]e have third-generation machines and first-generation minds' (Westrum 1988). This view was strengthened by accident reports at that time emphasising that accidents did not have purely separate technical or human causes but tended to combine these into interactive or synergetic errors. Time pressure could not be blamed; humans could simply no longer understand the complexity of the systems they were operating, and traditional feedback learning was no longer sufficient to achieve optimal process control. The breakdown of traditional feedback learning was part of the HRO philosophy; optimal process conditions were a utopian condition. That left two basic options: either the system had to be simplified, which Kletz proposed, or the operators had to become smarter. When the complexity outgrew the capacity of a single human, teams or networks had to do the job. These teams had to be composed of people from diverse backgrounds to enhance reliability. An example was the Diablo Canyon reactor, in California, which was built near four tectonic faults (including the San Andreas Fault and the Hosgri Fault) and had to be able to withstand earthquakes with a strength of up to 7.5 on the Richter scale. An elaborate seismic detection system was used that could shut down the reactor automatically. At shift change, process control was handed over as a group process in which people with different backgrounds bore responsibility: a supervisor, an engineer and some hands-on operators informed the next shift about the previous one. To Weick this represented a good example of dealing with complexity and an efficient delegation of responsibility and expertise with a minimum of hierarchical structure (Weick 1987).

Another example central to HRO reasoning was the almost flawless operation of aircraft carriers. Operating these carriers defied standard ways of working safely. Organisationally, there was a horizontal structure of squadrons, a vertical structure of maintenance and operations, and a cross-cutting command structure to deal with military units at sea and in the air. On top of that, all staff were replaced every 40 months. And the hazards were huge. An aircraft carrier stored vast amounts of high explosives and aircraft fuel; it controlled complex air manoeuvres, was propelled by a nuclear reactor and it often operated in hostile waters. An officer of an aircraft carrier explains it as follows:

> *So, you want to understand an aircraft carrier? Well, just imagine that it's a busy day, and you shrink San Francisco Airport to only one short runway and one ramp and gate. Make planes take off and land at the same time, at half the present time interval, rock the runway from side to side, and require that everyone who leaves in the morning returns that same day. Make sure the equipment is so close to the edge of the envelope that it's fragile. Then turn off the radar to avoid detection, impose strict controls on radios, fuel the aircraft in places with their engines running, put an enemy in the air, and scatter live bombs and rockets around. Now wet the whole thing down with salt water*

and oil, and man it with 20-year-olds, half of whom have never seen an aeroplane close-up. Oh, and by the way, try not to kill anyone.— Senior Officer, Air Division.

(Rochlin et al. 1987)

The only way to achieve reliability was to build in redundancy in several ways. Some responsibilities were delegated to the experts in the front line, meaning someone lowest in hierarchical rank could have the authority to cancel flights on the flight deck, or land planes. Officers had learned to trust the authority of these front-line men, on whom their safety depended. Young officers were trainers and learners at the same time, as if the ship were one big school. Fail-safe redundant systems were introduced: operations were supported by several teams that could take on one another's responsibilities, and double-checking was operational in nearly all safety-critical decisions and actions. The organisation was competent in dealing with process deviations, and all staff could be used multidisciplinarily (Rochlin et al. 1987; Roberts 1988). Organisational culture was a dominant feature of such highly complex high-risk operations. People accepted the way it functioned, shared preconceptions and assumptions, identified with personal fulfilment of the activities and tasks, and respected the way decisions were made. In contrast to organisations in which traditional feedback learning was prevalent, HRO organisations hardly required any oversight (Weick 1987). These specific high-reliability organisations did not appear to require a separate safety management system, whereas the majority of organisations that dealt with less extreme safety concerns trusted and used safety management systems.

Safety management

Accident reports, time and again, demonstrated that management failure was the fundamental cause of accidents, especially major accidents, and thus adequate safety management systems were desperately required. Scientific literature amounted to pretty much the same by concluding that management decisions were often haphazard, leading to unsafe situations. This was found in both high-risk organisations and occupational safety. Safety simply did not get the attention it deserved from contemporary managers. The only safety indicators used were sick leaves after accidents, but these provided no insight into the basic causes of accidents (Grondstrom et al. 1980; Kjellén 1982, 1984a, 1987; Robinson 1982; Kletz 1985b; HSE 1985; Fischhoff et al. 1987; Harms-Ringdahl 1987a). By focusing on occupational incidents, managers were assumed not to make errors, whilst front-line workers were seen as consistently faltering. This prolonged the popularity of the accident proneness theory, and it was still a prevalent theory in the design of accident prevention methods (Tombs 1988). The disparity between science and practice became almost impossible to reconcile (Purswell and Rumar 1984; Kjellén 1984a). Another complication was the rising importance of litigations: after major accidents new laws were adopted, which made safety laws very detailed and almost impenetrably complex. This led to an unclear relation between the law and safety performance (Kletz 1984b, 1986; Benner 1975, 1985). Kletz (1984a) went as far as to claim that criminal prosecution was counterproductive for safety; prevention was much better.

Wildavsky (1988), a political scientist, reconciled the streams in safety management. Wildavsky postulated that for well-known hazards in a stable environment, a trial approach without error would prevail. In these situations, hazards were predictable, and controlled with safety barriers, protocols and simulations. This approach was risk avoidance aimed at guarding the stability of a system. Risk avoidance was dominant amongst companies. Inevitably this stimulated an unbridled growth of legislations, protocols and rules amongst organisations and governments. Wildavsky postulated that stable situations were practically non-existent, and a different strategy was required in a dynamic environment: resilience. Resilience was an organisation's capacity to adapt to, and react to, dynamic conditions and hazards, before serious problems were caused. These two fundamentally differing approaches concerning hazards and risks illustrated how little was known about the functioning of complex organisations. Yet, most high-tech-high-hazard industries should be classified as such (Daft and Weick 1984).

The huge differences between organisations and their responses to safety issues led sociologist Ron Westrum (1988) to a classification of organisational reactions towards safety in his contribution to the Safety Control and Risk Management conference of the World Bank in 1988. His classification was similar to the one suggested by Petersen 13 years earlier (Petersen 1975):

1. *Pathological organisations: Even under normal operational circumstances, pathological organisations could not deal with hazards effectively. Significant economic pressures forced these organisations to bypass safety regulations. Safety protagonists were ignored, suppressed or punished in such organisations. Examples were Union Carbide in their bullying tactics in the Bhopal investigation (Shrivastava et al. 1988; Shrivastava 1992), and Occidental leading to the Piper Alpha major accident.*
2. *Calculative organisations*

   Such organisations worked 'by the book' when it came to safety management. They were better than pathological organisations and could survive during relatively placid time periods. However, this approach was no longer functional in times of change; when unforeseen events occurred, the organisation was not able to respond adequately. The incident at Three Mile Island fitted that management category (Kemeny 1979). During periods of calm, safety management functioned reasonably well, but during the four-day crisis, operators could not comprehend the seriousness and magnitude of the disturbances occurring inside the reactor.
3. *Generative organisations*

   Generative organisations were resilient and in line with Wildavsky's approach. They were characterized by strong leadership and creativity, which stimulated the entire organisation. High-reliability organisations were examples of such an organisation.

The Netherlands

Some major industrial accidents in the 1960s and 1970s

Some major accidents, one at Shell Pernis in 1968, another at DSM Beek in 1975 and the oil blowout at NAM Schoonebeek in 1976, instigated quite a few research initiatives and social debates in the Netherlands. Apart from research on the devastating effects of gas explosions and the fighting of blowouts, the discussions centred on the quantitative risk assessment and more fundamentally the risk concept. Being a nation that is largely below sea level, the Netherlands was a leader internationally in terms of its approach to risk. Research on the risks of flooding started in the late 1930s and the experience influenced risk discussions in the 1970s on nuclear energy and chemical process industries. From the late 1970s, a series known as 'coloured books' were published on quantitative risk assessment. These books presented an international standard of models of effects (e.g. the blast wave of an explosion of LPG), of assessing failure probabilities, of transmission models (e.g. of a toxic cloud through air) and of damage models. Detailed research after incidents (such as the major LPG accident in Mexico) was used to improve the damage models (Pietersen 2009). The models and fault trees formed an important basis for quantitative risk assessment. But the uncertainties were large, as shown by the criticism of the Canvey Island report discussed above (Cremer and Warner 1980; Warner 1981; TNO 1983). In the early 1980s, the Brede Maatschappelijke Discussie (broad societal discussion) on the introduction of nuclear energy fuelled the national discussions on risk acceptance. The criticism of QRA never ceased. In an advisory report to the Dutch government, the Adviesgroep Gevaarlijke Stoffen (AGS; Hazardous Substances Advisory Council) criticized the legally prescribed software program used for the calculation of individual risk and societal risk (AGS 2010).

Shell Pernis, 1968

The Shell refinery in Pernis was, and still is, the largest in Europe. Many kinds of petroleum products were produced from crude oil. The cleaning of the installations and pipes was usually done with hot water and steam. The water, mixed with oil residues, was called 'slop' and collected in slop tanks. On 20 January one of these tanks was completely full, with 1575 cubic metres of slop. The slop was heated by 130° C steam from a supply pipe and came to the boil. An overflow occurred and a hydrocarbon vapour cloud formed over the adjacent area. One or two smaller explosions and a particularly violent one, equivalent to 20 tons of TNT, took place. The blast caused extensive damage and a large fire. The fire covered an area of about 250–300 metres. Two people were killed and 85 were injured, mainly by flying glass. Because of the cold weather the slop had to be heated and it was believed that a layer of water-in-oil emulsion had built up. It covered the coils and reduced the heat transfer from them into the oil layer above, so that a substantial temperature difference had developed near the coils. Vapour formation initiated mixing, causing further vapour production so that the tank overflowed, the hydrostatic pressure at the bottom of the tank was reduced and a violent boil-up occurred, causing the condition known as 'stopover' (Roolvink 1968; Mannan 2005; Zwaailichten 2016).

DSM Beek, 1975

Early in the morning of 7 November, the start-up of Naphtha Cracker II on the 100,000 tons per annum ethylene plant at the Dutch State Mines (DSM) works at Beek was underway. Compressed gas was sent to the low-temperature system and an unexpected escape of vapour was observed. Shortly after, the cloud formed found a source of ignition and this resulted in a massive vapour cloud explosion. The explosion caused extensive damage and started numerous fires. The explosion killed 14 and injured 104 people inside and three outside the factory. The evidence suggested that the escape occurred due to low-temperature embrittlement. The rupture had been caused by an explosion of organic peroxides that might have been formed from material trapped in a section of flush pipe that was isolated. The amount of flammable material that escaped prior to the explosion was estimated at about 5.5 tons of hydrocarbons, mainly propylene (Mannan 2005; Zwaailichten 2016).

NAM Schoonebeek, 1976

On Monday, 8 November 1976, something went wrong at one of the NAM wells at Schoonebeek. A valve broke down in the early morning and fluid, mixed with sand and oil, spouted from the well for two days (Figure 6.21).

A large part of Schoonebeek was covered with a layer of oil-bearing sand. Only the next evening did the technicians close the well. NAM had everything cleaned by cleaning teams and paid compensation. It took three months for Schoonebeek to be completely clean again (Nammogram 1976; Gool 2016).

Impact of these major industrial accidents

Impact of vapour gas explosions

The major accidents at Flixborough in 1974 and Beek in 1975 had a huge impact in the Netherlands. DSM, a Dutch company, owned the factory in Beek and co-owned the company in Flixborough. Both major accidents surprised experts and policy-makers because the damage from the explosion was far worse than expected (Marshall 1987). This led to many questions about the explosion severity of released flammable gases and vapours (Cobben et al. 1976). Research on vapour cloud explosions had already started after the Shell Pernis accident, but

Figure 6.21 Schoonebeek blowout
Source: © Bert Verhoeff Fotocollectief (1976)

the explosions at Flixborough and Beek led to a heightened interest in fundamental research (Groothuizen 1976). Field experiments in the US (Nevada, Texas) and the Netherlands (Mosselbank, TNO) in the 1970s gave a first insight into the chemical and physical mechanism of a vapour cloud explosion and the conditions that could worsen a mild deflagration, or explosive combustion, into a much more violent and destructive deflagration, and even into a powerful detonation (Siccema 1973; CPR 1982). A detonation is an explosion with supersonic velocity and is much more powerful than a deflagration. The presence of obstacles such as piping was found to be a major factor increasing the destructiveness of gas and vapour explosions (CPR 14E 1979; Berg 1985).

Fighting blowouts

After the Schoonebeek blowout, the illustrious American blowout fighter Red Adair came to the Netherlands, invited by the Nederlandse Aardolie Maatschappij (NAM; Dutch Petroleum Company). NAM is owned by Shell and ExxonMobil. Red Adair provided NAM with further advice regarding blowout situations. This led to the updating of existing plans and elaboration of organisation schemes for emergency failures and to the purchase of special equipment to combat a calamity. His lessons proved to be very decisive for oil companies in preventing such disasters as a blowout. He was the man who was hired if there was something big to fight, up to and including the fires in Kuwait (Nammogram 1980).

Loss prevention

The major accidents in and outside the Netherlands in the 1970s led to the development of a more systematic, comprehensive and technical approach to safety that contributed to the creation of loss prevention (Kletz 1999). The Koninklijke Nederlandse Chemische Vereniging (KNCV; Royal Dutch Chemical Society) and the Koninklijk Instituut van Ingenieurs (KIVI; Royal Institute of Engineers) organized a conference called Inherent Hazards of Manufacturing and Storage in the Process Industry in 1969. Together with the 1971 symposium Major Loss Prevention in the Process Industries, of the European Federation of Chemical Engineers, in Newcastle, these events were the predecessors of the first European conference on loss prevention in Delft and The Hague, the Netherlands, in 1974 (Pasman 1974; Pasman and Snijder 1974). Issues like the comparison of risks were not raised during the Delft – The Hague Loss Prevention Symposium. The problems addressed were guidelines for safe design in the process industry and studies on the hazards and risks attached to static electricity, gas and dust explosions, the transportation of hazardous materials and the reliability of system components. Oddly, safety management was not explicitly addressed either.

In the opening speech at this first European congress, the Dutch Minister of Social Affairs Boersma made a plea for the development of methods for quantitative risk assessment (Boersma 1974). The scaling up in process industries required new insights and methods. New methods were needed to estimate the size and probability of potential major accidents and devise the necessary measures to avoid these accidents. Boersma stressed that safety regulations for the process industries were similar to the regulations of nuclear power plants where, at that time, the debate centred on system safety, on the maximum credible accident and the resulting risks. Buschmann, a chemical consultant working at the Dutch Ministry of Social Affairs, spoke about new challenges for the process industry and referred to the report 'The Limits to Growth' by the Club of Rome in 1972 (Buschmann 1974). Far more efficient and cleaner production processes needed to be developed; new materials and alternative fuels needed to be applied. At this time, society was becoming more critical towards the introduction of new technologies, which created a need for in-depth analysis methods and clear guidelines as to whether acceptable risk levels were being exceeded (Gezondheidsraad 1975). In addition to that, it was important to demonstrate that the benefits of new technologies outweighed the risks they created. Buschmann translated this into a challenge for science by asking for quantification of risk. His call had immediate effects. Without delay, a research programme was started for determining the probabilities of failure and effect modelling. In the end, this effort led to the so-called coloured books in the CPR series, and the COVO and LPG studies, as discussed in one of the next sections.

The 1974 loss prevention symposium ended just a day before the Flixborough accident (Buschmann 1974). This symposium was the very first of a successful series of loss prevention symposia, organized every three years throughout Europe. These symposia focused on the process industry and process safety and therefore were attended largely by safety-oriented engineers and scientists from companies, government and universities. Later, quantitative risk assessment also became a central topic in the congresses of the Society for Risk Analysis (SRA) and the European Safety and Reliability Association (ESRA), founded, respectively, in 1980 and 1991. This stimulated further development of risk analysis in various fields of science.

Publications on system safety, such as those of Pope (1976), Wijk (1977) and Wansink (1976), could count on criticism. In line with Fischhoff, the main argument was that safety science should focus on a human approach. Man was not, after all, a component of a system and did not fit into mathematical formulae, and this was consistent with criticism of the failure probabilities of human behaviour in the United States and the United Kingdom (Rigby and Swain 1971; Kirwan 1994; Pasmooij 1979). Studies of human behaviour in automated process plants, following the taxonomy of behaviours and the skill-rule-knowledge model of Rasmussen and Reason, were discussed in the *Dutch Safety Journal* (Kolkman 1980, 1981; Stassen 1981). This was also the topic of the Professorship on Safety Science at Delft University of Technology (Hale 1985). The chair was taken by Andrew Hale, who with co-author Ian Glendon published a textbook called *Individual Behaviour in the Control of Danger* (Hale and Glendon 1987). They argued that, generally, the mental information processing of operators and maintenance staff was excellent. Only occasionally were errors made, leading to system deviations. However, operators possessed good correction skills, and therefore could put a process back into its intended operation mode. But systems could also move too far out of control. According to the authors, and in line with Rasmussen, this was, however, not usually caused by human error. It was a consequence of insufficient system knowledge, information overload or management decisions that had insufficiently considered potential risks, and major accident scenarios. The system approach first led to the quantification, assessment and evaluation of risk, followed by all kinds of decisions made on acceptability, risk acceptance and the resultant measures and interventions or risk control. An extended argument concerning risk quantification and acceptance was published in *De Veiligheid* and in *Risikobulletin.* Probability calculations were often, it was thought, based on guesswork and the uncertainty surrounding probabilities was seriously underestimated (Wetenschap en Samenleving 1978; Reijnders 1979; Andreas 1979; Cremer and Warner 1980; Stallen 1980; Stallen and Vlek 1980; Warner 1981; TNO 1983) (Figure 6.22).

Furthermore, a comparison of the risk figures, in which non-equivalent activities were compared, made little sense. Not all risks were taken voluntarily, and risk acceptance was ultimately a political issue presented as a scientific problem (Boskma 1977; Leij and Mutgeert 1977; Boesten 1978, 1979).

Origin of the Dutch risk concept

In the Netherlands, developments in flood safety had a major impact on the risk-based and probabilistic approach in safety science. Van Veen and Wemelsfelder, civil servants in the government agency Rijkswaterstaat (Department of Public Works), had already pioneered this approach in the 1930s. The department was responsible for the design, construction, management and maintenance of the main infrastructure facilities, including sea dykes. Their work led to a fundamentally different approach to flood control.

Figure 6.22 Special issue of *Science and Society*, risk acceptance

Source: Wetenschap en Samenleving (1978)

At that time, it was customary to increase the height of dykes to slightly above the level of the last flood. But in the statistical approach, dyke heights were determined by statistical analysis of tidal flood levels. Statistical analysis dictated that there was no such thing as a risk-free dyke, but that there was a probabilistic relationship between the height of sea dykes and flood risk: the higher the dykes, the lower the risk of flooding. This work paved the way for the warning issued to the government in 1939 of the inadequate protection of the dykes in the south-west provinces of the Netherlands in extreme weather conditions. The warnings were repeated shortly after the Second World War (Horn–van Nispen 2001), but to no avail. In 1953, the Great Flood in the south-west provinces of the Netherlands, the west coast of Belgium and the east coast of the United Kingdom caused 1,836 fatalities in the Netherlands, 326 in the United Kingdom and 28 in Belgium. The Dutch started the Delta Works project and policy decisions on flood safety and dyke heights were based solely on risk assessments. For a vulnerable part of the Netherlands, the village of Hoek van Holland on the west coast, the dyke exceedance probability was calculated to be $4*10^{-5}$/year. This was translated into a maximum permissible exceedance probability of 10^{-4} per year, defined as the probability that the water height was equal to or higher than the dyke height (Dantzig 1956; Dantzig and Kriens 1960). Flood risk was defined as the probability of flooding each year multiplied by the amount of damage, defined as the number of deaths (Brinke and Bannink 2004). All other policies on flood defence were abandoned from that point onward. This successful risk concept – there have not been any seawater floods from that point onward – would influence safety assessments in industrial domains in the rest of the world.

Starting from 1975, the Commission for Prevention of Disasters (CPR) of the Dutch Ministry of Social Affairs and Public Health published standards for the safe storage of hazardous substances. The predecessor of the CPR was the Gasoline Commission from 1927, which was transformed into the Commission on Storage of Hazardous Substances in the 1960s. In these commissions sometimes the word 'risk' was used, but without a clear definition, or any quantification. The CPR standards of 1975 marked the beginning of a 'risk thinking' within the ministry.

Already in the late 1960s, in the Netherlands a broad national debate had emerged on the calculation of risk as a measure of safety. The debate attracted both the media and the scientific press. The focus of these discussions was on dangerous substances, particularly when it came to low-probability and high-impact events, which were later designated 'major hazards'. But the discussion went much further. Social issues were at the heart of the discussion on large-scale technological developments, such as the mooring of LNG in Rotterdam harbour, the construction of LNG pipelines to Germany, the use of LPG in cars and the development of the chemical industry in Eemshaven, in the north-east of Holland. The proposals for LNG landing led to spectacular images of vapour cloud explosions with the power of a nuclear bomb. These exorbitant projections drove the need to 'put some numbers in the arguments' to get a better grip on the problem and facilitate sensible safety decision-making. The heated debate about LNG abated after the discovery of a huge natural gas field in the north of the Netherlands, but the debate about the introduction of the risk paradigm and the acceptability of risks continued.

Resolving these social issues called for a new approach to safety (Ale 2002, 2003). The new approach was aimed at delivering a better understanding of the dispersion of hazardous substances in the atmosphere, the type and severity of the potential of such dispersions and all forms of damage they could cause. In addition, there was a need for a clear definition and quantification of the concept of risk and for a criterion for an accepted level of safety. The scientific debate about the long-term health risks from prolonged low exposure to hazardous substances took place mainly in the medical, environmental and occupational hygiene disciplines (see Chapter 5) and paralleled a similar discussion on the acceptability of risks of hazardous substances in industry. Thus, the stage was set for change.

In this context, the discussion about adverse consequences for local residents provided the motive for risk research in the 1970s. The limited land space in the Netherlands forced consideration of safe separation distances between the (process) industry and residential areas. This development emphasised the central role of the Dutch risk-based approach. As risk-based research progressed, the definition of risk as a combination of the probability and the effects of incidents became increasingly accepted. This definition had a structuring effect on the public debate on risks, and reduced the debate to realistic proportions (Pasman 1999). From the 1970s onwards, risk was frequently placed on the political agenda. The scientific debate on risk as a basis for safety decision-making was not restricted to the field of safety science, and is still ongoing (Vlek and Stallen 1979; Gezondheidsraad – Dutch Health Council 1996, 2006; WRR 2008; Ministry of the Interior and Kingdom Relations 2012).

The Dutch Labour Inspectorate was also very productive in this period of time. It produced guidance for guidelines for Hazop, process safety analyses and risk detection systems with their V- and R-series (V- for *veiligheid*, or safety, and R- for *risico*, or risk) (DGA 1979, 1981, 1984a, 1984b, 1989). Also, the professional press, like *De Veiligheid*, continued the tradition of reporting on the quality, consequences and results of risk analyses (Hanken and Andreas 1980; Leeuwen 1982; Dop 1981, 1984, 1985). But only a handful of experts seemed to understand the numbers of QRA (Bjordal 1980) and it was suggested that QRA should be labelled as a scientific approach to forestall public fears (Irwin 1984; Eindhoven 1984). Further comments came from Hofstede (1978), who, like Weick, voiced fierce criticism of systems thinking. The systems approach, with its clear goals, input-process-output scheme, risk calculations, comparison with standards and the feedback and feed-forward loops, was also prevalent in management literature as a model for decision structure. However, the managing of companies was primarily a social process in a socio-technical environment. Any systems approach remained flawed because there were no clear defining goals, only certain general remarks, derived from business visions. The claims, or the quality of management activities, were only partially measurable or not measurable at all, which renders feedback information either unusable or not desirable.

COVO study, LPG study

The turbulent development in the process industries greatly influenced the focus of research in safety science. This applies in particular to quantitative risk analysis. The Canvey Island study of the HSE had a follow-up in the Netherlands (HSE 1978; Gezondheidsraad 1978). Following this British study, Cremer and Warner (1982) conducted a risk assessment study at the industrial cluster of the Rijnmond area near Rotterdam harbour. The name of this study, COVO, was an abbreviation of the name of a safety committee of residents living in the Rijnmond area. It was the first quantitative risk assessment in the Netherlands that identified residents' industrial risks. The European Seveso Directive – named after a major accident near the Italian town of Seveso (1976) – was for a large part based on research on quantitative risk assessment as a follow-up of these studies, the coloured books.

In 1978, the TNO was commissioned to carry out a study on 'The Safety of Current and Future Activities Using LPG in the Netherlands', also called the LPG integral study (TNO 1983). Different methods were used in conjunction, including incident analysis, hazard and operability studies (Lawley 1974), fault tree analysis and event tree analysis. The report of the LPG integral study mentioned that due to the complexity of the concept of risk, three different measures were used to assess both the expected consequences of incidents and the likelihood of the occurrence of these consequences:

- *The average number of expected deaths;*
- *The probability that an individual person was killed at a particular location or in a certain area;*
- *The probability that a certain number of people were killed in the total area affected.*

Much later, in 2004, the last two measures became the basis for defining the criteria for the personal (individual) risk at a given location (PR) and the group risk (GR) in the Dutch Besluit Externe Veiligheid Inrichtingen (BEVI; Decree on External Safety of Industrial Plants). These three types of risk were calculated for different types of major accidents and scenarios. The most serious scenario was the BLEVE, a 'boiling liquid expanding vapour explosion' (TNO 1983) (Figure 6.23).

This is an explosion due to failure of a vessel containing a pressurized liquid at a temperature well above its atmospheric boiling point. With flammable liquids such as propane, the initial explosion is usually followed by an intensely burning fireball. Thus, the individual risk of a BLEVE followed by a fireball was calculated as the probability of occurrence of an incident BLEVE (10^{-2} per year) multiplied by the probability of a fatality resulting from exposure to the intense heat radiation of the fireball. The researchers extensively discussed the uncertainties of the risk analysis, including uncertainties related to the design, to the construction of engineering systems, to assumptions about failure probabilities, to the probability of occurrence of consequences and to the influence of human factors. The authors of the TNO report did not include the social acceptability of the activities examined. Their analysis was limited to a technical risk analysis concerning public safety and thus did not include an analysis of other aspects, such as environmental and economic ones. These aspects played a role in political decision-making on the use of LPG in the Netherlands. Nevertheless, the hard numbers infused the social debate.

Coloured books

After the publication of the LPG integral study, there was a growing need to pool the knowledge about risk analysis of hazardous substances. Commissioned by the Dutch government, this knowledge was put together in the so-called Yellow Book (CPR 14E 1979) by experts from ministries, research institutes and industry. The models presented were based in part on reliable scientific knowledge. White spots or parts of the risk analysis process for which sufficient reliable scientific information was lacking, or essential data were missing, were filled with the available rules of thumb. The reliability of the effects models was tested in laboratory and field tests. For many years, the Yellow Book was internationally used to calculate the consequences of hazardous substance releases in risk analysis studies. The methods were mainly used by designers and engineers of industrial plants and in quantitative risk analysis aimed at assessing 'safe distances' around industrial

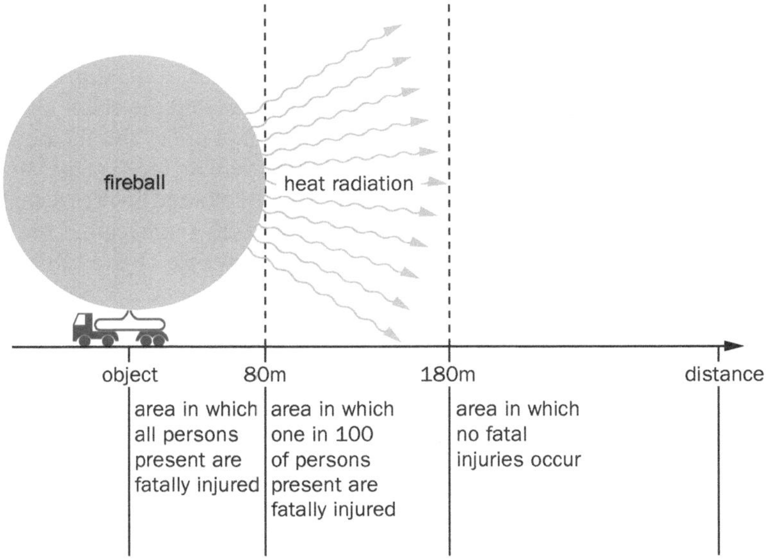

Figure 6.23 Effects of a BLEVE followed by a fireball

Source: TNO (1983)

plants and clusters. Later, other publications in the same series of 'coloured books' publications appeared. These treated specific aspects of risk assessment.

The Green Book was the second coloured book (CPR 16 1989). It described damage models for the determination of possible damage to people and property caused by the release of hazardous substances. The distinction between effect and damage was based on the sequence of the physical effects of the release of hazardous substances (e.g. an explosion producing a destructive pressure wave, a blast) and the probability of damage caused to the persons or properties exposed (e.g. exposed to blast resulting from an explosion). The terms 'effect' and 'damage' thus had different meanings in quantitative risk assessment compared to the same terms in industrial hygiene and occupational safety. Instead of the term 'effect' the term 'exposure' was used in industrial hygiene. And the term 'effect' in occupational safety and industrial hygiene was used for damage to health and property. The Green Book described effect models for calculating the transmission of released substances in different major accident scenarios. Thus, for a toxic gas release in a specific scenario, the expected gas concentration as a function of time at a certain distance from the source could be calculated. In the damage models these data were used to determine the risk, the probability of death for persons exposed to given concentrations for a given time duration. Damage models were also called 'vulnerability models'. These damage models were based on laboratory animal tests and military injury data from explosions and data on the effects of poison gas during and after the First World War.

The third book in the series, the Red Book (CPR 12E 1988), focused on the analysis of failure probabilities, such as the failure rates of individual technical components, e.g. valves and pipes, in order to determine the risk of failure of a technical system (Figure 6.24).

This analysis began with an analysis of the incident probability. This was based on hazard identification and potential accident scenarios and the chain of events leading up to an incident. In addition, both the potential effect and the potential damage had certain probabilities. These probabilities were included in the calculation of the risk. Perhaps surprisingly, the models in the coloured books were rarely used to assess risks for employees. In fact, the risk-based approach did not enter the occupational safety field until years later (see Chapter 5). The fourth book in the series, the Purple Book published in 1999, will be discussed in a next chapter.

The risk-based approach was discussed at the safety congress in Amsterdam in 1975 (Trier 1975; Leij 1977; Leij and Mutgeert 1977). In the introduction to Van Trier's lecture on risks, he stated that 100% safety did not exist. He urged the development of a systematic method for risk assessment that incorporates the loss of life, property damage, and social and ecological costs. The degree of voluntariness should be included in the acceptable risk criteria. Van Trier mentioned 10^{-6}, the probability of accidental death per year, as a generally accepted criterion for voluntary risks. The criteria for involuntary risks should be a factor 1000 lower. Van Trier uses the same factor published in the study of Starr (1969) in which a criterion for involuntary risk should be comparable to the risk of death by natural disasters, namely 10^{-9}, the probability of death by natural disaster per year. Buschmann

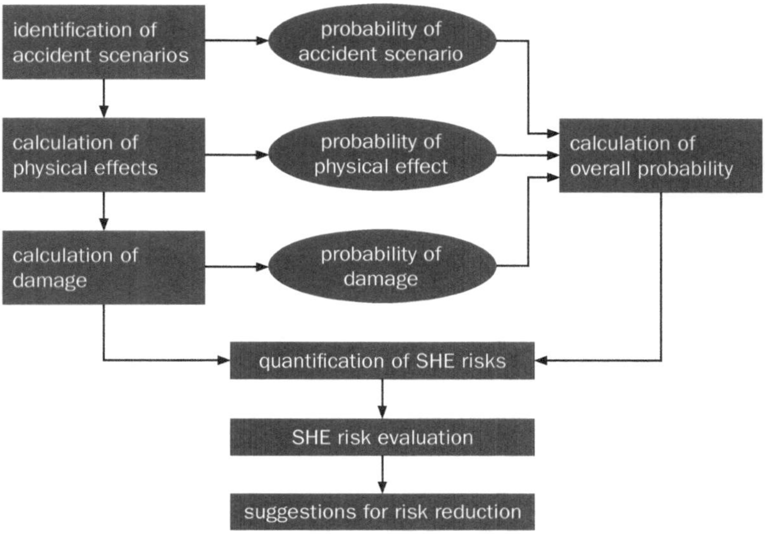

Figure 6.24 Schematic representation of the quantitative risk assessment. SHE: Safety, Health and Environment

Source: After CPR 12E (1988)

(1972), on the other hand, considered a mortality rate of 10^{-7} per year negligible. He based this on an accepted probability level for flooding achieved by Dutch coastal protection, a topic under discussion at that time in the Dutch parliament. Balancing the costs of raising the dykes to different levels resulted in an accepted probability of flooding in the coastal areas of once in 10,000 years (10^{-4}). At a mortality rate of about 1:1000 (10^{-3}) in a flood, this gives a mortality rate per inhabitant per year of 1×10^{-7}. This is 1/10 of the probability of death due to natural disasters in the United States. This probability figure differed by a factor of 100 from figures mentioned by Van Trier. The introduction of risk analysis would require a lot of explanation in the years following Van Trier's 1975 lecture in Amsterdam. Even with a seemingly tiny calculated risk, a particular incident could nevertheless occur tomorrow. Risk information based on a probability x effect calculation was incomplete if other information was missing, such as the type of risk, who ran the risk and controlled the risk, and how the risk could be managed. Something that seemed rationally acceptable was in fact emotionally often not acceptable. The discussion following the reading of Van Trier illustrated this:

> *Once we start doing risk analysis, and we have assessed the risk of a particular activity or a particular machine or a certain substance, and subsequently we know that something bad can happen with a certain probability, this often raises the question: so can it also happen tomorrow? And that*

question actually creates a kind of emotional atmosphere in which the risk suddenly becomes unacceptable.

(Hendrickx 1991)

Broad societal discussion

It is sometimes said that safety is a feeling. All kinds of considerations play a role in the perception and acceptance of risks, which differ from person to person, and the Dutch government at that time was struggling with the perception of risk in terms of its population. It organized a broad societal discussion between 1981 and 1983, in which safety and the perception of risk with regard to nuclear energy were prominent on the agenda. It was a popular consultation on the future energy supply and the possible use of nuclear energy. When the final report was issued, the conclusions caused so much debate that the social debate almost began anew and it was concluded that the broad social discussion ended in failure (Wolsink 1985). Despite that, in the 1980s, safety slowly grew into an applied science. It addressed individuals' experiences of risks, which received a lot of attention, and society slowly learned to deal with risks through trial and error. The era of the risk society (Beck 1986) had arrived (Zwaard 2007).

Risk perception determined the extent to which groups in society were willing to accept risks. The interest in risk perception and risk acceptance led to a great deal of research and numerous publications in the 1980s. This quickly yielded an image of the factors that were important. For example, the degree of voluntary participation played an important role in the perception of risks. Involuntary risks were generally more difficult to accept. The manageability, the influenceability and the extent to which people thought they could control the risks themselves and the importance of the risk-bearing activity were also important. In addition, the nature of the effect had an influence. The risk of an immediate effect, for example, and a fatal accident, was experienced differently to the risk of a delayed effect, for example cancer that led to death after a longer period of time. Foreseeability was another issue: unknown risk types were often weighted more heavily than known risks. Finally, the alleged knowledge about risks played an important role in the perception of risk – not only the knowledge of those directly involved but also the knowledge of experts and the image of those directly involved. Risk perception appeared to be largely determined by the confidence people had in experts and decision-makers. The acceptability of risks was also related to ideological background and political disposition. All in all, risk perception and risk acceptability were complex and unruly phenomena. As the broad societal discussion showed, discussions about safety could not exclusively be conducted in terms of probabilities and effects. Risk was more than a number.

Research on risk perception

Alongside the rise of quantitative risk assessment, interest grew in the fields of psychology, sociology and public administration concerning the public debate on how to deal with increased risks. The Dutch researchers Stallen and Vlek, who presented their findings alongside international researchers, gave various definitions of the concept of risk and discussed factors determining the risk perception of local residents (Vlek 1990). Right from the start, the social sciences debate strongly influenced the technical-scientific debate on quantitative risk assessment. The debate between the more technical-scientific approach to risk and the social-science approach is sometimes referred to as the debate between objectivists and constructivists.

Research in the field of risk perception, risk acceptance and decision analysis flanked the development of QRA. Stallen and Tomas (1985), for example, described the development of social science research aimed at distinguishing between objective and subjective risks in the public debate on industrial risks and nuclear power. The scientific debate on risk perception in the vicinity of industrial plants was conducted by a wide range of social scientists, economists and psychologists. The definition of risk was an important aspect of the debate. In 1977, the central question for a social scientist studying risk was not how risky something was but how acceptable a risky activity was (Rowe 1977). This was in contrast with the more technical-scientific view of QRA, which was, and still is, based on a distinction between risk quantification: how big is the risk? (with risk expressed in terms of adverse effects and probabilities – both quantitatively), and risk assessment: is the risk acceptable? However, social scientists questioned this distinction.

Psychological field studies showed that actually different risk aspects determine the risk perception of citizens (Slovic et al. 1984; Starr 1969; Vlek and Stallen 1979; Rowe 1977). The relevance of a comparison of risks, based on QRA, was therefore questioned, that is, as far as the view of the general public was concerned. Risk turned out to be a multidimensional concept with dimensions that were weighted differently by different groups of stakeholders (Vlek 1990). The most important psychological dimensions are:

- *Potential harm or lethality of a release or spill;*
- *Controllability by safety and control management;*
- *Number of persons affected simultaneously;*
- *Awareness of effects and consequences;*
- *Voluntariness of exposure and risk;*
- *Type of harm caused and relative fear, e.g. the probability of cancer was highly feared.*

According to Vlek (1990), personal risk acceptance was mainly determined by the interest in the intended or expected benefits, the severity of a maximum credible accident and the supposed process controllability of the activity.

The Shell casus

Another important contributor in the Netherlands was the psychologist Wagenaar (1941–2011). In the 1970s he worked at the Human Sensory Studies section, an institute of TNO. His work focused on decision-making and factors that influenced decision-making. In 1982, he accepted a professorship at Leiden University, where he studied the origin of human errors. His inaugural speech analysed a historic Dutch disaster, namely the huge explosion in Leiden in 1807, where a ship filled with gunpowder destroyed much of the city and killed hundreds of citizens. Wagenaar concluded that it was probably erroneous to completely blame the captain who lit a cooking fire on the gunpowder-loaded ship. It would be more correct to consider the wider web of decision-makers and the decision-making that had occurred (Wagenaar 1983). The Maritime Research Institute granted TNO and Leiden projects to investigate maritime safety. They used a checklist for air transport safety by Feggetter (1982), which was used to reanalyse 100 maritime accidents. The accidents were classified into three factors of subsystems: cognitive, social and situational. In 28% of cases, human errors were the main contributing factor, but even with technical factors there was always a human element present (Wagenaar and Groeneweg 1987). The conclusion of the research was that there are many contributing causes and that they were already present before the accident occurred. Furthermore, it was concluded that the person who committed the error at the end of the causal chain was not a *guilty person* and that punishment was an ineffective strategy to prevent future accidents. The focus should shift to the management of an organisation as responsible for the day-to-day working conditions. This finding justified further scrutiny of the human contribution to the accident causation process. The conclusion of the subsequent research into the causes and backgrounds of small and fatal accidents was that most accidents were too complex to blame on humans because it was often too hard for them to completely oversee the consequences of their actions, on the grounds that these same actions had not led to accidents on previous occasions. The study also concluded that a *blame culture* stood in the way of learning from incidents (Groeneweg 1988). The concept of *impossible accident* was introduced to reflect the fact that the occurrence and results of accidents were too unpredictable for directly involved humans to assess the consequences of their actions; their mistakes could be explained only after the accident. The researchers found that it was much more sensible to find the factors that caused people to make mistakes; those causes were much more manageable than the consequences of accidents. By eliminating disturbances caused by humans, entire accident classes could be prevented. Also, the primary cause of such disturbances was found to be due to poor decision-making by management. Analysis of accidents at NAM, the Dutch Natural Gas Corporation, and cases of misuse of force by the Dutch police supported these findings time and again (Groeneweg 2002).

The maritime safety investigation attracted the attention of Royal Dutch Shell. According to Van Engelshoven, the top manager at the Exploration and Production Division of Shell, the number of people killed in Shell's industrial activities was completely unacceptable, and he challenged local manager Koos Visser to come up with solutions to lower the number of casualties. In an organisation of a predominantly technical nature it was quite a challenge to introduce a more social science approach. Visser was convinced by the work of both Vlek and Stallen on risk perception and Wagenaar's work showing that there was an important role for social scientists to play in improving safety. He assigned a research grant to Wagenaar (Leiden University) and to Reason (Victoria University in Manchester, UK) to investigate industrial safety in Shell's process plants. At Shell there was some concern that the scientists, although very good at telling gripping stories and presenting compelling models, lacked knowledge about the day-to-day operations in the petrochemical world and that this would negatively influence their credibility. The first two years of the project involved an in-depth familiarisation of the scientists with a wide range of operations activities and locations. This project, 'From Jungle to Boardroom', had to develop a sensible and practical applicable model on accident causation and tools for identifying 'local triggers' and 'resident pathogens'. Additionally, better safety management techniques had to be developed. To make sure that the project had sufficient scientific basis, a workshop with a number of leading scientists was held in Gabon. This location was chosen to attract these scientists as it was thought that being exposed to an exciting environment like the jungle of Gabon was an opportunity they would not like to miss. The team consisted of renowned experts like Rasmussen, Fischhoff and Turner. After a week of wading through swamps and discussing the most effective way forward, a green light was given to the multi-year project. It was named 'Tripod', after a three-legged dog in Gabon that had been put down due to rabies that he turned out not to have. The concept of the safety management system was developed and based upon accident research at NAM (Groeneweg and Wagenaar 1989). The next chapter will discuss the Tripod model in detail.

At that time, in 1987–1988, Reason had already started writing his seminal work, *Human Error*, which appeared in 1990 (Reason 1990). It is a loss for modern safety scientists that Chapter 7 of the original manuscript, which dealt with a history of safety science from the ancient Greeks to modern times, was discarded because one peer reviewer thought it was too dull. Reason decided to write an alternative chapter, 'Latent errors and system disasters' (Reason 2013), in which he illustrated the new human error concept by applying it to modern incidents, such as the sinking of the *Herald of Free Enterprise* and the nuclear meltdown at Chernobyl. In the Netherlands, Wagenaar became a local celebrity and the first media star in the field of human error and safety. What this painfully demonstrated was that modern thinking about organisational factors in accidents, which is quite popular today, was quite well developed before Piper Alpha blew up in the North Sea.

University training and research in safety

In 1978 a symposium was organized by what was at the time the Technical High School of Delft (now Delft University of Technology), the Foundation for Road Safety Research (SWOV), the Safety Institute (VI) and the Directorate General of Labour (DGA). The University Teaching and Research Symposium in Delft discussed academic education and the research continuing into safety (THD 1978).

This symposium was much wider in scope than the 1974 European conference on loss prevention in Delft and The Hague. At the 1978 symposium the safety domain was not merely restricted to occupational or process safety. It was extended to include private safety, safety at home, safety in sport and road safety. The presentation topics were broader than the Delft–The Hague loss prevention symposium, which was restricted to experts from universities, industry and government. At the Delft symposium there was also space for presentations from unions, chemistry and science shops and action groups, which, alongside the technical aspects, also highlighted the social aspects of safety. The Delft symposium marked the start of the Safety Science Group, which was established at the Technical High School in 1979 (Goossens 1981). Safety Science thus became an academic discipline in the Netherlands, following earlier initiatives undertaken by the University of Wuppertal (1974), the Catholic University of Leuven (1975) and the University of Aston, Birmingham, UK (1976). In reality, process safety in industrial plants had become so complicated that academic training in chemistry and chemical engineering was required for adequate safety analyses (Lemkowitz and Zwaard 1988; Lemkowitz 1992; Kletz 1988b). Delft University of Technology and the University of Amsterdam followed up on these developments in 1989 by starting postgraduate courses on Management of Safety, Health and Environment (MoSHE) and *Veiligheid, Gezondheid en Welzijn in de Arbeid* (VGWA; Safety, Health and Well-being at Work). In the Delft chemical engineering curriculum, chemical process safety was added as a required course (Sectie Veiligheidskunde 1983; Hale 1987; Scholte 1993; Swuste and Sillem 2018).

In western Europe and Nordic countries, in North America and the Netherlands there were major contributions to the development of Safety Science. A series of major accidents in different sectors of industry had a decisive impact on this development. The consequence of these major accidents, the death toll, was in many cases the motive for extensive media attention. In western Europe and Nordic countries, the first models of accident processes in the nuclear sector were published. Major accidents in the process industries were a starting point for the loss prevention movement. And like the nuclear sector, this movement stimulated quantitative risk analyses and inherently safe designs of installations. On the one hand, engineers were the main driving force behind loss prevention and the accident model for the nuclear sector. This explained the preference for technical solutions and quantification, as in QRA. In this period the human influence was studied as part of a man–machine interaction, leading to the skill-rule-knowledge model. On the other hand, sociologists were more concerned with organisational conditions leading to major accidents. This led to the disaster incubation theory, pointing at delayed effects of organisational focus and decision-making on accident processes of major accidents. One of the major accidents in Italy and Spain initiated European legislation to regulate companies storing significant amounts of hazardous substances.

Also, in America, the nuclear industry was the first to initiate a model for accident processes, but here the model was integrated with a management model: MORT. The major accident at Three Mile Island was the impetus for the normal accident theory of sociologist Charles Perrow. He supported the notion that major accidents were inherent to the technology used and these accidents occurred in normally functioning organisations and companies. Major accidents in Mexico City and Bhopal fuelled the discussion on quantitative risk assessments. This was not without a heated debate. Comparison of industrial risks with commonly accepted risks was seriously questioned, as were the initiatives to introduce databases with human error probabilities. Finally, another theory emerged. Social scientists developed the high reliability theory, which was actually not a safety theory, but an organisational theory on how organisations could cope successfully with highly dangerous processes.

In the Netherlands, there was a focus on risk and risk assessments. The first international loss prevention symposium started a few initiatives. Risk assessment studies were initiated on LPG, and on risks for residents of the highly industrialized area of Rijnmond, near the city of Rotterdam. And finally, a series of 'coloured books' on risks (CPR books), probabilities and risk assessments were produced. These books would gain much international appreciation and were used as a source for the Seveso Directives. Also, in the Netherlands there was a lively debate on risk, and risk perception.

7 Occupational safety, safety management, culture

1990–2010

Knowledge development on occupational accident processes was rather limited in this period. Only three models were published, all reflecting a system safety and socio-technical approach. But quite a lot of attention was given to safety management and safety culture. Safety management was heavily influenced by developments in quality management. Safety management was heavily influenced by developments in quality management. The section on North America is extended to cover both North America and Australia, because publications from Australia are entering the safety domain.

DOI: 10.4324/9781003001379-7

The 1990s was a decade with wars and amazing technological developments.

In 1999, the world prepared for the **millennium bug** and the 'Year 2000 problem'. Problems were anticipated because many computer programs represented four-digit years with only the final two digits – making the year 2000 indistinguishable from 1900. Fortunately, the vast majority of problems had been fixed when the clocks rolled over into 2000. The situation was essentially one of preemptive alarm.

In the early 1990s, the idea that a computer was for everyone began to take shape. In 1995, **Windows 95** introduced a redesigned shell based around a desktop metaphor; the desktop was repurposed to hold shortcuts to applications, files and folders. Windows played a major role in both Internet Explorer's success and bringing the web to the masses.

In 1990, the **Hubble Space telescope** was launched into low Earth orbit. Hubble produced numerous remarkable images. The telescope has helped resolve some long-standing problems in astronomy, while also raising new questions. Some results have required new theories to explain them. The Hubble data had a strong positive impact on astronomy.

The **Yugoslav wars**, a series of ethnic conflicts and wars of independence, were fought from 1991 to 2001, leading to the break-up of the Yugoslav state. Often described as Europe's deadliest conflicts since the Second World War, they were marked by many war crimes, including genocide, crimes against humanity and rape.

In 1989, the **World Wide Web** was invented. A browser was released to the general public in 1991. Connected by the Internet, many websites were created around the world. The World Wide Web has been central to the development of the Information Age and is the primary tool billions of people use to interact on the Internet.

In the **Gulf War,** coalition forces from 35 nations fought against Iraq in response to its invasion of Kuwait. The 1990–1991 war marked the introduction of live news broadcasts from the front lines of the battle. Apart from the impact on Arab states of the Persian Gulf, the resulting economic disruptions after the crisis affected many states.

Nelson Mandela was a South African anti-apartheid revolutionary who was imprisoned for 27 years during the apartheid regime. In 1990 he was released from prison, and four years later he became the first black president of South Africa after the first free non-racial elections of the country. He retired in 1999 and died in 2013.

In 1994, the **Channel Tunnel** opened, a 50-kilometre railway tunnel that connects England with France. It is the only fixed link between the island of Great Britain and the European mainland. The tunnel is considered one of the seven modern wonders of the world. It offers an alternative transportation mode unaffected by poor weather.

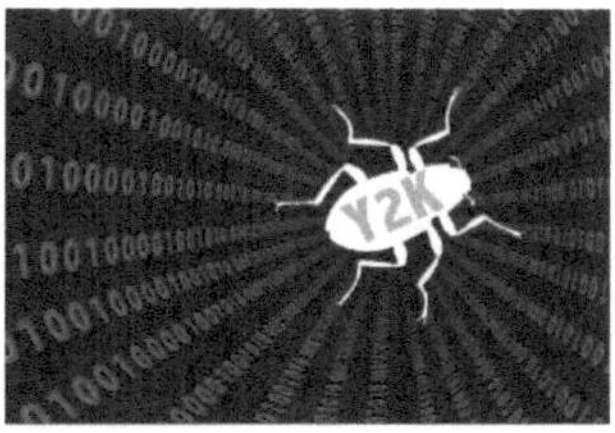

Millennium bug

Windows 95

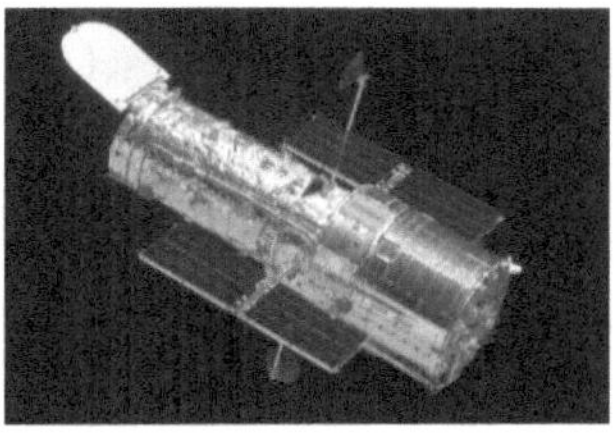

Hubble Space telescope

Yugoslav wars

World Wide Web

Gulf War

Nelson Mandela

Channel Tunnel

Sources:

a. Y2K.psd. Picture composed by JH van der Schrier. Public domain. (background: https://pixabay.com/photos/binary-binary-system-data-dataset-2728117/)

b. https://commons.wikimedia.org/wiki/File:Windows_Logo_1995.svg

c. https://en.wikipedia.org/wiki/Hubble_Space_ Telescope#/media/File:HST-SM4.jpeg

d. https://nl.wikipedia.org/wiki/Dutchbat#/media/Bestand:Dutchbat.jpg

e. https://pixabay.com/illustrations/monitor-binary-binary-system-1307227/

f. https://commons.wikimedia.org/wiki/File:Operation_Desert_Storm_22.jpg

g. https://en.wikipedia.org/wiki/Nelson_Mandela#/media/File:President_Bill_Clinton_with_Nelson_Mandela.jpg

h. https://en.wikipedia.org/wiki/Channel_Tunnel#/ media/File:Fresque_25_ans_Eurotunnel.jpg

In the new millennium, the world was connected with phones, money and commerce. Also, the environment and terrorism became new challenges.

The **September 11 attacks,** in 2001, a series of coordinated terrorist attacks by the Islamic terrorist group al-Qaeda, it resulted in 3000 deaths, seriously harmed the economy of New York City and had a significant effect on global markets. As a result of the attacks, many governments across the world passed legislation to combat terrorism.

The year 2001 introduced the **mobile phone** with monochromatic display technology. In 2007, the touch-screen 'smartphone' followed. This phone enabled users to run applications (apps) for specific purposes. Smartphones changed the way we communicate, do business, entertain ourselves, socially interact with others and learn.

In the 2000s, **social networking sites** bloomed on the Internet. Initially they were meant to give users the ability to interact with people with the same interests, and to publish their own content. LinkedIn, established in 2003, met considerable success. Companies use social media as a way to learn about potential employees' personalities and behaviour. Web-based social networking services connect people who share interests and activities across political, economic and geographic borders. Through e-mail and instant messaging, online communities are created.

The **financial crisis** of 2007–2008 was the most serious financial crisis since the Great Depression of the 1930s. The crisis began in 2007 with a depreciation in the subprime mortgage market in the United States, and it developed into an international banking crisis.

In the 2000s, **global warming,** also known as climate change, dominated the public debate. Through global warming, the Antarctica and Greenland ice sheets melt and sea levels rise. Documentaries such as *An Inconvenient Truth* and even science fiction films such as *Avatar* raised international public awareness. Their environmental message is clear: continue to treat the planet with disregard, and we're in for it big time.

In the 2000s, **globalisation** has grown due to advances in transportation and communication technology. Proponents of economic growth, expansion and development view globalising processes as desirable or necessary to the well-being of human society. Opponents see the phenomenon as a promotion of corporate interests.

In 1999 the **euro** came into existence. Notes and coins began to circulate in 2002. The euro rapidly took over from many European Union member nations' national currencies. Travel was made easier by removing the need for exchanging money, and more importantly, the currency risks were removed from European trade.

In 2003 the **Human Genome Project**, the world's largest collaborative biological project, was completed. Now all base pairs that make up the human DNA are known and all genes of the human genome are identified.

September 11 attacks

Mobile phone

Social networking sites

Financial crisis

Global warming

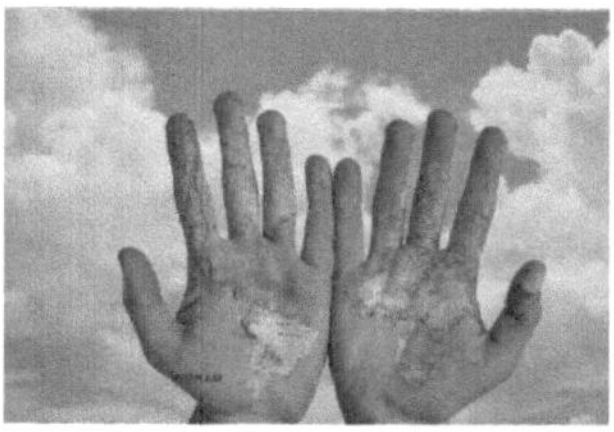

Globalisation

Euro

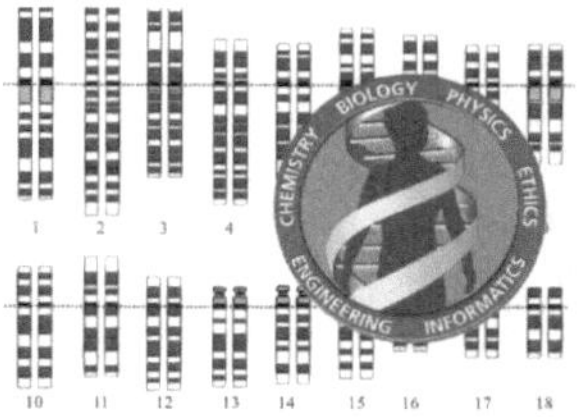

Human Genome Project

Sources:

a. https://commons.wikimedia.org/wiki/ File:WTC_smoking_on_9-11.jpeg
b. https://en.wikipedia.org/wiki/Nokia_3310#/me¬dia/File:Nokia_3310_blue.jpg
c. https://en.wikipedia.org/wiki/YouTube#/media/ File:YouTube_Logo_2017.svg
d. https://commons.wikimedia.org/wiki/File:Leh-man_Brothers_Times_Square_by_David_Shankbone.jpg
e. https://pixabay.com/photos/life-beauty-scene-arctic-iceberg-863034/
f. https://pixabay.com/photos/hands-world-map-glob¬al-earth-600497/
g. https://nl.wikipedia.org/wiki/Chartaal_geld
h. https://en.wikipedia.org/wiki/Human_genome#/me¬dia/File:Karyotype.png / https://en.wikipedia.org/wiki/Human_Genome_Proj¬ect#/media/File:Logo_HGP.jpg

Western Europe and Nordic countries

Basically, safety management is modelled after quality management. In the period 1990–2010, the combination of safety management and safety culture was expected to control occupational accidents. This period also saw the birth of three models of occupational accident processes, the Swedish model of the Occupational Accident Research Unit (OARU), the Occupational Risk Model (ORM) of the Dutch, British and Greek Workgroup (WORM) and the Tripod model. The WORM workgroup was one of the initiators of the biannual European conference Working-on-Safety (WOS).

Quality management

The Allied victory ending the Second World War and the fall of the Berlin Wall, on 9 November 1989, boosted the influence of the Anglo-Saxon management model on economy and organisation in Europe. Maximisation of profits through detailed planning and control was one of the main features of this approach, along with rationalisation of organisational processes, making them more easily manageable. This included standardisation, protocols and transparency as success factors, creating an extensive bureaucratic accountability. The European alternative to the Anglo-Saxon model was the Rhineland model, which focused on continuity, customer experience and worker participation. The Rhineland model was based on decentralisation and delegated responsibilities to teams and individual workers (Brouwer and Moerman 2005). The Anglo-Saxon view, with its focus on centralisation, was almost diametrically opposed to Rhineland's views. Economists characterised these differences as control versus commitment, and organisation management experts as control versus involvement oriented (Vinkenburg 2010). Despite the apparent disagreement, the views were mixed in distinctly European systems.

Quality management theories like Total Quality Management (TQM) had a major impact on safety management (Zwaard and Groeneweg 1999a). TQM, which was developed just after the Second World War in Japan, found its way to important management schools in the Western world in the 1980s. TQM attracted a lot of attention in Europe with the installation in 1988 of the European Foundation for Quality Management (EFQM), a non-profit membership foundation. In that particular year, 14 presidents of European multinationals met. Philips, KLM, Ciba-Geigy, Fiat, Nestlé and Volkswagen were amongst the driving forces behind the formation of this foundation. Together they searched for ways to make their companies ready for the future, meaning they would be resistant to global competition. Four years later, the first European Quality Award was issued. TQM was the leading management system for the EFQM. Two of the key features of TQM are its focus on customer requirements and workforce participation as an essential input to optimise (production) processes (EFQM 2019).

Another relevant development was 'corporate social responsibility' (CSR). This term, originally from the 1970s (Hanekamp et al. 2005), had a strong comeback in the 1990s. CSR was a reaction to the Anglo-Saxon approach to value shareholders and focus purely on corporation profits. CSR focused on three areas: how corporations treated their staff, suppliers, funders and other dependent parties; the impacts a corporation had on external factors, such as job creation and the environment; and the effects of a corporation on its community and society as a whole (SER 2000; Sluyterman 2004).

The European Foundation for Quality Management (EFQM) in Brussels developed a management model based upon Rhineland's approach, incorporating CSR (Figure 7.1).

In the Netherlands, the Instituut Nederlandse Kwaliteit (INK; Dutch Quality Institute) translated the EFQM model into a Dutch version. Both the INK and EFQM models assumed that the competitiveness of organisations was improved by balancing the interests of their stakeholders. Both models required constant adaptations, fuelled by constantly changing market forces, technology and customer demands. These adaptations were based on the well-known Deming cycle: 'Plan-Do-Check-Act (or Adjust)' (Deming 1982).

Via the Deming cycle there is some connection between EFQM and ISO standards, in particular the ISO 9000 series of standards dealing with quality assurance and quality management. The ISO series focused only on managerial aspects through standardisation and certification, while the EQFM presented a more

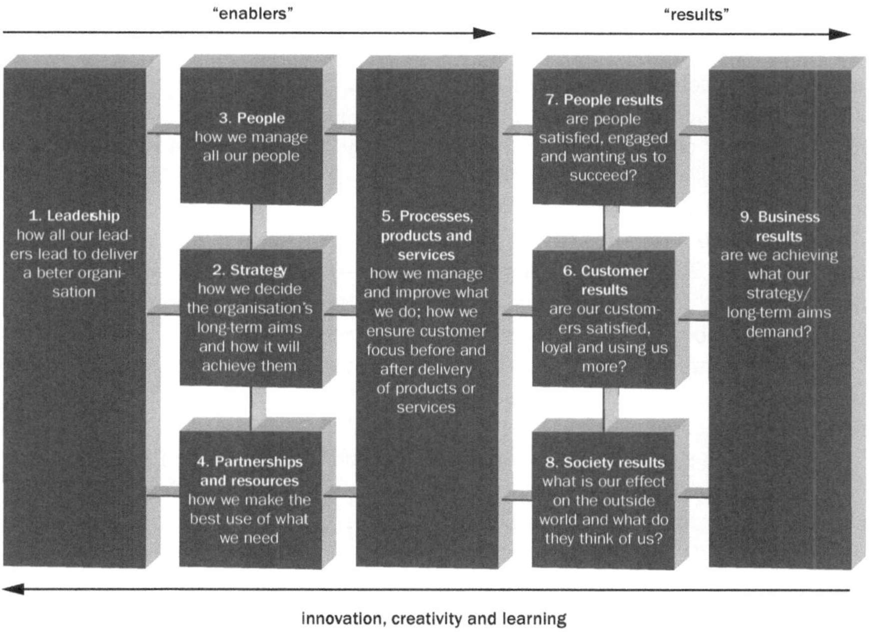

Figure 7.1 The EFQM management model

Source: EFQM (2019)

holistic model. The ISO, or the International Organization for Standardization, is the world's largest developer of voluntary international standards. The purpose of the ISO was to facilitate international trade and cooperation in commercial, intellectual, scientific and economic endeavours by developing international standards. The ISO originally focused on industrial and mechanical engineering standards but later, with great success, it ventured into setting standards for organisations' processes, policies and practices.

The roots of the ISO 9000 series, standards for quality assurance and quality management, dated back to the Second World War, when the Ministry of Defence in Britain decided to implement a set of standards to reduce mistakes following accidents in the manufacturing of munitions. The standards focused on the management of procedures rather than the actual manufacturing. The Ministry of Defence would first inspect the procedures used during manufacturing, and then inspect the manufactured product to ensure consistency of quality. The ISO 14000 series on environmental management was similar to the 1994 ISO 9000 series and was published two years later. Both series required third-party certification, and the certification business boomed, thanks in part to these two standards. The European Union, with the 1993 Maastricht Treaty, created a single internal market. This led to the perception that certification was required in order to do business within and with Europe, and consequently to the widespread acceptance of the ISO 9000 series (Hall 1995).

Safety management

In this period, detailed models of safety management were developed but without any solid scientific justification. Safety management was defined as the sum of all activities conducted in a coordinated way by an organisation to control the hazards presented by its production process and technology. These hazards could be potentially harmful for its workforce, to its assets, including damage to buildings and structures, to its customers or to those living around the company site (Hale 2003). Rasmussen (see Chapter 8) characterized safety management as keeping the organisation within a more broadly defined safe 'envelope', away from the boundaries of uncontrollability (Rasmussen 1997). According to Rasmussen, companies were constantly being pushed towards these boundaries by competitive pressures and financial restrictions or might drift towards them through complacency.

The attention given to safety management had already started in 1950 with Heinrich's safety ladder, which was also when the first publication appeared of a study relating psychological climate to accidents in the automotive industry (Keenan at al. 1951). The basic principles of management – feedback and control – had already been known for many years. Research on the structural view of management, and on the driving force behind safety performances, resulted in concepts like *safety climate* and *safety culture*. Safety management received full attention after the British 1972 Robens Report, which introduced self-regulation of safety by companies, and again after the 1988 Piper Alpha major accident. Safety climate and culture followed somewhat later.

Concurrently, two forces reinvigorated the attention given to safety management systems. The first was based on progressing insights. It was found that the traditional way of working with safety management systems should be changed. It should no longer be an add-on to an organisation but an integral part of it. There should not be one management system that controls production and another for safety. The task of a separate safety department was to coordinate, monitor and evaluate the safety management system and its performance (Hale 2003). Another force was the drive to self-regulation or internal control, as an alternative to governmental control. Norway did a lot of work in this area, and the general attitude towards internal control was positive. One of the reasons to pursue this route is that a high degree of internal agreement was desired: the internal fights between line management and safety experts and quality assurance were found to be very ineffective. One condition for successful implementation of internal control was a high degree of consensus on basic values between public safety and health executives and company managers, and between employers and employees and their organisations. This could be considered a 'social-democratic' model of society and safety management systems that might support that aim (Hovden and Tinmannsvik 1990).

The basic elements and structure of safety systems were known, inspired by TQM, the EFQM, the ISO 9000 series and INK (HSE 1991; Gun 1993; NPR 1997; Visser 1998; Redinger and Levine 1998; BSI-OHSAS 18001 1999; ILO 2001). This generic safety management structure was originally developed for high-tech-high-risk activities, such as the nuclear sector, or large bureaucratic organisations with a division structure (Mintzberg 1983). In these sectors and organisations, primary processes were governed by rules and procedures, as was safety management. In such companies, rules were cornerstones of risk control, marking the boundaries of safe production, at least in theory, though often not in practice. Too much dependence on rules and procedures created a dehumanised atmosphere, negating human initiative and creating resistance to rules and to rule control. At that moment rules became the problem rather than the solution (Elling 1991; Maidment 1993; Lawton 1997; Hale and Swuste 1998; Heertje 2009; Knudsen 2009).

The management task can be broken down into subtasks, as shown in Figure 7.2, with a clear allocation of tasks within the organisation (Hale 2003). This problem-solving process takes place at three interlocking levels of functioning.

The primary control of hazards takes place through the actions of those directly in contact with these hazards: the workforce, operators, drivers, maintenance fitters, and so on. They exert hands-on control within a short timescale of minutes or hours before the direct results of their actions become manifest. This is the execution level, the lowest-level box in Figure 7.2. In organisations with only a rudimentary safety management system, these actions will be improvised by the people concerned, based on their competence, or will be learned implicitly from

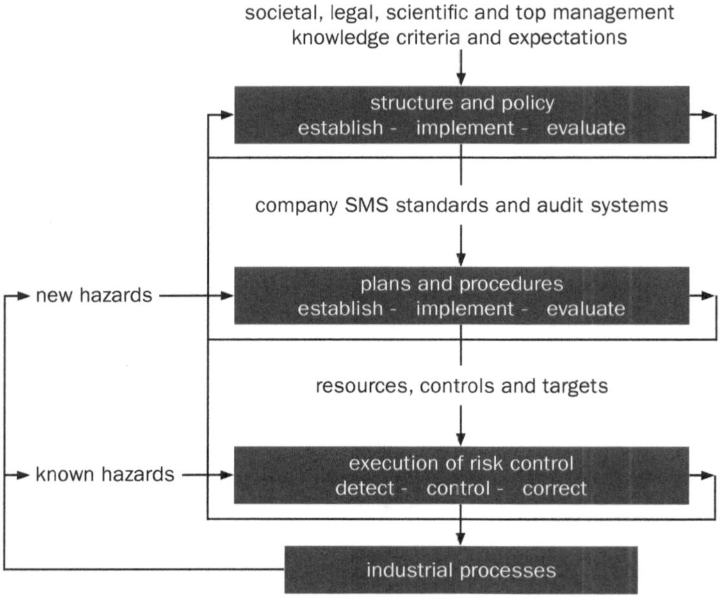

Figure 7.2 Safety management as a problem-solving activity operating at three levels
Source: Hale (2003)

their peers during job training. If this is all that exists within an organisation, one cannot really speak of a safety management system.

The first step in developing a safety management system is to make these control actions explicit, to document them in training manuals, safety handbooks, procedures or guidelines, and to assess their logical compatibility and completeness (see the middle box in Figure 7.2).

This need for completeness demanded a systematic way of predicting and cataloguing the risks to be controlled. This required a clear understanding of the company's primary production processes, their activities and all possible scenarios leading to significant harm or damage. This risk inventory and evaluation, required by the European Framework Directive (European Commission 1989), had to anchor the safety management system to the specific hazards of a specific company. The task and job safety analysis had to be rooted in a functional analysis of the processes so that the deviations in the flow of those processes, which led to accidents, could be traced to their origins and linked to barriers (Hale 2002). This 'plans and procedures level' also had to provide the resources and criteria for the 'execution level' to operate effectively. However, problem solving could still be quite reactive and ad hoc. A top level of problem solving could be found in more advanced organisations. They subjected this level of plans and procedures to a periodic review and improvement process, triggered by, for instance, a plateau reached in safety performance, or a large difference in safety achievements compared to rival companies, or a major and unexpected incident, or new theoretical or practical advances in safety management internationally. This review process was the structure and policy level of problem solving (see the top box in Figure 7.2). It could result in a major rethinking of the philosophy of risk control, in challenging new targets for performance or in a better adjustment of the risk control system to a change in the organisational structure or culture.

Although the proposed safety management model was quite straightforward, various sectors still had great difficulty in managing safety, the building sector being one of them. The available literature on construction safety, as one of the more dangerous sectors, was not very optimistic about the chances of safety management in the construction industry exerting a positive influence (Swuste et al. 2012b). Many articles indicated that the structures and processes that were designed to ensure safety in the industry were poorly developed (see, for instance, Niskanen and Saarsalmi 1983; Laukkanen 1999; Spangenberg et al. 2002, 2003; Spangenberg 2010; Sertyesilisik et al. 2010). Safety management systems were not efficient, and the business processes executed were fragmentary. It was not clear who was responsible for safety and parties lower in the construction hierarchy tended to be saddled with the consequences. Safety detracted from the primary production process and was seen as a bureaucratic burden. With the slogan 'Manage the risk, not the paperwork', the HSE called for a return to the controlling of hazards and risks at construction sites (HSE 2009; Donaghy 2009). But there were some positive developments as well. Lists of prevalent accident scenarios and significant events were available, and information was published on barrier failures. What was missing was a reliable exposure gauge of the relative importance of scenarios and the identification of pivotal events. The more clearly the cause-effect chains of accident processes could be recorded, the more specific the measures, solutions and interventions for avoiding or reducing the effects of accident scenarios. Audit methods were developed, such as the Safety Index, which assessed safety not only negatively but also positively (Mattila et al. 1994; Laitinen and Ruohomaki 1996; Laitinen et al. 1999). Finally, an approach that could best be described as *frappez toujours* (be persistent) seemed to yield noticeable results. In such cases it did not really matter what precautionary safety steps were taken, as simply highlighting the issue of safety was a factor that could, in itself, have an effect.

The accident process

The accident process remained a complex phenomenon that was poorly understood. It was difficult to capture the combined influence of organisational culture, management, the individual behaviour of front-line workers and the rapid development of automation in a comprehensive accident process description. It was expected that new technologies would lower accident frequencies by distancing the operator from hazards through automation, remote control and robotics. However, contemporary research showed a shift towards other accident scenarios rather than to a decrease in accident frequency (Laflamme 1993; Menckel and Kullinger 1996). Accidents related to maintenance and impact with moving parts of machines and with robots would become prevalent over hands-on production. New production technologies also changed management styles and safety management. Organisational processes became more complex. Lower levels of control were initiated by quick-reaction-just-in-time management. Organisations with a flexible management style, teamwork and human-centred labour processes adapted their safety systems more easily than traditional hierarchical organisations (Harrisson and Legendre 2003).

A change in dominant accident scenarios also influences the validity and usefulness of frequently used metaphors, like Heinrich's 300:29:1 occupational accident triangle, or iceberg, the relation between serious accidents, minor accidents and near misses (see Chapters 2 and 3). For instance, Salminen and co-workers showed in Finland an occupational accident triangle of 1:100:1200 for fatal accidents, major accident injuries and minor injuries, with an unknown number of near misses (Salminen et al. 1992; Hale 2001a). A rigid ratio was expected only if scenarios of serious and minor accidents and near misses had some amount of overlap, which in general was not the case. Whether an accident was serious or not was dependent, not on the incidence of minor accidents or near misses, but on the energy content of its hazard. More energy meant more harm.

Worldwide figures on occupational accidents were reported in 1999 by the International Labour Organisation (ILO). Annually, 335,000 lethal accidents and 254,000,000 non-lethal accidents were estimated. The report recognised three geographic areas. The first area included the developed countries in the West, the established market economies with an overall occupational mortality rate of 4.2 per 100,000 employers per year. The second included India, China and the former European socialist countries, scoring about three times higher. The third area was the rest of the world, including South-West Asia, Sub-Saharan Africa, South America, the Caribbean and the Middle East, with a rate seven to eight times higher (Hämäläinen et al. 2006). These numbers changed after the investigation was repeated in 2006. Established market economies and China and India showed a drop in lethal accidents but an increase of 20% in non-lethal ones. In South-West Asia and Africa, the mortality and morbidity rates increased sharply, especially the latter (Hämäläinen 2010). The drop in lethal accidents in the established market economies was explained by better occupational safety management, but alternative explanations were also possible. One such explanation comes from a British study, which showed a clear shift in employment away from primary industries, utilities and manufacturing towards the service sectors, which employ a large percentage of workers in relatively safe office environments. There was also a relation with employment. Workplace injury declined rapidly as employment on a permanent basis increased (HSE 2005). Another argument came from the Dutch occupational hygiene domain – not so much that the hygiene control strategy reduced exposure, but that the mechanisation and automation of production processes were predominant (Kromhout and Vermeulen 2000; Vermeulen et al. 2000; Tielemans et al. 2004). One could expect a similar effect with occupational injuries. Also, dangerous production processes were exported to low-wage countries in Asia, which lowered mortality rates in developed market economies (Swuste and Eijkemans 2002; Benavides et al. 2005; Takala 1999). Most likely, a combination of factors mentioned above had caused the reduction in occupational mortality and morbidity. Despite these developments, dangerous and lethal sectors had remained the same in five areas: agriculture, fisheries, transport, construction and mining.

Accident models

No comprehensive theory on occupational accidents was published in literature in this period, just two accident models, coming from different scientific disciplines: the model from the Swedish Occupational Accident Risk Unit (OARU) and an international workgroup that developed the Occupational Risk Model (ORM), a barrier-based model founded on the bowtie metaphor. In Sweden, a model was developed based on accident research in mining, construction, steel production and railways (Kjellén 1996). The Occupational Accident Research Unit was a unit of academics with an engineering, natural sciences and social sciences background. This unit presented the OARU model, a systems model using a socio-technical approach. Hazards and risks were not considered generally; instead, the focus was on deviations and process disturbances as causes of accidents (Figure 7.3).

Disturbances led to injuries, and system factors determined accident sequences, as previous research had already shown (Kjellén and Larsson 1981; Kjellén

		Deviations in the Process	*Loss of Control*	*Type of Loss*
human	occupational safety	common	always present	sudden damage to the body
	occupational disease	may be present	minor acute events may contribute	injury resulting from long-term exposures
environmenten-vironment	acute emission	common	always present	sudden major emission
	'planned' emission	may be present	minor acute events may contribute	continuous emission
material	material damage	common	always present	sudden damage to material
	product quality	may be present	not relevant	product fails to meet quality standards
	production regularity	common	not relevant	production loss

Figure 7.3 The OARU model

Source: Kjellén and Larsson (1981)

1984a–c, 1996). The OARU model was first published in 1981 (Kjellén and Larsson 1981). This model was quite holistic and combined occupational accidents, work-related disease, environmental emissions and quality of production and products (Kjellén 1996). Central in this model were deviations from normal operations. These deviations, leading to a loss of control, could occur in various formats, such as deviations from a planned production process, in communication to workers, in unplanned activities, in the physical environment of workers or deviations in man–machine interfaces. The broad definition of deviation made it a container concept and it was therefore difficult to pin down (Kjellén and Hovden 1993). A short list of structural factors included management tasks, worker tasks, work instructions, planning, characteristics of machines and tools, and management factors. These factors were always present in manual labour organisations and thus could always initiate accident scenarios (see Figure 7.4) (Menckel and Kullinger 1996; Laflamme 1990; Kjellén and Hovden 1993).

Such factors were seen as relatively stable, present during accident processes, and referring to physical and technical conditions of a work process, workplace and work environment. Additional and situational factors were organisational, like decision-making, maintenance, training, and so forth, and social-individual, such as management instructions, communication and competency. There was a shared belief that accidents could originate from a relatively small number of these structural and situational factors. This finding is corroborated by overseas studies in Australia (Williamson et al. 1996) and Canada (Shannon et al. 1997). Case studies also illustrated similar examples of risk underestimation in various industries all over the world (Takala 1993; Larsson and Rechnitzer 1994; Gardner et al. 1999), as well as in the Netherlands in a waste recycling plant (Zwanikken and Swuste 2002). Other examples in line with the OARU came from marine transport and construction, where a lack of feedback on safety issues lowered safety performance (Saarela et al. 1989; Saari and Näsänen 1989), or an almost absent management system, as in construction, put this sector in the top four of the most hazardous sectors (Pekkarinnen and Anttonen 1989; Aneziris et al. 2010; Swuste et al. 2012b; Hale et al. 2012).

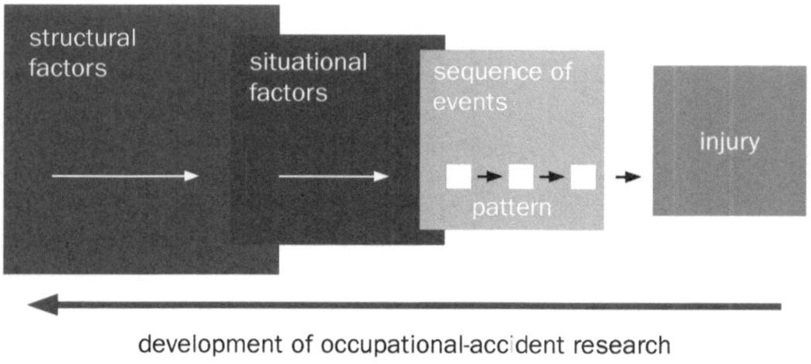

Figure 7.4 Factors of the Swedish deviation model based on OARU

Source: Menckel and Kullinger (1996)

The ORM model was developed by an international collaboration of Dutch, British and Greek researchers, with a similar background to the OARU group (Hale et al. 2007a; Papazoglou and Ale 2007). The ORM model shared commonalities with the OARU model through its focus on socio-technical aspects of an organisation. This model was based on the bowtie metaphor. This metaphor had precursors in the 1970s (see Chapter 6) and was given its final form in the late 1990s after a brief flirtation with the *hourglass* or *percolation* metaphor (Visser 1998). Maintaining barriers is the core of the ORM model. By turning the bowtie 45 degrees, the idea was that failures to control hazards at the top would percolate down to consequences at the bottom. However, the leading human error researcher Wagenaar from Leiden University always wore a bowtie and decided that the model had to be reverted to reflect his preference for this garment (Figure 7.5).

The centre of the bowtie represents the central event, a condition when a hazard actively becomes manifest, uncontrollable and changes into a risk. The arrows between hazards and the central event depict scenarios leading to the central event, and the arrows between the central event and consequences represent damage processes. In complex accident processes, more than one central event is possible. The bowtie metaphor implicitly contains a time axis. But the speed of events can vary greatly from one scenario to another. Furthermore, due to latent factors, anything from minutes to years is possible on the left-hand side of the bowtie. On the right-hand side, however, timescales tend to be much shorter – seconds or even less.

The barriers are the instruments of control, and can stop scenarios, reduce or slow down their effects. The ORM was a dynamic model of barriers, distinguishing different barrier types. Passive hardware barriers are always present without

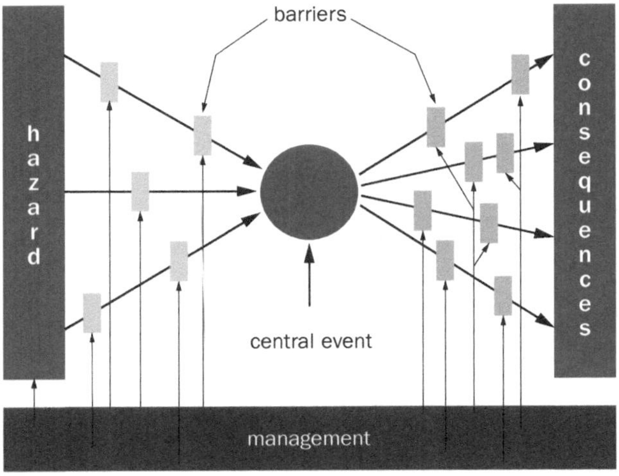

Figure 7.5 The bowtie metaphor for supporting safety management

Source: Visser (1998)

additional management action. Active barriers require some kind of activation by humans, technology or electronics. Behavioural barriers depend on an intervention by a person (Schupp et al. 2006; Bellamy et al. 2007). Another important aspect of the barrier model is its relation with the safety management system (the vertical arrows in Figure 7.5). The arrow linking management and hazard directly reflects an inherent safe design of an installation, equipment or tools.

In the European ARAMIS project, nine 'management delivery systems' were postulated, which served as a starting point for an audit (Guldenmund et al. 2006) (Figure 7.6). Management was responsible for the correct operation of barriers.

This basically meant that management was responsible for the quality of performance of these nine management delivery systems: risk identification, design and installation of adequate barriers, maintenance and monitoring of the barriers, and training of staff and conflict resolution, which together formed an integral part of a safety management system. Efficient management processes were required to guarantee current and future barrier quality.

The ORM model was based on Dutch and British national accident registration databases. These databases were updated by factory inspectors when notified of a serious occupational accident and contained information on more than 9,000 occupational accidents from 1998 to 2004. The database was also the source for identifying failing barriers and failures in the management delivery systems. Because the inspectors' reports contained narratives of accidents, the tool developed to analyse these was called *story builder* (Bellamy et al. 2008a). The investigation revealed patterns in the occurrence of accidents and safety management tasks (Bellamy et al. 2010).

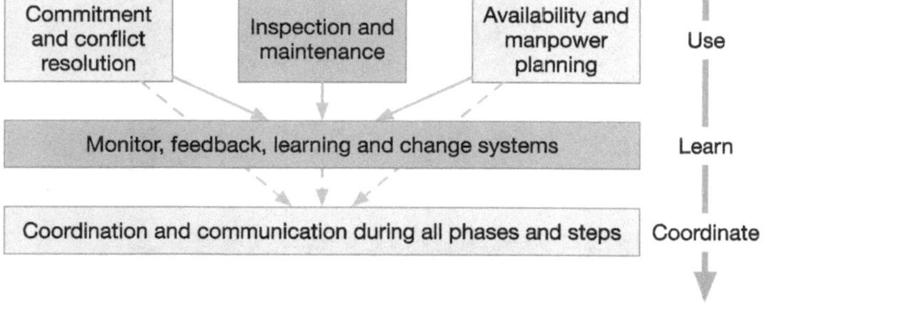

Figure 7.6 The nine delivery systems
Source: Guldenmund et al. (2006)

ORM included risk exposure. The inclusion of exposure required expertise on occupational hygiene, since it was not a traditional safety science research topic. For ORM, however, the traditional units of risk, such as accidents per 100,000 persons, were too coarse to use in this model. WORM included exposure to risk per hour worked, such as hours working on a ladder, when for instance a risk was taken by climbing a ladder (Aneziris et al. 2008). Combining detailed exposure data with probabilities allowed a quantification of occupational risk (Ale et al. 2008; RIVM 2008; Damen 2011). Some risks were broken down into even smaller subtasks, this to demonstrate the necessity of safe design to designers and architects (Frijters and Swuste 2008). Others suggested defining machine-specific exposures, such as the amount of time parts of a body are in close proximity to a machine (Raafat 1989; Gardner et al. 1999). But this level of detail, unfortunately, was not included in ORM's exposure measurements.

Neither the OARU nor the ORM model referred to human factors or to human errors or unsafe acts. These models were not focused on blaming the victim, as a human error approach would imply. This interpretation of suboptimal acts and their causes was in line with an ergonomics approach. This view was shared by the Swedish investigators. Errors and suboptimal actions of workers occurred in dangerous environments where abnormal work was carried out: for instance, during maintenance, or interventions during process disturbances. Furthermore, in research on international football matches, a workplace with a high risk of injury where some players sometimes carried out spectacular tackles, no evidence was found of the existence of personal traits explaining the occurrence of injuries (Fuller 2005). The researchers blamed top managers, stating that a top manager knew – or should know – the impact of these conditions on occupational risks, but did not act on them (Döös et al. 2004).

The accident proneness theory, however, reappeared in the safety literature, even right up to the 2010s. Defining an accident-prone worker was still a point of discussion, as in the period before the Second World War (see Chapter 4). A French biomechanic and a Dutch human movement scientist persisted in their view that personality was a primary consideration in preventing accidents (Gauchard et al. 2006; Visser et al. 2007).

Working-on-safety

The Occupational Risk Model initiative was one of the driving forces behind the European Working-on-Safety (WOS) Network, which aimed to establish a permanent network of experts in occupational accident prevention, including researchers, policy makers, safety professionals, labour inspectors and others in the prevention of occupational accidents. The network was an extension of what was until then a predominantly Nordic expert network and started with a four-day biannual conference. The first WOS conference in 2002 was held in Elsinore, Denmark, with strategies for prevention as an overarching theme, followed by the WOS-2004 conference, in Dresden, Germany. Here the theme was prevention and inspection. WOS-2006, in Zeewolde, the Netherlands, focused on three areas:

(1) research and knowledge development; (2) intersection of research and policy; and (3) policy. The theme of WOS-2008, at Crete, Greece, was 'Prevention of occupational accidents in a changing work environment'. And the WOS-2010, in Røros, Norway, was themed 'On the road to vision zero?'. These conferences established an informal platform for the exchange of new findings, on the scientific foundation of safety initiatives, on alternatives to (soft) regulations and successful best practices for the prevention of accidents at work. Usually the WOS conferences were held in relatively isolated places, to ensure mutual discussion between participants. The conferences also had a relatively small number of participants, between 200 and 300, most of whom presented their work. This conference design stimulated discussions, cooperation and sharing of knowledge between attendees. From 2006 onwards a selection of presentations was published in *Safety Science* (Swuste 2008a, 2008b; Swuste et al. 2010, 2012a). Many contributions dealt with safety management issues, frequently using 'Does it work?' as a leitmotiv. From the first conference, there was a growing number of presentations addressing safety culture and safety climate.

Safety culture and safety climate

According to Hale and Hovden, safety management combined with safety culture represented the third era in the development of safety and the main areas of attention, both in science and organisations, after 1990 to lower accident frequencies (Figure 7.7) (Hale and Hovden 1998).

Central to the first era of safety were technical solutions and safety engineering concepts, which lasted until after the Second World War (Chapters 2 and 3). The second era had already started before the Second World War, and it peaked in the 1970s. From the accident proneness theory onwards, human interactions, human errors and human factors were seen as dominant factors. This resulted in insight into the limitations of training and of teaching workers to work safely with dangerous machines. This concept was the basis of ergonomics, adapting machines and workplaces to workers, from the notion that context determined human acts and behaviour (Chapter 4). Also, in the United States, the National Safety Council considered the focus on human errors and unsafe acts to be outdated and proposed

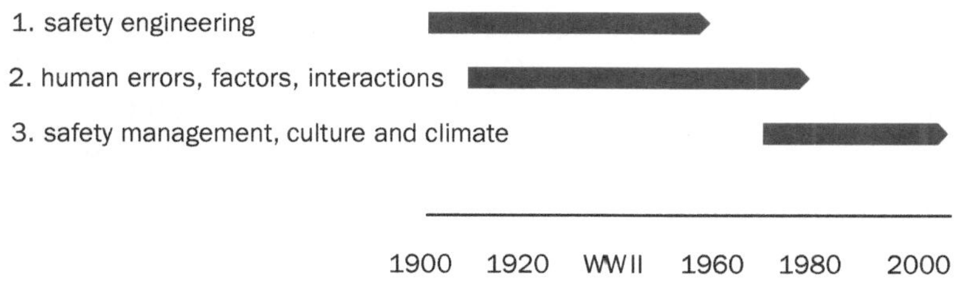

Figure 7.7 Three phases in safety science

Source: Swuste, Gulijk, Groeneweg, Guldenmund et al. (2020)

replacing them with causal factors and unsafe conditions (Manuele 2002; Wachter and Ferguson 2013). Safety culture seemed to be a way to combine many of the approaches with human factors.

Safety culture and safety climate became a topic of interest for an increasing number of researchers. At the beginning there was some confusion as to whether or not safety culture as an independent phenomenon had a right to exist, and the difference between culture and climate remained ambiguous for some time. Also, some authors addressed the issue of power relations and safety management, a topic neglected by most culture researchers.

Early articles on safety climate in the scientific press included the previously mentioned publication on psychological climate of Keenan and colleagues from the 1950s and Zohar's seminal publication on safety climate (Zohar 1980a). In the second half of the 1980s, organisational culture became a topic in contemporary reports on major accidents such as Chernobyl (1986), King's Cross (1987) and Piper Alpha (1988). These reports explicitly referred to organisational culture. These major accidents and their aftermaths created a strong link between culture and safety. The International Atomic Energy Agency (IAEA) and the International Nuclear Safety Advisory Group (INSAG) played a key role in forging this link (INSAG 1988; IAEA 1991; Choudhry et al. 2007). Safety culture and safety climate were both popular research topics at conferences, in special issues and in many journal papers, in both scientific and professional journals, from 2000 onwards. Figure 7.8 shows the distribution of articles on safety culture over the years (Glendon 2008). Meanwhile, social scientists still disagreed on whether a safety culture should be considered a distinct entity,

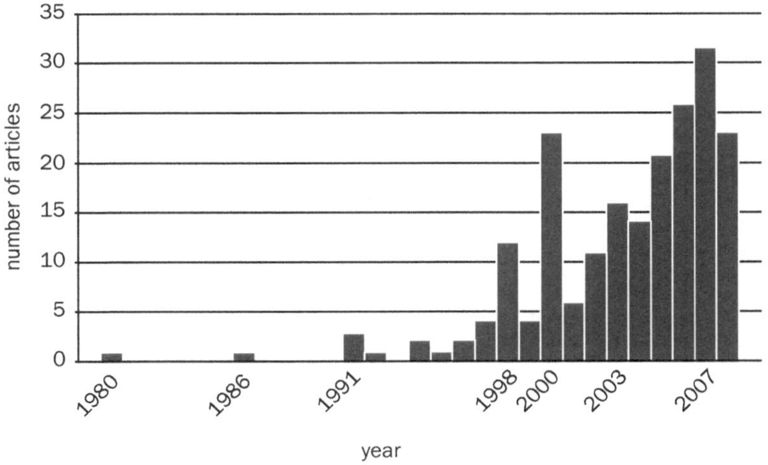

Figure 7.8 Distribution of safety culture, climate articles during the years 1980–2008, 2007 is the second last bar

Source: Glendon (2008)

open to scientific scrutiny. This finally led to three approaches to safety culture (Guldenmund 2010):

1. The academic approach

   The academic approach pertains to the qualitative and interpretative approach to the study of safety culture. Its purpose is to understand the culture under scrutiny through the interpretation of various empirical data sources. According to Schein, at the heart of an organisation the basic assumptions operate, which are accepted by the majority of the employees beyond doubt. These assumptions are neither explicit nor directly observable, and therefore need to be deciphered from observable artefacts and espoused values where they are, or at least should be, at work. By investigating work processes and documentation, and through observing decision-making processes, results, conflicts, communications, symbols and so on, the basic assumptions will ultimately reveal themselves to the researcher. Such data have to be augmented further with personal and group interviews with organisation members. Reactions to the results of the analysis can provide even more valuable information on the basic assumptions as well as conflicts within or between various data sources. The academic approach to safety culture is fundamentally qualitative in nature and requires active and extensive fieldwork, hence the adjective 'academic', in the organisation or company under study to understand the organisation's dynamics.

2 The analytical approach, or the measurement of safety climate

   The analytical view on safety culture originates from social psychology and sociology. Its purpose is to measure the concept by taking it apart, hence the adjective 'analytical', and operationalising these parts separately. This approach provides a mere 'snapshot' of safety culture in the sense that a momentary insight into the perceptions, attitudes and behaviours of the workforce towards safety is captured. Psychometric questionnaires quantify workers' perceptions of, for example, safety, attitudes and behaviours towards hazards and risks, workers' participation in safety decisions and the quality of safety management. Therefore, not so much safety culture but safety climate is captured with these methods, which do not require immersive investigations and still provide useful insight (Glendon and Stanton 2000; Flin et al. 2000; Zohar 2008). Different research groups developed different scales for measuring safety climate; these scales all focus on the perception of the priority of safety and the effectiveness of safety management in an organisation. However, in this view, safety climate is only a reflection of psychological attributes based on the perceptions of individuals rather than basic assumptions (Neal et al. 2000; Rundmo 2000). The analytical approach is quantitative in nature and does not require active fieldwork in the organisation or company under study.

3 The pragmatic, normative approach

   This approach is not based on empirical research but rather on opinions and experiences of managers and experts. The approach is also normative because it links the quality of the safety management to safety culture. In contrast to the academic view, in this approach a distinction is made between good and bad cultures and companies falling somewhere in between. Basically, a 'good' safety culture leads to 'good' safety performance and a 'bad' safety culture leads to 'bad' safety performance, or so the reasoning goes. The aim is to improve the safety culture of an organisation, so the organisation will reach an overall higher level of safety according to this approach. This approach is also known through Shell's 'Hearts and Minds' programme and the 'Safety Culture Maturity Model' from the Scottish Keil Centre (Fleming 2001). 'Hearts and Minds' was initiated by Visser from the Exploration and Production department at Shell after he visited the World Bank Conference in 1988. Visser was impressed by Westrum's presentation in which he introduced the three levels of safety culture (see Chapter 6). The pragmatic approach addresses factors that coincide with the structure and processes of an organisation (Westrum 1993; Lawrie et al. 2006; Parker et al. 2006; Hudson 2007). Group assessments score behaviour in issues like leadership, conflicting goals, subcontractor management and the organisation's response to accidents. These scores yield a position in a safety culture hierarchy (like a ladder) that varies, in the case of 'Hearts and Minds' from 'pathological' to 'generative' (Figure 7.9).

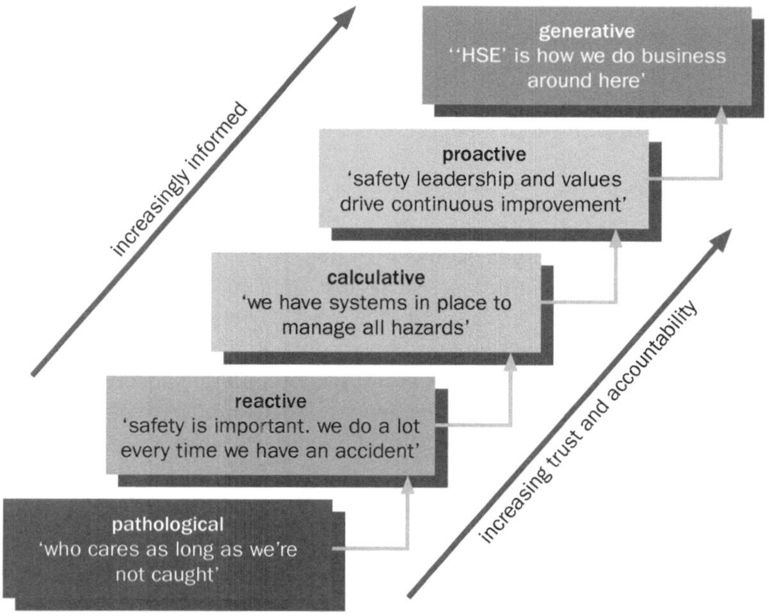

Figure 7.9 Safety culture ladder

Source: After Lawrie et al. (2006)

This view on safety culture is similar to an American safety management characterisation method published by Dan Petersen in 1975 (see Chapter 6). According to one of its developers, the safety culture ladder is actually not a culture assessment but a tool to initiate discussions on safety (management) in companies (Hudson 2010a).

The academic, the analytic and the pragmatic views on safety culture provide information on aspects of an organisation such as culture, safety management and perceptions of the priority of safety. In the literature, the distinction between safety management and safety culture in particular led to some confusion. These terms were not defined unambiguously, and some authors seemed to confuse them (Choudhry et al. 2007). The three views on safety culture and climate do not contradict but rather complement each other. However, if an assessment is limited to a safety climate investigation or a safety culture ladder assessment alone, then the step towards a full appraisal of safety culture is still a long way ahead. Safety climate investigations and safety culture ladder assessments were mostly limited to desktop research and lacked fieldwork input, and hence they are not able to capture organisational dynamics and the context within which the answers given should be interpreted.

Safety culture analyses were often used in high-tech-high-hazard industries, like the nuclear and petrochemical sectors, and after major accidents. Safety climate analyses seemed more appropriate for occupational accidents, because accident or incident frequencies could easily be related to questionnaire results. In the literature it is sometimes questioned whether safety culture assessments could predict major accidents and safety climate assessments occupational accidents. Meta-analyses of safety climate studies, however, seemed to show a relationship between safety climate and occupational accidents (Clarke 2006; Johnson 2007; Zohar 2008, 2010), while convincing results for safety culture have not been published yet.

Another interesting finding in this period of time is the relation between safety management systems and power (Antonsen 2009). Antonsen explained that most researchers assumed that organisations operated as harmonious workplaces and did not consider conflicts and power struggles. Both culture and power are hard to capture with scientific methods and their relation remained elusive. Some of the accident models discussed contain aspects dealing with power issues. Tripod, a safety model that will be discussed in the section on the Netherlands, included 'conflicting goals' as a basic risk factor, and the bowtie metaphor considered rules for conflicts between safety and production. These aspects deserved further elaboration in research into organisational aspects of occupational safety, and safety culture/safety climate surveys could incorporate these aspects into their investigations too.

North America and Australia

Just as in western Europe, in North America there were also major developments in quality management and in organisational learning. There was a growth in safety audit techniques for assessing the quality of safety management systems, and behaviour-based safety as a safety intervention started to emerge, together with initiatives to introduce best practices in companies. Finally, an ethnographic approach to organisational culture was developed, which did have a profound effect on the safety culture research tradition.

Quality management

A typically American model of quality management was Six Sigma, which embodied the Anglo-Saxon model for statistical control. Six Sigma (Figure 7.10) refers to a statistical concept. Sigma is the standard deviation of a normal distribution, being the unit indicating the deviation from the desired or optimal state.

Quality management deviating with an error less than six sigma referred to a quality of 99.99966%, which translated to 3.4 errors per million products. The prestigious Six Sigma label was awarded only to management and engineering processes achieving this high level of quality. Motorola applied Six Sigma to all of its production processes from 1986 onwards. Jack Welch introduced the method at General Electric and popularized Six Sigma further in 1995 (Welch 1995).

The DMAIC circle around the triangle in Figure 7.10 is an adapted version of the Deming cycle, used for continuous improvement and adaptation to change conditions, as in the EFQM and INK models. The ISO quality system at the top of the triangle required a third-party certification, as mentioned in a previous paragraph. Questions were raised regarding the fairness of this certification process. The ISO 9000 series was demanding in its documentation requirements and required a significant amount of time and personnel. The proponents of the

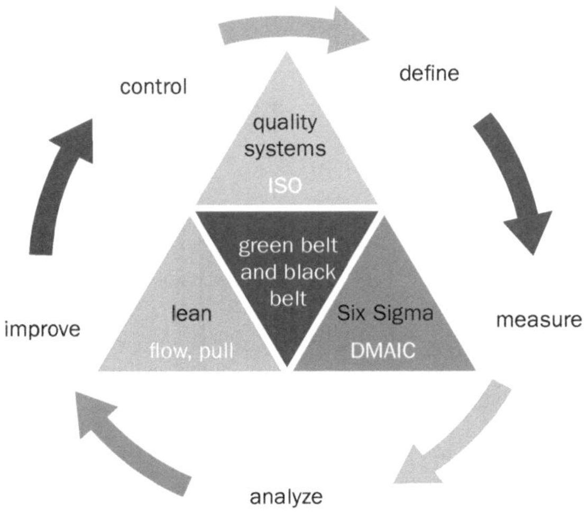

Figure 7.10 Six Sigma

Source: After Welch (1995)

series point to the economic benefits, such as the opening of new markets and the development of streamlined procedures, which could lead to increased profits. There were also intangible benefits of ISO certification, such as improved corporate image, and the feeling of doing the right thing. Although the basis for the management processes was different to the one in Europe, it had similar aims and used similar tools.

Organisational learning

Continuous adaptation to changing environments and learning was central to the quality management models. Learning loops found their way into safety management models as the previous section showed. Safety audits were tools for assessing the quality of safety management, and the development of behaviour-based safety was a major safety intervention. But the most important contribution in this period was the ethnographic concept of organisational culture.

Organisational learning required an organisational memory (Senge 1990). Organisational memory was the most important feature for preventing repetitive errors. But an organisation could not learn by itself. People working in the organisation, on the other hand, could. Senge, who had both an engineering and a management background, described key organisational parts and a right organisational attitude towards learning. Feedback loops, so prominent in the systems approach for engineering and safety, were also key elements for organisational learning.

The psychologist Argyris proposed a mechanism for organisational learning, by introducing the concepts of single- and double-loop learning (Argyris 1976, 1992; ILO 1992). Single-loop learning was characterised by adapting or fixing situational elements in particular processes (Figure 7.11).

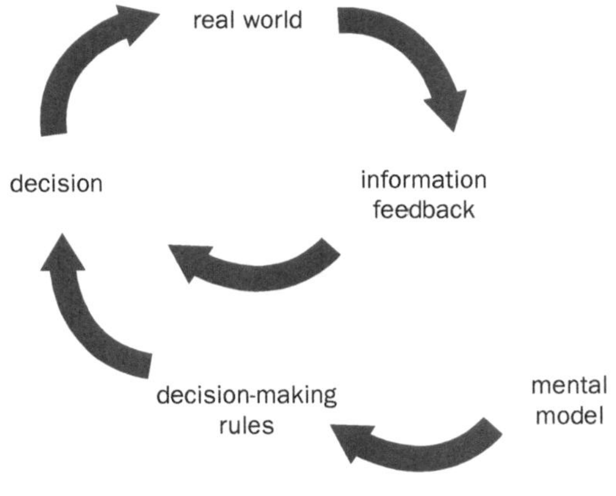

Figure 7.11 Single-loop learning

Source: Argyris (1976)

Single-loop learning is a basic type of learning that can take place within a system. Also described as 'incremental learning', single-loop learning focuses on fixing a problem within the present organisational structure so the system will function better, without altering the basic structure of the system. An example of single-loop learning would be to adapt procedures or training requirements after an incident has happened. In double-loop learning, deeper, underlying causes behind the problematic action are changed or corrected (Figure 7.12).

Double-loop learning, or reframing, contrasts with single-loop learning and questions the 'mental model' of the organisation, its basic rationale. This mental model includes the purpose, structure and function of the organisation and of work being done, on which decisions depend. It thereby changed the decision-making rules, which will have a huge impact on organisational functioning. Double-loop learning does not take existing organisational structures for granted; on the contrary. Organisational learning encompasses all activities within an organisation, but is particularly useful for safety purposes, as it is already familiar with feedback loops.

Safety audits

The development of safety management systems stimulated the interest in safety audits. Safety audits were planned – systematic investigations to assess an organisation's administration and operation with respect to safety. Audits were key instruments in determining whether a given safety management system was 'delivering' sufficient safety to make it worthwhile, as described in detail above. The safety diagnosis was made by analysing parts of the safety management system and conducted by an experienced safety staff member (Chaplin and Hale 1998; Tinmannsvik and Hovden 2003). Deriving from research done in the 1970s

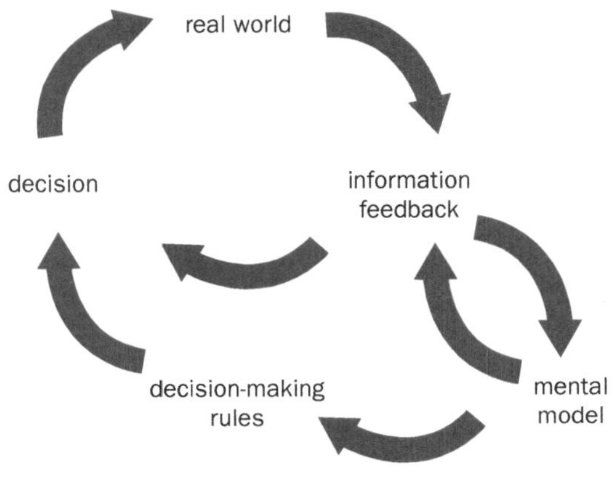

Figure 7.12 Double-loop learning

Source: Argyris (1976)

(see Chapter 5), many audit methods originated in the early 1990s. Widely known ones included the International Safety Rating System (ISRS) (Bird and Germain 1990) and the audit system of the British Health and Safety Executive (HSE 1991; Lindsay 1992). The ISRS audit consisted of 580 audit questions covering 20 key elements and yielded a safety score (Figure 7.13).

This ISRS audit was developed for large companies, for both occupational and process safety. Because of its generic character, it was applicable in different sectors, opening up possibilities for benchmarking. Smaller companies, though, did have problems with the bureaucratic approach, red tape and paperwork involved (Hale 2005).

The cost of safety has always been an issue. Again, Heinrich was the first to introduce a rule of thumb, the ratio 4:1. Direct costs of occupational accidents were much lower (1) than indirect costs (4) (see Chapter 2). However, estimating safety costs was difficult (Linhard 2005). Governments were trying to convince corporations that safety investments were economically sensible. But due to the lack of accurate cost estimations, that argument fell on deaf ears.

Even if safety investments provided long-term benefits, managers who might hold their position for only a few years did not see much return on investment. The Australian sociologist Andrew Hopkins suggested that governments switch arguments away from economics to ethics and the law. Hopkins advised governments to emphasise the metaphorical 'costs' of ignoring safety culture: reputation damage, emotional and psychological stress, and potential law enforcement (Hopkins 1999a).

1. Leadership
2. Planning and administration
3. Planned inspections and maintenance
4. Task analysis and procedures of critical tasks
5. Accident and incident investigation
6. Task monitoring
7. Emergency procedures
8. Rules and permits
9. Accident and incident analysis
10. Knowledge, skills and training
11. Personal protective equipment
12. Occupational health and well-being
13. System analysis
14. Control and change management
15. Communication between individuals
16. Communication in groups
17. Safety promotion
18. Recruitment and placement
19. Supplies and services
20. Safety outside work

Figure 7.13 ISRS elements

Source: Bird and Germain (1990)

Safety interventions

In essence, safety management was a problem-solving activity. Several surveys looked at the effectiveness of safety interventions (see, for instance, Saari 1998). Two kinds of interventions were prominent, namely behaviour interventions and accident prevention interventions, the latter being based on an exchange of experiences between companies. Behaviour interventions were thought to improve organisations' safety performance through behaviour-based safety (BBS) programmes. Prevention interventions through information exchange were based on the exchange of best practices; the exchange of predominantly technical safety solutions could make safety more cost effective.

Behaviour interventions were revived after the international attention given to safety climate and normative safety management. A key publication on behaviour-based safety is the one by Krause and colleagues (Krause et al. 1990). Publications on this topic had already started to appear in the 1970s. Fox and colleagues, working on open-pit mining in Utah starting in 1972, showed that by using a token economy, improvements in safety results were maintained for more than 12 years (Fox et al. 1987). Another author is Judi Komaki, from the Georgia Institute of Technology. Komaki was part of a small group of applied behaviour analysts in the academic community working on safety applications, showing positive effects of feedback and reinforcement on the safety of bakery workers (Komaki et al. 1978). In the 1990s, more publications on BBS appeared. In the Netherlands, this reversal was not received with open arms. In the Dutch *Maandblad voor Arbeidsomstandigheden* (Monthly Journal for Occupational Conditions), a paper entitled 'Return of the Human Error' mocked the return of an old subject (Zwaard 1990). However, some research showed a positive effect of BBS programmes (McAfee and Winn 1989; Guastello 1993; Lingard and Rowlinson 1998; Krause et al. 1999). BBS programmes work only when management is supportive of the idea and when the programmes include both obvious and less obvious risks (Swuste and Jongen 2007). Without clarity and sufficient management support, BBS would fail and a predictable negative reaction of the workers subject to the programme could be expected (Harper et al. 1996a, 1996b), although some researchers claimed that managerial support was good to have, but not always necessary. Some BBS processes were run entirely by the workforce, without any management input, apparently with great success (Cooper 2017). The main criticism of BBS was the dominant position of human behaviour as a causal factor for accidents (Hopkins 2006). In contrast to phases introduced by Hale and Hovden (Figure 7.7), advocates of BBS foresaw the importance of behaviour, as shown by Hopkins in Figure 7.14 (Hopkins 2006).

The second type of intervention, information exchange on accident prevention, was not very successful, due to what was termed a knowledge-application gap. The inability to control safety and health risks adequately was due not to a lack of knowledge but to a lack of applying existing knowledge (Beaumont and Else 1990; Abeytunga 1990). Kogi from the International Labour Office was one of the first to actively collect and spread 100 examples of successful health and

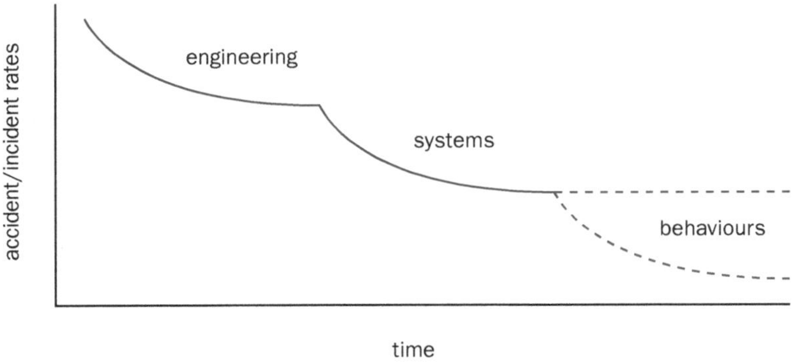

Figure 7.14 An alternative view on the development of the safety domain

Source: Hopkins (2006)

safety solutions, developed in Asia, for workplace problems related to occupational hygiene, ergonomics and hazard reporting (Kogi et al. 1989). The aim was to demonstrate that significant safety improvements did not need to be very costly and were even affordable for small companies.

Organisational culture

Around the same time, the psychologist Edgar Schein proposed an ethnographic concept of organisational culture (Schein 1992, 1996, 1999). Ethnography was a domain unfamiliar to most safety scientists. It is a social-anthropological method to study differences between groups and societies. According to Schein, three levels of culture could be distinguished inside each other like the layers of an onion (Figure 7.15).

On the surface are the visible and tangible artefacts, which can be observed in company or plant slogans, meetings, reporting systems, posters, dressing codes and behavioural patterns. Underlying these at the second level are the so-called espoused values. These values are easily espoused, yet the extent to which they actually guide people's behaviour in the organisation is unclear. They might show themselves, for instance, in training materials, procedures, instructions and other official expressions of the company. The core of the culture consists of the 'basic assumptions', which encompass a small set of implicit and largely unconscious premises. They manifest themselves in the typical ways in which an organisation conducts itself, responds to its environment and solves its problems, including those related to safety. They define what is 'normal practice' and the right way of acting and are often so deeply ingrained that those in the organisation cannot imagine other ways of responding. Only an outsider can see, after a while, what they are. In Schein's view, culture is established both by the founder(s) of the organisation or company, and during the first major stress test for its ultimate

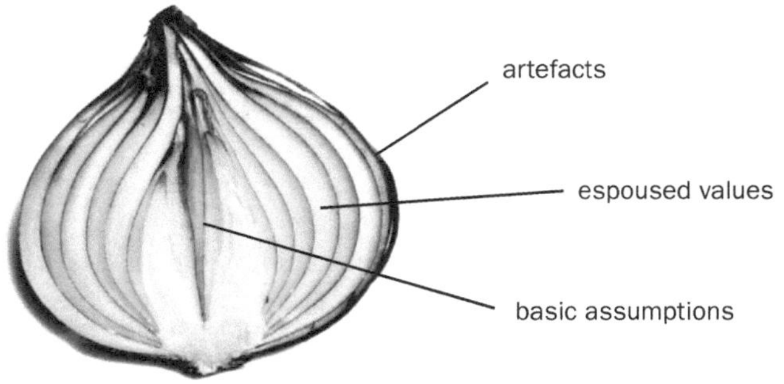

Figure 7.15 Organisational culture as an onion

Source: Guldenmund (2000)

survival. The personal relationships, work patterns and organisational structures which are successful during this crisis, will (re-)define the soul of the organisation, its basic assumptions. These basic assumptions – the shared views on time, space, reality and human nature – define its culture for years to come, and the relation between people and business partners. In Schein's approach, the culture of an organisation is a value-free concept. 'Good' or 'bad' cultures do not exist. The visible expressions of culture are the official communications, artefacts, symbols, espoused values and behaviour patterns. Conflicts between artefacts and espoused values can provide information on the implicit and largely unconscious basic assumptions, which will clarify these conflicts, according to Schein. An example of such basic assumptions will be presented in the next section.

The Netherlands

In the Netherlands, there were quite a few changes in safety thinking in this period. *Well-being* disappeared from the expert jargon as well as from the legislation and *risk* finally reached a central position in the field of occupational safety. The wave of certification systems drove many safety experts to despair, and knowledge development was slow. Only one model was developed, and an ethnographic approach to culture gave insight into cultural dimensions at a Dutch steelworks.

Organisational learning

An increased interest in organisational learning was present in this period. In the Netherlands, this was sparked by an investigation into the survivability of corporations (Geus 1997). The average life expectancy of corporations was less than 12.5 years, irrespective of their size and profitability. This seemed to contradict the inaccuracy of these two indicators (size and profitability) of organisational health. Nurturing human resources by attracting talent and retaining competence as well as a flexible response to market changes were much better indicators. This boils down to an organisation's capacity to learn and adapt.

The concept of well-being

Well-being, which was embedded in law by the 1983 Arbowet (Working Conditions Act), addressed the topic of the humanisation of labour (see Chapter 5). But well-being was a difficult subject to grasp, as became clear from publications. Companies seemed obsessive in their pursuit of efficiency. With as few people as possible, they aimed for maximum production in the shortest possible time. This led to a high workload (Bergen 2000). The instrument for measuring well-being was the WElzijn Bij de Arbeid-WEBA method (Well-being at work). This method measured the degree of complexity of tasks, as well as the possibilities of adjusting work to personal needs and having contact with other workers. WEBA was not very popular with companies, nor amongst experts in the domain of working conditions. The method was too complicated, too subjective, and the terms used were not clear (Peperstraten 1992). In the 1999 Dutch Arbowet, the triple concept of safety, health and well-being disappeared. Instead, the term 'working conditions' became commonplace in the law.

ISO madness

Due to increasing Anglo-Saxon influences on its economy, the Dutch government adopted the system of self-regulation, based on corporate safety management systems, around 1990. Under the European Framework Directive of 1989, the Dutch Working Conditions Act had changed, leading to a major transfer of governmental duties to third-party certification. From that point onward, the Dutch government increasingly focused on the presence of safety management systems in companies and no longer on technical safety prescriptions. To still be able to exert some influence on self-regulating companies, the Dutch government, as well as governments in other European countries, had introduced a certification scheme in 1994, for safety and occupational health services, and a personal certification for experts in health, safety and hygiene. Apart from the advantages of personal certification, like continuing education and maintenance of knowledge, certification had some serious drawbacks. Certification could be interpreted as a system of formalised mistrust. The focus on transparency created a widespread bureaucracy and did not encourage cooperation between experts in the health, safety and hygiene disciplines (Oortmans and Hale 1991; Hohnen and Hasle 2011). Also, there was a real threat of certification upon certification, for instance in education: not only educational institutions but also teachers, educational materials and examinations had to follow the certification regime. In literature, certification was provocatively described as 'ISO madness' (Hale and Storm 1996; Swuste 2008c, 2008d). In smaller companies and organisations in particular, these governmental interventions or regulations were perceived as a bureaucratic burden (Holmes et al. 1997; Larsson 2003; Hasle et al. 2009). Such shortcomings are found over and over again in Dutch research that focused on businesses in the construction and chemicals sectors (Ziekemeyer and Nossent 1990; Nossent et al. 1990; Swuste et al. 2012b).

Risk assessment and evaluation

In 1989, the concept of risk was formally introduced in the Dutch occupational safety domain with the publication of the EC Framework Directive on Safety and Health at Work (EC 1989). In the Framework Directive, the main elements of risk were defined. Anticipating the implementation of the Framework Directive in the Netherlands, methods for risk ranking and semi-quantitative assessment of risks, as described in Chapter 5, were being increasingly used by safety engineers (Henstra 1992). The Framework Directive was anchored in the 1994 Dutch Working Conditions Act and introduced both the risk assessment and evaluation (Dutch: RI&E) and a workplace sickness absence policy. Many safety experts believed that the emphasis on absenteeism due to illness was far too strong in practice and diverted attention from preventive health and safety at work. The Framework Directive required every company, small or large, to commit its risk to paper by performing and reporting risk analyses. A risk assessment was the basis for all safety and health activities in a company. The obligation for a risk assessment and evaluation was a determining factor (Zwaard et al. 1996). After a long period of discussion (see Chapter 5), the term 'risk' had become the central theorem in occupational safety, even though the text of the Dutch Working Conditions Act was still formulated in terms of 'hazards' (Willems and Kam 1993; Zwaard 1994, 1999; Zwaard and Veld 1994). The inventory and evaluation of risks could be interpreted as two steps in a risk management process, first an assessment, then an evaluation (Zwaard 1993a, 1993b, 1995; Zwaard and Passchier 1995). With the RI&E, the concept of risk became central in virtually all methods and practices of occupational safety professionals.

Often the RI&E was performed using a checklist in combination with a ranking method. But there were questions about the quality of the outcomes. There were many factors that caused differences in outcomes in comparable situations, such as estimating the probability of an effect. The government policy was characterised by self-regulation and development and the choice of RI&E methods and content were left to the market. It was not surprising that the 'no-fuss RI&E' was emerging. Companies did not want to have any hassle with the Labour Inspectorate, with working conditions services or with certifiers (Zwaard 1996b, 1998; Nissen et al. 1996).

As in many other European countries, with the new act, the government moved away from highly detailed standards to frameworks of essential safety objectives. Technical inspections by the Factory Inspectorate gradually reduced in numbers, and in 1997 the inspectorate stopped the production of publication papers (known as P-papers) that contained sector-specific technical safety measures and served as a source of information, and as a measuring rule for auditors, safety managers and inspectors (SER 1997; Zwaard 2007). The main reason to abolish these papers was the change in the governmental approach mentioned above. Another argument was the high speed of technological change, which was difficult to follow and to capture in the P-papers. The safety vacuum it left created a new research domain of safety indicators to help safety inspectors decide in which company, or department, inspections were mostly needed (Zwaard 1996a).

Accidents and accident models

In the first decade of the twenty-first century, the yearly number of lethal occupational accidents in the Netherlands was around 75. Agriculture, fishing, transport and construction were the sectors with the highest incidence rate. For non-lethal accidents, under-reporting was a problem. Despite a legal obligation, unlikely low accident frequencies were found in the Netherlands. Employers' ignorance of this obligation seemed to be the cause of under-reporting (Andriesen and Swaan 1991; Hollander and Vliet 1992; Schouten and Faas 2007). A proper estimate for the Netherlands was derived from German statistics, a country with similar criteria for the recognition of occupational accidents, as well as occupational diseases. To account for differences in risky activities in occupational sectors between the two countries, a standardisation of the numbers from these sectors was performed. Results suggested that the Netherlands ought to have had about 266,400 lost-time occupational accidents and 3,700 cases of occupational disease in 1993, much more than the official number suggested (Gorissen and Schroder 1996). However, the 2005 and 2010 Dutch national labour survey again suggested a serious level of under-reporting (Venema et al. 2007; Klauw et al. 2012).

Another problem that was completely separate was the influx of migrant workers into sectors such as agriculture, construction and production. Most of these workers held low-wage and dangerous jobs. Here, language could become a problem and a hazard (Starren et al. 2009). Migrant workers without full command of Dutch were simply put to work without clear instructions and were given rules and protocols they could not understand (Corvalan et al. 1994; Döös et al. 1994; Lindhout and Ale 2009; Lindhout et al. 2010; Hovden et al. 2010; Swuste and Jongen 2013). The effect on accident statistics was not known yet.

Risky behaviour at work and the dualistic approach towards this topic was another theme of publications. If this behaviour saved production or made it run smoothly, then it was not considered dangerous behaviour, but accepted and even encouraged. Safe and unsafe behaviour were therefore relative concepts and dependent on the effect (Kerckhove 1991). This put the emphasis much more on the organisation, which stimulated certain behaviour and punished other behaviour. The behaviour of workers was thus drawn into an organisational context, something that had been propagated by British and Scandinavian researchers since the second half of the 1970s (Zwaard and Groeneweg 1999b).

The Dutch model on occupational accident processes known as Tripod was developed by the psychologist Groeneweg (Groeneweg 1992; Groeneweg and

Roggeveen 1998). In line with psychological practice, it addressed perception, and cognitive and psychological processes influencing the suboptimal behaviour of workers at the onset of an accident process. The Tripod model assumed workers were practically unable to anticipate the consequences of their possibly suboptimal behaviour until it was too late. This behaviour was highly dependent on the context and conditions of their work, and the managerial style of the organisation. Consequently, it was not the worker that had to be changed but his or her working conditions. Accidents were negative outcomes of a suboptimal organisational delivery process, a view similar to earlier work in Nordic countries (see Chapter 5). According to Groeneweg, prevention changed over a period of 50 years from investing in safety equipment and personal protection devices to the design and organisation of labour processes (Figure 7.16).

In Figure 7.16 a Murphy margin is present. This margin relates to Murphy's law: 'If anything can go wrong, it will.' This 'law' relates to the idea that even when all measures are taken and all control systems are in place, there will always be things going wrong. The Murphy margin was also called the 'zone of creeping entropy', where factors beyond the control of management and workers would influence and initiate accident scenarios, leading to accidents.

In the Tripod model, the decisive factor was an *operational disturbance*, followed by incomplete barriers. The barrier concept referred to Haddon's hazard-barrier-target model from the 1960s. Working conditions leading to accidents could be reduced to a limited number of underlying causes. These latent factors were called 'basic risk factors', or BRFs (Figure 7.17).

These latent organisational factors could be seen as a further specification of Turner's incubation period (see Chapter 6). Some BRFs were branch- or sector-specific, and some were more general (Table 7.1). A separate BRF dealt with

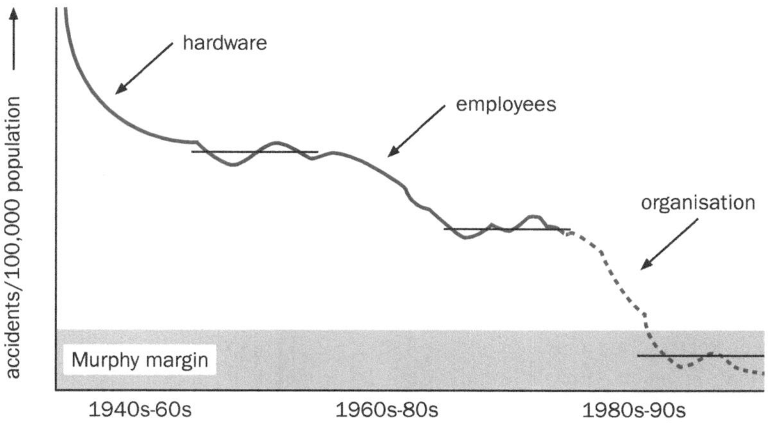

Figure 7.16 Three eras of safety improvement

Source: Groeneweg (1992)

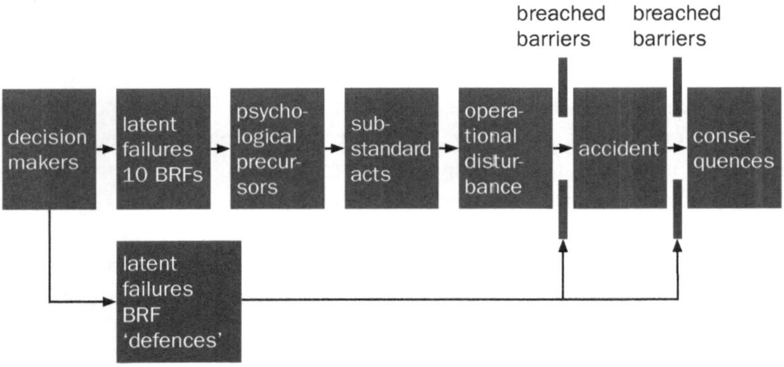

Figure 7.17 The Tripod model of accident causation

Source: Groeneweg (1992)

Table 7.1 Branch/sector specific and general basic risk factors

Branch/Sector Specific BRFs	*General BRFs*
1 design of installations, equipment, tools	7 comprehensible instructions and procedures
2 quality of equipment and tools	8 training of workers
3 physical working conditions	9 incompatible goals
4 housekeeping	10 communication
5 management of maintenance	11 structure of organisation
6 presence, quality of barriers	

defences and the quality of safety barriers. Tripod included 'substandard acts', but these were seen as consequences and not as causes of organisational shortcomings. The Tripod model could be used to map effects of safety management systems on operational conditions and safety (Cambon et al. 2006). This model found its way into industry with many applications, and into scientific literature. For instance, Tripod was used to generate indicators of train passenger safety for train conductors (Wielaard and Swuste 2001) and risk scenarios for an oil and gas exploration laboratory (Bruin and Swuste 2008).

The management delivery systems of the bowtie metaphor are compared with the basic risk factors from Tripod in Table 7.2. The similarity between the two classifications is striking. While the delivery systems are focused specifically on barriers, hardware and behavioural aspects, the basic risk factors are formulated more generally.

No new theories on accident causation and processes of occupational accidents were developed between 1990 and 2010. Rather, the focus in science was on the development of aspects of safety management systems such as safety culture and

Table 7.2 Comparison of management delivery systems and Basic Risk factors

bowtie, management delivery systems	*Tripod, BRF's*
barrier systems, hardware	
1 scenario and risk identification, barrier selection and specification	6 presence of, and, quality of barriers
2 monitor, feedback, learning and change	
3 barrier design, purchase, construction, installation	1 design of installations, equipment, tools
4 barrier inspection, testing, maintenance	5 management of maintenance
barrier systems, behavioural aspects	
5 procedures, plans, rules, goals	7 comprehensible instructions and procedures
6 availability, manpower planning	
7 competence	8 training of workers
8 commitment, conflict resolution	9 incompatible goals
9 coordination and communication	10 communication
barrier systems not matching	
	2 quality of equipment and tools
	3 physical working conditions
	4 housekeeping
	11 structure of organisation

The numbers refer to the original numbers of delivery systems, and BRF's (Figure 7.6 and Table 7.1)

the Tripod model, the latter representing a bowtie-based system approach for barrier control. Both the bowtie metaphor and the Tripod model reflected the dominant view of analysing these processes within socio-technical systems, including context of work and organisational factors. But some authors questioned the necessity of new theories or models and approaches to occupational accident prevention. As a caveat, Hovden had rejected the systems approach because he claimed that the boxes-and-arrows approach was not supported by any scientific evidence and was impractical as well. He stated that the era of 'boxicology' had to end (Hovden et al. 2010). This was also true for safety management systems. Self-regulation of companies was a strong stimulus for research on safety management systems and audits (Hale 2005). Despite this research on elements of safety management systems, related audits and safety interventions, the theoretical understanding of safety had not advanced much.

New systems tended to be based on the dated Deming cycle and experts' opinions, creating lists of factors similar to the EFQM/INK model (Figure 7.1), the generic safety management cycle (Figure 7.2) and the ISRS list of crucial management factors (Figure 7.13) (Hale and Hovden 1998). Unfortunately, the empirical evidence to support the validity of this research was limited (Eisner and Leger 1988; Robson et al. 2007; Molen et al. 2007; Hale and Guldenmund 2010; Hale et al. 2010). Most investigations used a pre- and post-measurement approach as evidence for success but tended to ignore potential biases. It simply seemed too difficult to prove that a reduction in numbers of accidents was actually due to the introduction of management systems (Shannon et al. 1999).

Reason et al. (2006) raised the question of whether the search for organisational factors had prevented organisations from taking actions directed at making sure that key safety barriers would not be breached. After reviewing work-related fatalities, Shell found that failure to comply with safety rules was a significant factor in the majority of these incidents. In response, and as the next step in the Goal Zero journey, Shell launched its '12 Life-Saving Rules' in 2009. They were not new rules, but were promoted in such a way as to increase awareness of the importance of following the rules and to drive compliance in the areas with the highest risk of fatal injury. They set clear expectations about what employees and contractors and their supervisors must know and do to prevent injuries or fatalities. Compliance is mandatory for the entire Shell group, including for all Shell employees and contractors while on Shell business or sites. Following the launch in 2009, Shell's lost-time injury frequency (LTIF) reduced between 2008 and 2011 significantly, as did the fatal incident rate (FAR), compared to the rest of the oil and gas industry (Peuscher and Groeneweg 2012). The rules have since been adopted by many international organisations and the effect on safety performance has also been significant in the long term (Bryden et al. 2016).

All models discussed refer to the control of accident processes, either as operational disturbances in the primary process (Tripod) or as deviations in the primary production process (OARU) or as central events (barrier based). Scenarios led to such conditions, and both Tripod and ORM include barriers, BRF defences and management delivery systems to prevent these conditions. One line of research can be the impact of safety management (systems) and interventions on the development and status of these scenarios. Another line of research is the direct or causal relation between these management delivery systems and the actual status of barriers. Recent research at the Delft Safety and Security Science group has been conducted along this line of thinking, using falling loads from cranes as an example, and the required management deliveries to prevent such drops (Li 2019). Furthermore, a first attempt has been published to use the bowtie metaphor to monitor the progress of scenarios, enabling accident incidences to be predicted. One article discusses occupational accidents among pallet movers in a manufacturing company (Nunen et al. 2018).

Safety interventions

One of the safety interventions promoted in many countries was best practices. This posed two interesting questions: 'Can a solution be transferred from one company, or location, to another, and if so, under what conditions?' and 'How can safety experts and managers with specific problems be guided toward useful solutions?' Research showed that such transfer is possible only if technologies are similar or industries have similar production processes

(Hale and Swuste 1997). This finding was based on prior research in Delft, where safety aspects and exposure to shocks and vibrations were investigated (Drimmelen et al. 1989; Swuste et al. 1997b). By classifying production processes into production functions, like 'unit operations' in the process industry, similarities between production processes were revealed, a major condition for an exchange of solutions. The exchange of ideas and solutions for safety was possible even if its practice was not widespread (Swuste and Hale 1994; Swuste 1996; Swuste et al. 1997).

Safety culture

Also, in the Netherlands, culture was a major area of research. One prominent publication on differences between the cultures of countries was by the Dutch organisational psychologist Hofstede. He studied IBM workers worldwide in the 1960s and 1970s. Results, however, were published much later (Hofstede 1991). According to Hofstede, differences between national cultures could be described using five factors: power distance, uncertainty avoidance, individualism versus collectivism, masculinity versus femininity and long-term orientation versus short-term orientation. The influence of national culture was also studied in a research project on French, Argentinian and American sites of a multinational company (Janssens et al. 1995), and the differences found matched Hofstede's national culture dimensions. A similar national culture effect was present during the implementation of ergonomic interventions in Japan and the United States (Liker et al. 1989).

Another study was carried out by the psychologist Guldenmund. He found a lack of agreement on a definition of safety culture (Guldenmund 2000). Therefore, he suggested a compromise: safety culture could be regarded as a reflection of those aspects of organisational culture that deal with safety (Hale 2000; Guldenmund 2000; Guldenmund and Swuste 2001). Next to organisational structure and processes, organisational culture is a basic organisational component, influencing human behaviour in organisations (Guldenmund 2007, 2009) (Figure 7.18).

This three-component approach shares characteristics with the EFQM/INK management model, and they are dynamically linked, both positively and negatively. Structure referred to the (formal) structure of the organisation, that is, the division of power and responsibilities, mechanisms of communication, coordination and control, but also the actual hardware, like buildings and installations; it determined how organisational objectives were met, who did the work and how it should be done. The influence of structure on safety was investigated in a Dutch steelworks (Guldenmund and Booster 2005). Processes represented both primary and secondary ones in a company, including their interactions. They posed requirements for management, support and strategy for the success of the organisation. The three components together determined the organisational context for human behaviour and, hence, also behaviour in relation to safety. The study of culture, as a set of shared, taken-for-granted basic assumptions, required extensive fieldwork

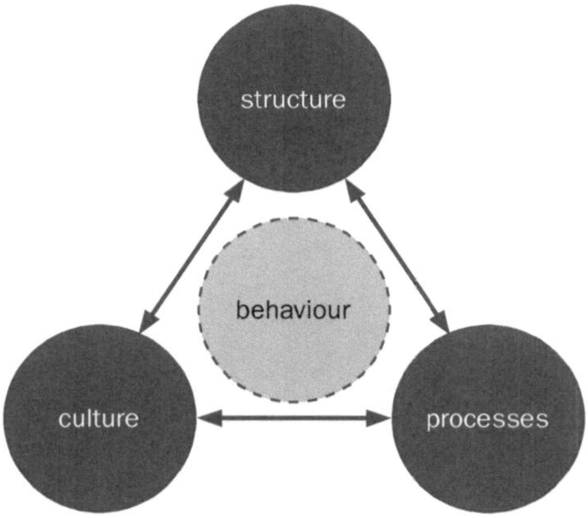

Figure 7.18 The organizational triangle

Source: Guldenmund (2007, 2009)

and much interpretative analysis afterwards (Guldenmund 2007). After their introduction into mainstream safety research, safety culture and climate investigations were conducted in many sectors. Manufacturing led the way, together with health care and transport. Process safety, construction and power plants came a good second. Research at the oxy-steel plant of a Dutch steelworks depicted a classical technocratic safety approach. The energy level of the material stream was high, with temperatures of 1,700–1,800° C and masses of 50 tons and more. There was a strong white-collar/blue-collar divide, a division not unique for this sector that is present in many more industrial sectors (Guldenmund et al. 1999; Swuste et al. 1993, 2002). Figure 7.19 presents some of the basic assumptions amongst steel workers in this plant.

An example of this white-collar/blue-collar divide was the action of one production team. This team generated solutions to prevent frequently occurring process disturbances when casting steel rods. Their white-collar managers were kept outside the loop. As a side effect, scenario accidents of these production disturbances were also substantially reduced. The solutions proposed interfered directly with accident processes. From a cultural perspective, freedom for blue-collar workers to solve their own problems was inherent in the culture of this steelworks at that time. It also shows that solutions fitting blue-collar culture had a high chance of success.

But safety culture and safety climate investigations did not yield clear conclusions on the overall safety level of a company. Methods and definitions differed, and it was difficult to distinguish between the effects of a safety management system and those of safety culture. Even the relatively straightforward safety ladder (Figure 7.9) did not create sufficient clarity. While the developers saw the

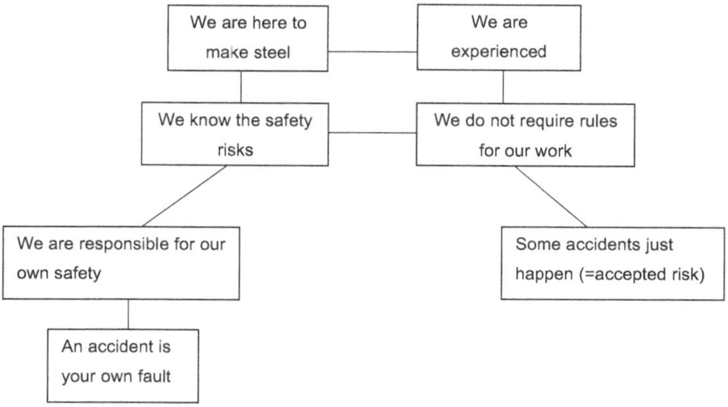

Figure 7.19 Basic assumptions found at the Oxy-Steel 1 steel factory

Source: Guldenmund et al. (1999)

safety ladder as a tool for opening discussions and deepening insights into safety, many practitioners took the ladder as a benchmarking tool. Also, hindsight bias after a major accident shaped the outcome of safety culture investigations. Such confusion prompted a special issue of *Safety Science*, opening with the editorial 'Cultural confusions' (Hale 2000). The effectiveness of safety interventions on the organisational/safety culture was more or less a black box in safety research (Dyreborg et al. 2010). The fragmented approach, focusing on different aspects of safety management, simply did not offer sufficient proof to support the notion that there were causal relationships. To find out whether a relationship existed, a longitudinal research approach was required and that was beyond the scope of researchers at that time, not without good reason. For one thing, funding of such research was probably a stumbling block. In itself, it was a painful finding that the quality of management systems and the quality of accident models, was not rigorously investigated. Models and metaphors were by definition representations of reality, without any form of validation. A validated model is a theory, making predictions possible. That was missing in the safety domain. The models and metaphors were good enough to make retrospective analyses, but predictions were not possible yet. In brief, it was a shortcoming of the safety domain itself, which was poor in terms of theories.

In this period no safety theory on occupational accidents was developed. Only three models existed, namely the Swedish OARU, the Dutch, British and Greek ORM, and the Dutch Tripod, all stressing organisational causes of accidents. Stimulated by developments in quality management, models

for safety management were developed in detail, as were safety audits. But managing safety was still a difficult and complex task, as the construction sector showed. Safety climate and safety culture received a lot of attention as a motor for safety performance, and there was a renewed interest in employee behaviour. Unfortunately, any scientific proof of a relation between safety management, safety culture and audits was still missing. The following chapter describes developments in the same period for the high-tech-high-hazard sectors.

8 High-tech-high-hazard safety, culture and risk

1990–2010

This chapter covers the same period as the previous chapter, but this time the emphasis is on high-tech-high-hazard safety. This is often called the 'golden years' of safety science. Quite a few metaphors and theories were developed during this period, which are briefly summarized in this chapter. Risk and safety culture dominate the two decades covered by this chapter.

DOI: 10.4324/9781003001379-8

General management schools

The management schools are the same as those discussed in the last chapter. Total quality management (TQM), the model of the Dutch Quality Institute (INK) and Six Sigma were the important management trends for the period 1988–2010. Corporate social responsibility (CSR), the business community's stated concern for society, was also an important concept during this period. CSR, which originated in the early 1970s, was adapted/incorporated by a few stakeholders, close at hand, to be more far-reaching and inclusive, eventually becoming global in scope (Carroll 2008). These developments included Anglo-Saxon and Rhineland management concepts, organisational learning and precautionary principles. This chapter ends with a separate section on the major accident involving the Deepwater Horizon oil rig in the Macondo oil field, in the Gulf of Mexico, which occurred on 20 April 2010.

Publications on the concept of risk, initiated by development in the nuclear sector and further developed by the Dutch series of 'coloured books', appeared around the same time in three geographical areas: western Europe and Nordic countries, North America, Australia and the Netherlands. Risk and the domino effect will be discussed as an overarching concept in the next section. This includes major accidents in high-tech-high-hazard sectors. Specific aspects related to risk perception, and risk and safety management, will be the subjects of separate sections.

Risk

In safety, a risk is a measurable uncertainty of loss, which has a negative connotation. Conversely, a certain degree of loss of control in risky sports gives practitioners an adrenaline rush. People learn by taking risks. Sometimes this is forgotten. People are not just *homo prudens* – the prudent man – seeking an environment with risks kept as low as possible; they are also *homo aleatorius* – the gambler, the risk taker (Breivik 2011).

The high-tech-high-hazard sectors have a different risk pattern to the sectors discussed in the previous chapter, dealing with occupational risks. The most pronounced difference is that exposure to risk and variations in that exposure are poorly developed concepts in high-tech-high-hazard industries, which is in stark contrast to occupational hygiene. DeBlois introduced this concept in the safety literature during the interwar period in 1926 (see Chapter 2). Following DeBlois, both Lees (2005) and Rowe (1988) argued that reduction of exposure to a hazard could be one of the possible prevention strategies (Rowe 1988; Mannan 2005).

There are many other different definitions of risk (Vlek 1990). A simple definition of safety risk is the one by Rowe (1988):

> *Risk is the potential for realization of unwanted, negative consequences of an event.*

Quantitative risk assessment (QRA) was introduced in the nuclear sector in the 1950s under the name 'probabilistic risk assessment' (PRA; Keller and Modarres

2005) with the famous WASH 740 and WASH 1400 reports of the American Office of Nuclear Reactor Regulation as landmark studies in risk assessment (WASH 740 1957; WASH 1400 1975; Wellock and Budnitz 2016). These reports were published after concerns about reactor safety in the Cold War period just after World War II and the rapid expansion of the nuclear sector. QRAs were developed for very rare, unacceptable, major accidents with a complexity so large the risk could not be approached deterministically, but only probabilistically (Pasman 2000; Rasmussen 1997; Hudson 2010b). Such was the philosophy in the period of the first major accidents with extensive media coverage (Kasperson et al. 1988).

National governments, including the Netherlands', played an important role in promoting QRA applications outside the nuclear sector. They saw results of risk calculations as an argument in political decision-making on land use planning issues. With the well-known series of coloured books published in the Netherlands from the late 1970s onwards, the magnitude of three risk types was calculated: individual risk, group (or societal) risk and collective risk to ecosystems. Internationally, these coloured books received a lot of international attention and were the basis for the European Seveso Directive (see Chapter 6). The first two types of risks, individual and group risk, were related to damage to human health and risks associated with industrial plants (Figure 8.1). The individual risk is the annual risk of death or serious injury to which specific individuals were exposed. The individual and external risk of an on-site activity is graphically represented using iso-risk contours (Pietersen and Veld 1992). The group risk was the probability that a group of ten persons or more per year would be victim of an accident. For non-site-specific activities, individual risks and societal risks are calculated differently: for example, by means of dose-effect relationships or dose-response curves.

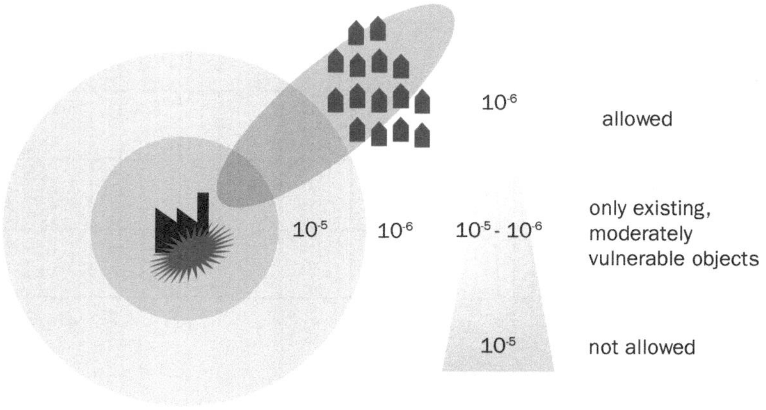

Figure 8.1 Individual risk contours around a hazardous establishment and the area affected by an individual accident scenario

Source: Pietersen and Veld (1992)

The collective risk to ecosystems was the risk of an adverse effect on the ecosystem from an exposure of one year to a particular substance. This risk group is shown using an F/N curve, in which the expected frequency of accidents, F, by at least N deaths is expressed as a function of N (ISO 2012) (Figure 8.2).

Due to the far greater complexity of ecosystems, with many interacting species compared to humans (who are only one species), and the corresponding larger lack of reliable, quantitative data (exposure-risk relations), the accuracy and precision of numerical results of QRA for studies of ecosystems were considerably more questionable than corresponding QRA studies concerning only humans.

Although it was well known that the human factor played an important role in the causation of incidents, attempts to systematically include quantitative estimates on the impact of human factor and organisational aspects on failures in a QRA proved to be quite a challenge (Skogdalen and Vinnem 2010). The human factor component is often estimated using expert opinions, which often lead to heuristics such as adding a factor of 10 or 100 to the failure probability of the technological components. This practice was indicative of a lack of implementation of existing knowledge concerning human error likelihood and human reliability assessment (Steijn et al. 2020), reducing the validity of the failure rate estimates even further.

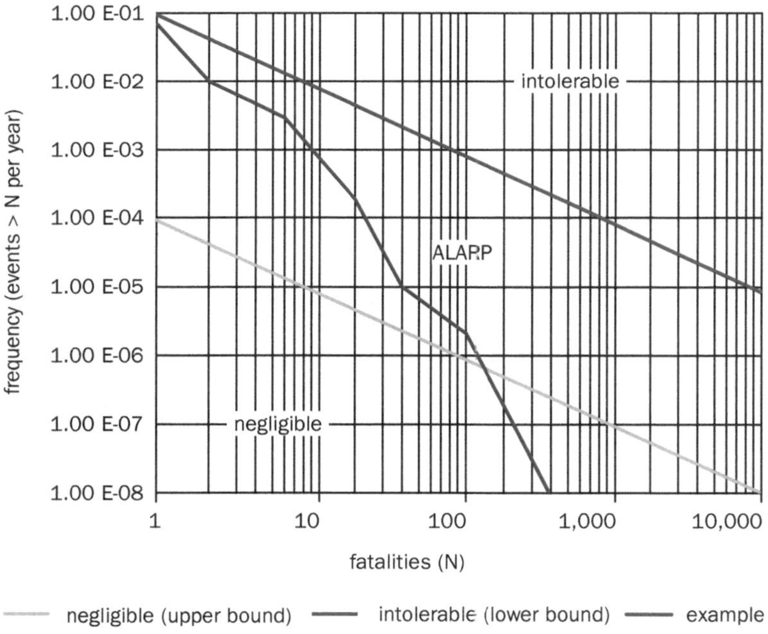

Figure 8.2 Example of an F/N curve, group risk

Source: ISO (2012)

Risk measures have numerical standards for acceptance of negligible risks and for unacceptable, inadmissible risks. Risks that were not negligible in the intermediate region required a further review. The ALARA (as low as reasonably achievable) principle attempted to reduce the risks as much as reasonably possible or feasible. The benefits were balanced against the remaining risks, and an activity with a non-negligible risk could still be allowed. Britain used the ALARP principle, the P standing for 'practicable' (Vlek 1990; Gezondheidsraad 1995; HSE 2001). At first, the influence of human interference was underestimated, for instance the knowledge and performance of employees to correct failures during production or transport. QRA used only generic failure data (Woods 1990a, 1990b; Pasman et al. 1992; Kjellén 2007). Just as with human recovery capabilities, the quality of safety management also had no effect on the outcome of a QRA (Hale 2001b).

The QRA approach had a fundamental shortcoming. QRA was used to compare alternative technologies, as part of a (political) decision-making process, but it did not allow a falsification of its predictions. QRA typically predicted events occurring with extremely low probability, like 10^{-6} per year. Testing and validating these results in practice, which related to such rare events, was simply not possible.

Another problem was the reliability of the results of a QRA. Results of a benchmarking study by the European Commission, carried out in the 1990s, showed a wide variation in the failure probabilities of a chemical plant. Among ten participating countries, this variation rose by three to five orders of magnitude, a factor of 1,000–100,000 (Lemkowitz et al. 1995; Pasman and Reniers 2013). The results of a QRA were shown as a number. Unfortunately, the uncertainties of the accuracy and precision of the outcome were seldom discussed. A final drawback was the fact that these calculations did not give additional information about the nature of the accident processes involved.

The 'precautionary principle' came up as a countermovement against the QRA approach. This was to the annoyance of policy makers who interpreted the results of risk calculations as a standard for political decisions (Smallman 1996; Vlek 2010a, 2010b). The precautionary principle dated from the 1970s and 1980s and emerged as a management approach for serious and uncertain environmental risks and risks to public health. The principle meant that proponents have the burden of proof that their operation is safe. When an activity may cause serious or irreversible damage while no scientific consensus yet exists on the safety of the activity, it should not be performed, or only under very strict conditions. This meant that a government did not have to wait to implement control measures until conclusive evidence of adverse effects had been delivered (Brundtland 1987; Rio Declaration 1992).

Accidents in high-tech-high-hazard sectors

During this period, the term 'impossible accidents' popped up again in literature. This term dated from the second half of the 1980s and referred to industrial accidents nobody expected to happen (see, for example, Wagenaar and Groeneweg 1987; Woods 1990a). At a detailed level, these impossible accidents were specific to the sector where they appeared, while nobody expected these major accident scenarios would ever occur.

Before exploring the accident processes in the 'high-tech-high-hazard' industries, this section will give a brief review of the frequency of incidents and major accidents in Western countries. Most countries report serious incidents and major accidents. These incidents and accidents occur not only in developing countries or in so-called intermediate countries but also in the technologically advanced West (Shrivastava 1992). Appendix 1 and the resulting Table 8.1 provide an overview of these incidents and major accidents occurring in the high-tech-high-hazard sectors between 1990 and 2010. Note that this table does not cover the world in its entirety. First, it is limited to only five sectors: aviation, rail transportation, process industry, upstream and downstream sections of the oil and gas industry and finally the nuclear sector. Second, only data from Western countries and Japan are included.

The nuclear sector is different from most industries. Nuclear technology is used outside Western countries and has been, unlike other sectors, under international scrutiny by a United Nations agency: the International Atomic Energy Agency (IAEA). The data in Table 8.1 were grouped into seven-year periods, which was an arbitrary choice. The last limitation was the sources of this table, which were restricted to publicly available information, with almost no information on non-Western countries in the sectors selected (Chemical Safety Board 1990–2010; Lees 1996; Rasmussen and Gronberg 1997; Khan and Abbasi 1999; Office of Nuclear Regulations 2000–2010; HSE 2003a; Mannan 2005; Kinnersley and Roelen 2007; Clough 2009; OGP 2010; Abdolhamidzadeh et al. 2011; Mihailidou et al. 2012.; Marsh 2012; Thomson 2013; Dort 2014; Kraaijvanger 2014; Wikipedia 2016; Zwaaiichten 2016).

Table 8.1 shows that only a limited number of countries, such as the United States, the United Kingdom and the Netherlands, report information publicly. Among companies there was and is a resistance against publishing these data publicly for fear of facilitating terrorism (Belke 2001). For instance, in countries where nuclear power provides a substantial part of the national energy supply, incidents will most likely be reported, but not necessarily to the public. This limitation also applies to other high-tech-high-hazard industries. France is an example of a country that rarely reports its nuclear incidents and major accidents publicly. Reporting of major accidents in the process industry in Belgium is another example of limited reporting in the publicly available literature (Swuste and Reniers 2017). The bias in the table varies by country, time and sector. The table includes an underreporting that is difficult to estimate. The numbers in Table 8.1 and Appendix 1 are difficult to interpret. After Chernobyl (1986) one would expect that the number of nuclear power plants worldwide would decrease and thus also the number of incidents. This effect, however, is not visible in Table 8.1. To a lesser extent, a similar effect was expected in offshore oil and gas exploration (upstream) after the major accident at Piper Alpha (1988). In the chemical processing industry, such an effect was not visible; in fact, the opposite occurred. There was an increase in risk sources, and the same was seen in the transport sector (Kirchsteiger 1999). It was argued that the number of industrial facilities putting large numbers of people at risk had in fact increased (Roberts 1990; Baram 2009). Or, as other authors stated, 'based on available literature, it is very difficult to draw unambiguous conclusions about an increase or decline in the number of accidents – occupational accidents and/or process safety' (Reniers and Khakzad 2017). Other sources, however, indicated an overall decrease in the number of major accidents in the West (Khan and Abbasi 1999; Khan and Amyotte 2002; Mihailidou et al. 2012), or that the number of explosions, fires and related incidents varied over time, but remained largely constant over a longer period (Bradley and Baxter 2002). However, due to uncertainties in both the numerator and denominator, there was no real basis for conclusions about an increase or decrease in serious incidents or major accidents. There was no information on exposure or an estimation of the number of active installations or activities in high-tech-high-hazard sectors. Therefore, no rates or time trends could be calculated (Kirchsteiger 1999; Reniers and Khakzad 2017).

Domino effects

The major accidents in process industries in Table 8.1 could be consequences from complicated domino-effect accident processes, which were mentioned in Chapter 6. A domino effect is a relatively complex event. In the past two decades these events have attracted increasing attention in scientific literature. The first documented domino accident was in 1947. In the port of Texas City, a ship carrying ammonium nitrate detonated due to a fire. This resulted in a chain reaction, and other ships and an oil storage facility on land exploded. Despite the fact that this major industrial accident in America was the largest, as measured by the number of fatalities (almost 600), it was a not a trigger for research into domino effects (Khan and Abbasi 1998).

There are a number of definitions of the domino effect. The simplest definition came from Lees: 'an event in one unit that causes a follow-up event in another unit' (Lees 1996). Reniers and co-authors described a domino effect as 'a cascade of events in which the consequences of a previous accident increase through successive events, both spatially and sequentially, and lead to a major accident' (Reniers et al. 2005a). The American Center for Chemical Process Safety (CCPS) of the American Institute of Chemical Engineers (AIChE) defined it as 'an incident that starts in one unit and affects nearby units through a thermal effect, an explosion or an impact of fragments' (CCPS 2000). In this definition, attention was paid to the mechanism of a domino effect, heat radiation, the pressure wave and the projection of debris. This was further elaborated in a definition by Cozzani and co-authors where four stages were distinguished (Cozzani et al. 2006, 2007).

1. *A primary accident scenario, the starting point of the domino effect;*
2. *The propagation following the primary event, caused by physical effects – the escalation vector factors – of the primary scenario and resulting in damage to at least one secondary unit;*
3. *One or more secondary accident scenarios, involving the same or another plant unit or establishment;*
4. *An escalation effect is the result, an increase of the domino effect in relation to the primary scenario.*

Table 8.1 number of major accidents in high-hazard-high-tech sectors from publicly accessible literature.

	Atlantic Ocean	*Australia*	*Austria*	*Belgium*	*Canada*	*Denmark*	*Finland*	*France*	*Germany*	*Greece*	*Hungary*	*Ireland*	*Italy*	*Japan*	*Luxemburg*	*Netherlands*	*New Zealand*	*North Sea*	*Norway*	*Portugal*	*Spain*	*UK*	*US*	*Sweden*	*Switzerland*	*India*	*Ukraine*	*Russia*	*South Korea*	*TOTAL*
’90 –’96 aviation	1							2		1			1	3		5	1	1	3	1		2	22	1	1					45
’97 –’03 aviation			1		2			2	1	2			2	1	1	1			1		2	3	17		2					38
’04 –’10 aviation	1	2			1	2		1		1			3	2		2			1		1	4	8							29
Total Aviation	2	2	1		3	2		5	1	4			6	6	1	8	1	1	5	1	3	9	47	1	3					112
’90 –’96 transport (train)					2	1	1	2	3					1		3					1	8	19		5					46
’97 –’03 transport (train)		5	1	1	4		1	1	5	1			2	1		5			1		1	5	21		6					61
’04 –’10 transport (train)		7	1	1	5		1	1	3	2		2	7	2	1	15			1		2	7	38	2	1					99
Total Transport (train)		12	2	2	11	1	3	4	11	3		2	9	4	1	23			2		4	20	78	2	12					206
’90 –’96 process industry									2			1		1		3				1		4	12		1					25
’97 –’03 process industry		5						2	1					1		3						3	28							39
‘04 –’10 process industry		2			1				1					1		4							30							39

Total Process Industry		7		1	2	4			1		3	10			1		7	70		1					109
'90 –'96 up-downstream		4			1	1	1			2	3	3	1	1		1	5	20							43
'97 –'03 up-downstream	1			4	2		2					5		1		1	3	35							54
'04 –'10 up-downstream		1	1		1							4		5			1	17							30
Total Up-Downstream	1	5	1	4	4	1	3			2	3	12	1	7		2	9	72							127
'90 –'96 nuclear				2							3						1	11			3		2		22
'97 –'03 nuclear				1	2			1			5						14	6			1	1	2	1	34
'04 –'10 nuclear				1	3						3						18	6	1						32
Total Nuclear				4	5			1			11						33	23	1		4	1	4	1	88

Source: Swuste et al.(2019)

This definition was more detailed in terms of mechanisms and made a distinction between an internal domino effect and an external domino effect. Process disturbances could start internal or external domino effects. The probability of these effects is increased in the clusters of high-tech-high-hazard companies. Clusters of chemical companies are present in industrialized countries and have the advantage of jointly organized support services and a reduction of logistical costs. Internal domino effects occur when a major accident in one installation spreads over several installations but remains within the geographically defined area of the company. The major accidents discussed in Chapter 6, at Feyzin, France (1966), Flixborough, UK (1974) and Piper Alpha, North Sea (1988), were all examples of internal domino effects. An external domino effect occurs when a major accident or an internal domino effect exceeds the geographically defined area of the company and affects one or several other companies. The major LPG accident at San Juan Ixhuatepec, Mexico City (1984), and the vapour cloud explosion and fires at the Buncefield oil storage depot, in Hertfordshire, England, were examples of external domino effects. In publications that analysed these domino effects in the chemical and process industries, characteristic patterns were found of various hazards: flammable liquids (oil, naphtha, gasoline, kerosene), gaseous hydrocarbons and toxic substances (Cl_2, NH_3, pesticides). The highest domino frequency was found with the gaseous hydrocarbons and the lowest with the toxic substances, although the consequences for toxic substances were the greatest. An overwhelming majority of domino effects were caused by flammable substances. But also, non-flammables, like CO_2 and Cl_2, and overheated water had created explosions and subsequent domino effects. The most important factors for domino effects were external events and mechanical failures. Accident processes that started with an explosion followed by a fire, or vice versa, were by far in the majority (Kourniotis et al. 2000; Ronza et al. 2003; Gómez-Mares et al. 2008; Darbra et al. 2010; Abdolhamidzadeh et al. 2011; Necci et al. 2015; Swuste, Nunen et al. 2019, 2020).

Table 8.2 shows an overview of the already known primary and secondary scenarios, including the escalating factors (Salzano and Cozzani 2012). Toxic emissions as the primary domino scenario are not held responsible in this overview for an escalation, although toxic release in combination. with a fire or a heat source might ignite (Necci et al. 2015).

Another point relates to domino effects in parallel pipelines. These effects are different to those in chemical plants. Corrosion is a very important factor here and a domino effect can occur if an adjacent pipeline lies in the hole or crater created by the primary scenario. Adjoining pipelines are protected by the ground, so that a distance of 10 meters between parallel pipelines appears to be sufficient (Ramirez et al. 2015; Silva et al. 2016).

Golden years of safety

In the 1990s, often called the 'golden years of safety science', new metaphors and theories of accident processes in high-tech-high-hazard sectors were

Table 8.2 Escalating factors and expected secondary scenarios (after Salzano and Cozzani 2012)

Primary Scenario	*Escalation Factor*	*Expected Secondary Scenario*[1]
pool fire	radiation, fire impingement	jet fire, pool fire, BLEVE, toxic release
jet fire	radiation, fire impingement	jet fire, pool fire, BLEVE, toxic release
fireball	radiation, fire impingement	tank fire
flash fire	fire impingement	tank fire
mechanical explosion[2]	fragments, overpressure	any
confined explosion[2]	overpressure	any
BLEVE	fragments, overpressure	any
VCE	overpressure, fire impingement	any
toxic release	concentration	none

(1) Expected scenarios are also dependent on hazards of chemicals.
(2) A primary failing reactor vessel can lead to other scenarios (e.g. pool fire, BLEVE, toxic emission).

developed, as well as revisions of previous theories. In western Europe and Nordic countries, the final version of the Swiss cheese metaphor appeared in 1997 (Reason 1997; Reason et al. 2006). In the same year, the 'drift to danger' metaphor, in combination with a socio-technical risk management system, was published (Rasmussen 1997), and it was followed a year later by the bowtie metaphor for industrial accidents (Visser 1998). The 'disaster incubation theory' finally received full attention in the safety science domain through a reissue of *Man-made Disasters* (Turner and Pidgeon 1997). Finally, 2004 saw the start of the biannual workshop of the Resilience Engineering Association (REA) (Hollnagel et al. 2006).

In North America and Australia, a reissue in 1999 of the 1984 book *Normal Accidents* appeared (Perrow 1999). Finally, there was also the high reliability theory of the Berkeley group where, during the late 1970s, Roberts began her research on nearly accident-free organisations (Roberts et al. 1978). Therefore, she is referred to as the 'mother of high reliability theory' (Weick and Sutcliffe 2001). During the same period, a number of important books appeared, which are now widely cited in the scientific safety literature. The first was by Sagan, a political scientist, on safety problems with nuclear weapons during the Cold War (Sagan 1993). The second one was from the sociologist Vaughan, with an analysis of a major accident involving a space shuttle (Vaughan 1996), and the last book was by the public administration scientist Snook, with an analysis of the shooting down of 'friendly' (non-enemy) helicopters (Snook 2000). In addition, the sociologist Hopkins (1999a, 2000a, 2000b, 2008, 2012) published a series of books on major accidents in Australia and the United States: the Moura mine disaster in central Queensland, the Esso Longford explosion in Victoria, and the major accidents at BP Texas and in the Gulf of Mexico. This last major accident will be discussed at the end of this article. All these publications refer to the high level of complexity within the high-tech-high-hazard sectors.

Western Europe and Nordic countries

Determinants of major accident processes

Sloppy management

A disturbing similarity in causes of major accidents was found during retrospective research at high-tech-high-hazard companies in various sectors. These accidents were extensively analysed, often based on case studies and information from legal proceedings related to these accidents. Conditions and contexts were different for each sector, as were the technical causes of the accidents. The similarities occurred in the organisational aspects of accidents: sloppy management of companies. This observation also applied to the Netherlands, where reports of the Dutch Safety Board repeatedly stressed the lack of management attention to safety. According to this board, employee training was too general, as were safety inspections. Within companies, safety paperwork was well organised but safety in practice had only a limited similarity to the safety paper trail. Measures and solutions imposed were mainly limited to technical changes. Knowledge of the effects and effectiveness of these solutions was virtually non-existent in companies (Kirwan 1998; Kirwan et al. 2002; Hovden 2002; Pasman and Baron 2002; Grote 2012; OVV 2013; Jongen and Swuste 2014; Kampen et al. 2014). Similar results were found for three major incidents in 2002 at the BP petrochemical complex in Grangemouth, in Scotland (HSE 2003b). In short, there was no focus on latent factors of the accident process (Groeneweg 2002). Critical events flooded the organisation. Employees were exposed to almost insoluble problems, and strongly relied on the quality of human repair activities and the design of the operating system (Woods 1990b).

Complexity and socio-technical systems

Complex technology required complex control, and therefore situations occurred that could not be predicted, or could be predicted but were not easily changeable. This was partly the consequence of larger distances between operators, workers or pilots controlling processes and the process itself. Automation limited their task to diagnosis, analysis and control of process disturbances in high-tech-high-hazard sectors (see Chapter 5), where processes were no longer directly visible or not even understood (Olsen and Rasmussen 1989; Reason 1990; Rasmussen 1991; Hollnagel et al. 1994; Nachreiner et al. 2006; Le Coze 2011). It could partly be the consequence of management's fixation on occupational accidents, assuming that a trend of these accidents was indicative of process safety, thereby implicitly or explicitly referring to Heinrich's safety pyramid or iceberg (see Chapter 2). Another reason was automation itself, which could be ill designed or fail to take into account of the tasks of drivers, process operators or pilots. These front men might understand automated functions insufficiently, recognising disturbances too late or not at all, leading to 'automation-driven scenarios' (Shepherd 1989; Amalberti 1993, 2001; Kirwan 2001). Automation did not decrease the incidence of major accidents but changed their nature. An example was the Turkish Airlines crash at Schiphol, the Netherlands, in 2009, caused by a conflict between the automated systems of the aircraft and pilots (OVV 2013). Another consideration was the operational reliability of processes. Companies investing in this topic, in the belief that a smooth and flawless production would also be a safe production, could not foresee major accident scenarios arising if no attention was paid to specific elements of the production (Rasmussen 1993a).

Complexity was also caused by the different timescales of departments within a company, which were essential for the process or production. For example, workers, operators, drivers and pilots had a time horizon of a few minutes in control rooms and cockpits. All operational problems and process disturbances at this level should be solved within a short period of time, adjusting process parameters and detecting failing process components. Decisions taken on these issues could only lie with the operational staff, and they should have all responsibility at this level. Another level was the (preventive) maintenance of plant and installations with timescales of one or a few months. A marketing department had a much longer time horizon; it determined annually what and how much should be produced. These differences in timescales complicated communication between departments on operational issues (Brehmer 1991; Manna 2005).

'There's not a damned thing on this site that cannot hurt you', was a statement by a Texas driller on an oil platform (Reason 1997). The company had a big sign at its entrance with an incredibly high number of accident-free days. The message suggested the company was safe, while the driller's statement suggested otherwise. One reason could be the generally little attention paid to complex man–machine interactions in technical organisations within the high-tech-high-hazard sectors (Kjellén 2007). The dangers of equipment and systems were detectable or limited with the use of various safety techniques. In contrast, these techniques missed organisational factors, such as weak skills, the age of equipment, increased changes in personnel and the growing complexity of processes (Knegtering and Pasman 2009).

In line with general management approaches, organisations are open systems, processing information. Increasing doubts were expressed in this period about the formal rationality of organisations. Problems are never fully explored, and actions are rarely preceded by an extensive problem analysis or a list of possible actions. Rationality is a facade when talking about goals, planning, analysis and intentions, and reality is a metaphor – a way in which people in an organisation try to understand the flow of information and experience (Weick and Sutcliffe

2001). In the literature, the hitherto poorly understood interactions between technical, social and organisational aspects of production and transport systems are referred to by the term 'socio-technical systems' (Wilpert 2002). In safety research the term 'socio-technical system' was introduced. Originally this term had come from the famous Tavistock Institute in London. Studies in the 1950s on the effects of mechanisation in British coal mines (Trist and Bamforth 1951) had shown that a newly introduced technology, mechanised conveyors, had actually reduced productivity rather than, as had been expected, increasing it. The reason for this difference was a change in work organisation, caused by the introduction of the new technology. Socio-technical systems and associated socio-techniques were introduced into the Netherlands by the sociologist De Sitter, who at the time was a professor at Eindhoven University of Technology. These studies of socio-technical systems had grown in the 1980s from an academic activity to a popular social movement, known as the 'Quality of Work' (Sitter 1975). This movement was spearheaded by significant employee participation and increasing human redundancy and fault recovery capability in complex systems, which also reduced labour needs (Trist 1981; Clarke 2005).

Gas clouds

As well as poorly understood organisational aspects of major accidents, technical aspects could also be badly understood, for instance the disastrous effects of gas cloud explosions as at Flixborough (see Chapter 6). Investigations were launched into the spreading and explosion of gas clouds, the formation and dispersion of toxic and/or flammable gas/vapour clouds and explosion of the latter. This research resulted in nationally and internationally widely applied methods, such as the 'coloured books' of the Dutch CPR and the concept of inherently safe design. These unwanted process-specific hazards causing major accidents did not only occur in companies with a low focus on safety. As Appendix 1 shows, companies with high safety awareness could also meet these conditions (Hovden 2002).

Human failure and human factors

Human failure, which played a prominent role in accident analyses in earlier periods (see Chapters 1–3), gradually moved into the background during this period. Human failures as the cause of major accidents were not explicitly mentioned in metaphors or theories. Behaviour was given a different interpretation in the high-tech-high-hazard industries, with their predominantly automated processes. Behaviour in this dynamic socio-technical environment was not understood from the perspective of a task sequence or the correct or rational execution of work. In terms of human factors, the behaviour of workers and managers was seen as a result of factors in a particular context (Rasmussen 1990, 1993b, 1997; Wilpert 2009). This context was not always favourable for operator performance. Automation of processes moved operators to control rooms, and they received a lot of processed information at a high level of abstraction. Controlling potentially hazardous installations seemed similar to playing a high-level computer game (Mostia 2010). Furthermore, operators did what they always do without having a complete overview: ignoring small deviations. Often, they had to react to situations not covered by procedures or which were in any case unclear. Automation is reliable but often difficult to comprehend, and requires a basic understanding by operators. The time to learn how to operate a complex system was always much shorter than the time needed to understand how the system worked under various conditions (Rasmussen and Vincente 1989; Brehmer 1993; Hollnagel et al. 1994). What used to be called 'failures' and 'errors' could also be attempts by operators to understand and intervene in a complex system.

Safety culture and inspections

In the 1980s, organisational culture became a topic in contemporary reports of major accidents (see Chapter 7). In reports of major incidents and accidents in the 2000s, a poor safety culture of the company involved was explicitly mentioned as one of the contributing factors to major accidents. Culture was seen as a *convenient truth*, which was an umbrella term, covering many aspects but without blaming any one individual. But it was unclear whether it provided a better understanding of accident processes. In the literature, several images emerged of safety culture (Guldenmund, cited in Jongen and Swuste 2014):

1. *Organisational culture as the characterization of a tribe, an anthropological (interpretive) approach. The only objective is to understand a culture without judgement;*
2. *Safety culture as the steps on a staircase, a moralistic perspective. The steps describe levels of maturity of an organisation. The well-known safety culture ladder is a prime example. The aim here is to improve culture, in order to attain the highest, generative level;*
3. *Safety culture as a number, an analytical perspective. This is actually not safety* culture *but safety* climate. *The aim is to provide a description, and possibly a diagnosis, based on a few overall means;*
4. *Safety culture as a process of making sense of ambiguities and arriving at consensus. This view is consistent with the first description and sees culture as a product of an organisation or group of people to successfully achieve a goal together.*

Safety culture in the domain of occupational safety was discussed in the previous chapter. During this period, many publications appeared on safety culture in high-tech-high-hazard companies. Despite all the research, no conclusive evidence was produced relating safety culture to a level of safety in a company (Groeneweg et al. 2010). All indications came from retrospective studies, where hindsight bias made results less reliable. Few concepts were as desired and as poorly understood as organisational/safety culture (Reason 1997; Hale 2003).

An important contribution to the cultural debate was made by Schein (Schein 1992). In one of his other contributions, Schein divided organisational culture into subcultures (Schein 1996). A distinction was made between the culture of operators, of engineers and of management. Schein defined culture as an attempt at sense-making. Culture is a set of basic, tacit assumptions about how the world works or should work, shared by a group of people. Their perceptions, thoughts, feelings and, to some extent, their overt behaviour are partly determined by culture. According to Schein, in the operator culture, the activities of humans were the basis. Knowledge and skills were organized locally. Given the design process, operators were required to deal with surprises and process disturbances. This implied a great level of interdependency and trust. The world was partly unpredictable, and innovative skills were necessary to cope with unforeseen circumstances. Rules and procedures were therefore obstacles (Dien 1998).

The technician culture was the culture of the engineers, who provided the basic design elements of the technology of the primary process of the organisation. This group was attracted to technology because it was concrete and impersonal. Engineers nevertheless recognised the human factor and were oriented towards safety. They took safety seriously during the design process, by automating many functions. They preferred a simple cause-effect relationship and ideally thought in quantitative terms.

Finally, the management culture was permeated with the need to maintain the financial health of a company and was mainly concerned with controlling, investing and dealing with competitive markets. The economic environment was one of constant competition and potentially hostile. Managers received no reliable data from their subordinates, so they needed to trust their own judgement. Intrinsically, the organisation and management were hierarchical. The organisation had to form a team, but responsibility lay with individuals. These differences in subcultures greatly complicated communication between the three groups; the differences also hindered the potential for learning from incidents and accidents.

The inability of companies and organisations to learn from incidents and accidents was reflected in almost all reports of major accidents. With the exception of high reliability organisations, companies and organisations were generally headstrong. They were not built to learn, primarily because they were action generators. Companies were more likely to dismiss and punish people in a blame culture, where questions of responsibility for process disturbances, incidents and accidents were paramount. This would impede careful reporting and analysis. Learning and blaming did not mix (Weick 1991; Pidgeon and O'Leary 2000; Amalberti 2002; CSB 2007; Mostia 2010; Grote 2012). According to Schein, a first step would be the awareness of the organisation that the three cultures mentioned above spoke different 'languages'. The second step was the assumption that the various presuppositions arising from the subcultures were equivalent. Only then could the process of learning be initiated, following from the analysis of incidents and accidents, and translated into future adjustments and improvements and the evaluation of the effect of these interventions. This translation was complicated, but in the literature the first steps had already been made (Drupsteen et al. 2013; Drupsteen 2014).

Did external inspections or governance have any effect, given the fact that the quality of the risk and safety management in companies is generally disappointing? Supervision and inspection by the authorities were a counterforce in a production system. This was about controlling, influencing, guiding, and was thus easily at odds with the independence and autonomy of a company or unit (Mertens 2011). Government inspections were based on legislation, and legislation could be a problem rather than a solution. A degree of over-regulation existed in the safety domain, caused by overzealous politicians who introduced new legislation after each major accident (Pidgeon and O'Leary 2000; Kirwan et al. 2002; WRR 2008). It could be argued that, in general, politicians were not interested in the effectiveness of their laws, because laws, inspections, monitoring and certification all escaped independent review of their quality and effectiveness (Amalberti 2002; Hale et al. 2002; Gundlach 2002; Mertens 2011). In addition, regulation through inspection and monitoring was a complex issue, given the great dynamism of technological changes in many high-tech-high-hazard industries. Governments could hardly keep up (Larsson 2002; Kirwan et al. 2002). The effect of supervision and inspection might be limited, but although companies were primarily responsible for safety as a core task, government had to guarantee safety for its citizens, according to some prominent authors in the Dutch safety domain (Vollenhoven 2008; Dik et al. 2008).

Design

The loss prevention initiative in the early 1960s was one of the first initiatives focusing on major accidents and prevention in the process industries. It was introduced by the British and American Institutes of Chemical Engineers (IChemE and AIChE) and the Dow Chemical Company, an American firm. At the beginning of the 1970s, loss prevention became a movement in the chemical process sector. It was a systematic, multidisciplinary and integrated application of physical sciences, applied mathematics, technology and management-relevant disciplines. Hazards and risks could be prevented by technical solutions and design options, leading to a so-called acceptable level. The movement has three guiding axioms, primarily directed towards top managers of chemical companies (Pasman et al. 1992):

1. A company that forgets its past repeats it;
2. Success in preventing emissions and loss of containment means anticipating the future;
3. A company is not in control if a loss has to occur before it is measured.

Inherent safe design was an important new concept in loss prevention (see Chapter 6). Inherent safe design would be useful for reducing the effects of specific process hazards, such as increases in pressure, temperature or volume, contamination from catalytic effects, adverse reactions that could cause process failures and emissions of toxic, flammable or explosive substances (HSE 1991; Pasman et al. 1992; Cates 1992; Lees 1996; Khan and Abbasi 1999; Dechy et al. 2004; Mannan 2005; Delvosalle et al. 2006; Pasman 2009; Kidam et al. 2010).

The design of a production or transport system could play a major role in any scenario. Small and financially weak companies often have mixed technologies. The production is partly manually controlled and partly remote-controlled and/or automated; having different control systems on different types of production frequently involves an endless period of redesign, and is a starting point for various accident scenarios (Poyet and Leplat 1993). In bigger companies, design is also one of the main causes of major accidents. Design failures occurred when proposed design specifications were not, or could not, be realized. Research revealed the proposed contribution of design factors to major accidents to be 50–60%, which also includes the quality of selected materials or poor design specifications (Hale, Kirwan et al. 2007; Kinnersley and Roelen 2007; Taylor 2007; Kidam et al. 2010). In aviation, during the 'bad years' (1920–1994), aircraft design was not challenged in any way (Amalberti 2002). As such, the high contribution of design to system failure in this period was not surprising.

In general, designers were often overly optimistic about shop floor reality, and intricate and complex designs disregarded human shortcomings of workers, drivers, pilots and operators (Roberts and Rousseau 1989). Furthermore, process disturbances were often not foreseen by designers, and design could not eliminate all hazards in high-tech-high-hazard sectors. It was usually not possible to design a process that was both economically sound and safe and that also met the highest standards of production and quality (Kjellén 2002). Safety was not a starting point for designers, and complying with safety demands and regulations generally occurred at the end of a design process (Fadier and Garza 2006; Kjellén 2007).

An important design feature was the concept of barriers, and defence in depth, by applying multiple barriers to stop a hazard from creating potential losses. As previously mentioned by Rasmussen, however, in reality, these layers of protection had their own weaknesses, depicted as holes in Figure 8.3 (Reason 1997) (see also Chapter 6).

If failures of a layer were not noticed by employees or operators, there was a problem: an unforeseen scenario developed after a process failure, becoming a major accident waiting to happen, or a 'wildness in waiting'.

The barrier layers were effective only if they were regularly tested for availability and reliability and replaced or repaired when one or more layers were 'out of service'. Apart from LOPA, a large number of indices were proposed in process safety, such as the PIIS (Prototype Index for Inherent Safety), ISI (Inherent Safety Index), HCI (Hazardous Chemical Index), HRI (Hazardous Reaction Index), TCI (Total Chemical Index), WCI (Worst Chemical Index) and WRI (Worst Reaction Index). However, the approaches based on these indexes had several shortcomings. The categorisation used was quite subjective, just like their weighting, which led to quantitative results that were difficult to interpret. In addition, coverage

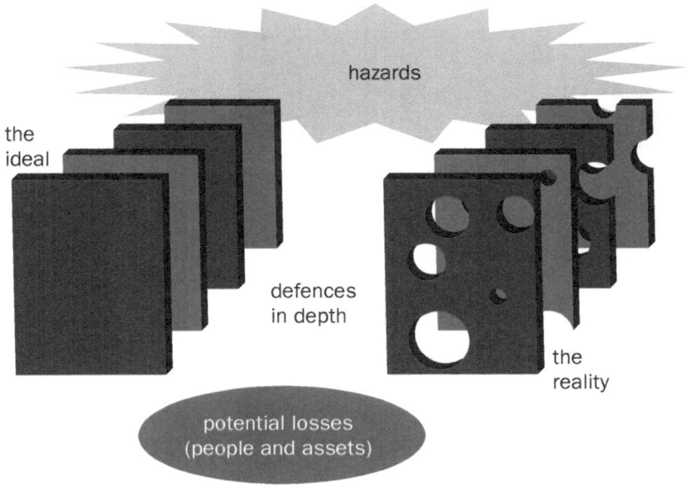

Figure 8.3 Ideal and actual state of defences in depth

Source: After Reason (1997)

of phenomena quantified with the index was generally limited (Sugiyama et al. 2008; Srinivasan and Natarajan 2012).

Metaphors

Swiss cheese

The 'hindsight bias' – a known tendency in retrospective studies – is a serious problem when detecting unexpected and hidden failures. In retrospect, these factors are always clear, but when a major accident takes place, they are far from clear for those involved, for whom the major accident is a mystery and chaos reigns. Even in the best-run organisations, decisions are deemed wrong retrospectively (Reason 1990).

During the reconstruction of major accident processes, several types of errors were discovered: big failures, small failures and latent failures. The big failures were obvious, but often a major accident had one or more small direct causes (Pasman et al. 1992; Weick and Sutcliffe 2001). Examples include the loose wires at a signpost at Clapham Junction, a leaking pump at Piper Alpha (both discussed in Chapter 6) and pollution that started as an exothermic reaction, like the ammonium nitrate explosion at the AZF chemical factory in Toulouse in 2001 (Dechy et al. 2004). Latent failures were factors that could create failure-generating conditions (Groeneweg 1992; Wagenaar 1998; Reason et al. 2006) (Figure 8.4). The origin of latent failures lay in the company's organisation and in its decision-making processes.

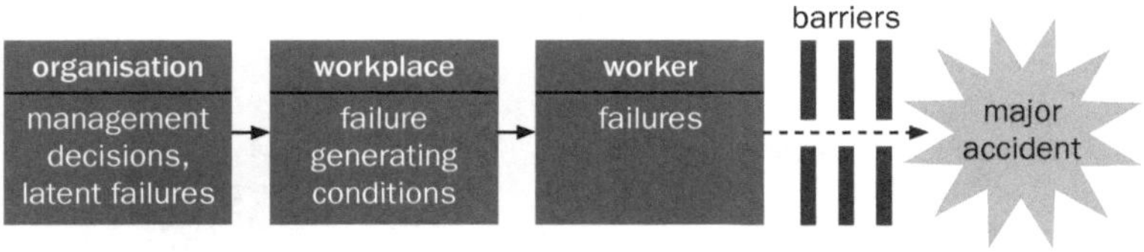

Figure 8.4 Route of latent failures

Source: Groeneweg (1992)

Decision-making within an organisation was determined by the context and limitations of the decision-makers, who tended to recycle known solutions for technical problems (Halpern 1989; Wagenaar et al. 1994; Rosness 2009). These latent failures were present in a system for a long period of time without causing problems, but were activated in combination with other system failures, breaking through barriers. The psychologist Reason described these latent failures using a medical metaphor: resident pathogens caused by designers, procedure writers and top managers represent the 'blunt end' of an organisation (Qureshi 2007; Reason et al. 2006) (Figure 8.5).

This was an important contribution by Reason, depicting human failures from a Freudian perspective as unconscious mistakes that were inherent to the brain's functioning and which were therefore unavoidable. The more complex a system, the more pathogens were present (Woods 1990a; Reason 1993).

Reason started his research with an analysis of reports on the London Underground station King's Cross St Pancras fire (1987) and the roll-on roll-off car ferry major accident of the *Herald of Free Enterprise* (1987) at Zeebrugge, Belgium. These major accidents were the result of a poorly understood interaction between technical and social aspects of the organisation and were called 'organisational accidents'. Human and technical failures alone were not sufficient to trigger the scenarios that occurred, but the combination of several latent factors had damaged the integrity of the barriers in place.

The cheese metaphor found application in many aviation accident investigations. A review, commissioned by Eurocontrol (Reason et al. 2006), summarised comments on this metaphor. This concerned the relationship between the causal factors and the position of barriers. The holes were insufficiently defined, the barriers were interdependent and the metaphor put too much emphasis on latent factors within the organisation. According to the review, manifest failures of workers or pilots could also be dominant factors in accident processes. The metaphor depicted the complexity of the accident process clearly, but it did not explicitly focus on the hazards concerned. Engineers commonly started with hazards. Authors outside the technical domain often disregarded the engineering aspects of major accidents and work processes (Rasmussen 1997). This was also evident in the barrier concept. Engineers defined barriers as physical entities stopping or slowing down scenarios. In the Swiss cheese metaphor, the barrier concept was expanded to include intangible barriers, such as procedures, human actions, work permits and other administrative routines.

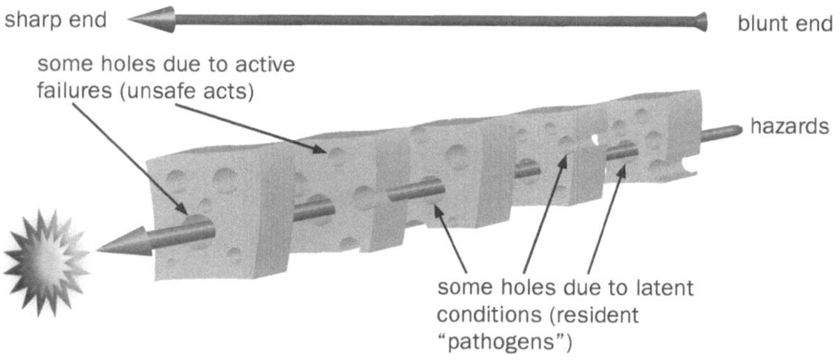

Figure 8.5 The Swiss cheese metaphor

Sources: Reason et al. (2006); Quereshi (2007)

Drift to danger

A major accident analysis looked for causal factors, that is, for dominant scenarios in accident processes. Usually these analyses were readable and revealed conditions for a major accident. The implicit assumptions of these analyses and the stop rules of the investigation, however, were less clear. Analyses stopped when an accepted cause was found, or when further information was no longer available. Another stop line was an abnormal but well-known condition explaining the accident. The final stopping rule was the solution. The dependence of stopping rules on information availability, awareness and solutions made results of analyses highly dependent on the bias of the investigator (Burggraaf and Groeneweg 2016). Most likely, a designer or an operator would come to different conclusions (Rasmussen 1991).

In 1997, the engineer Rasmussen published the 'drift into danger' metaphor, a metaphor that was quite different from the other safety metaphors. His approach to the accident process was holistic (Figure 8.6).

Like the Swiss cheese and bowtie metaphors, this metaphor discounted technical and human failures as main causes for major accidents. Rasmussen was suspicious about the purely psychological approach of the human reliability assessment analysts (see Chapter 6), where operators and workers were seen as output devices. In his opinion, quantifying human failures was not meaningful (Le Coze 2015). The working process, with its administrative, operational and safety constraints, limited workers' or operators' behaviours. But behaviour was not fixed: there were degrees of freedom, making variations of actions possible. These variations had a certain similarity with the Brownian motion of gas molecules, labelled in the centre of the figure. Market forces were another quite aggressive factor in many industrial sectors. The loss of knowledge and of process control increased the risk of major accidents. With staff reductions, operational knowledge disappeared and activities such as maintenance and repairs were outsourced to contractors and subcontractors (Baram 1998; Woo and Vincente 2003; Knegtering and Pasman 2009; Le Coze 2010). As a consequence,

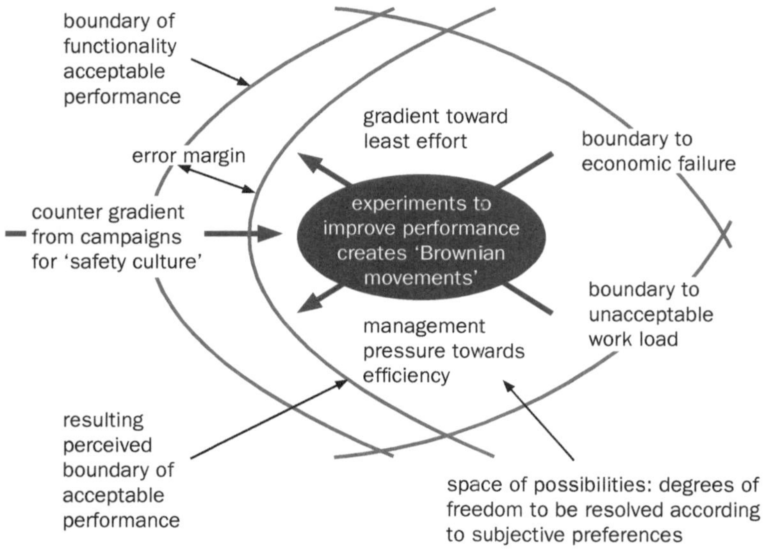

Figure 8.6 Drift into danger

Source: After Rasmussen (1997)

economic/financial risks for the company were spread out. However, companies could also lose control of these activities when many different contractor groups were active at the same time. This was the case during the major accident in Toulouse in 2001, and also played a role in the train accident at Clapham Junction (Dechy et al. 2004). Under the pressure of market forces, management emphasised the need for cost-efficient production and consequently increased the staff workload. This generated a negative efficiency gradient. The result was a systematic migration to the boundaries of acceptable functional performance of the work, which led to major accidents if limits were exceeded. Therefore, the metaphor was named 'drift into danger'. Major accidents in 1997 were not caused by human failures but by a systematic migration of organisational behaviour towards the boundaries of acceptable performance under the influence of cost efficiency and an aggressive market approach (Rasmussen 1997). The focus on costs led to uncertainty about the boundaries of safety (Svedung and Rasmussen 1998; Woo and Vicente 2003). First of all, these boundaries needed to be explicit, not only from a design perspective but also from a man–machine interactions perspective. Solutions included an increase of the margins of normal operation or the development of a counter-gradient, and training operators in adequate behaviour at these boundaries.

Rasmussen's second contribution was the presentation of the socio-technical system in which the risk management of a company had to operate (Figure 8.7).

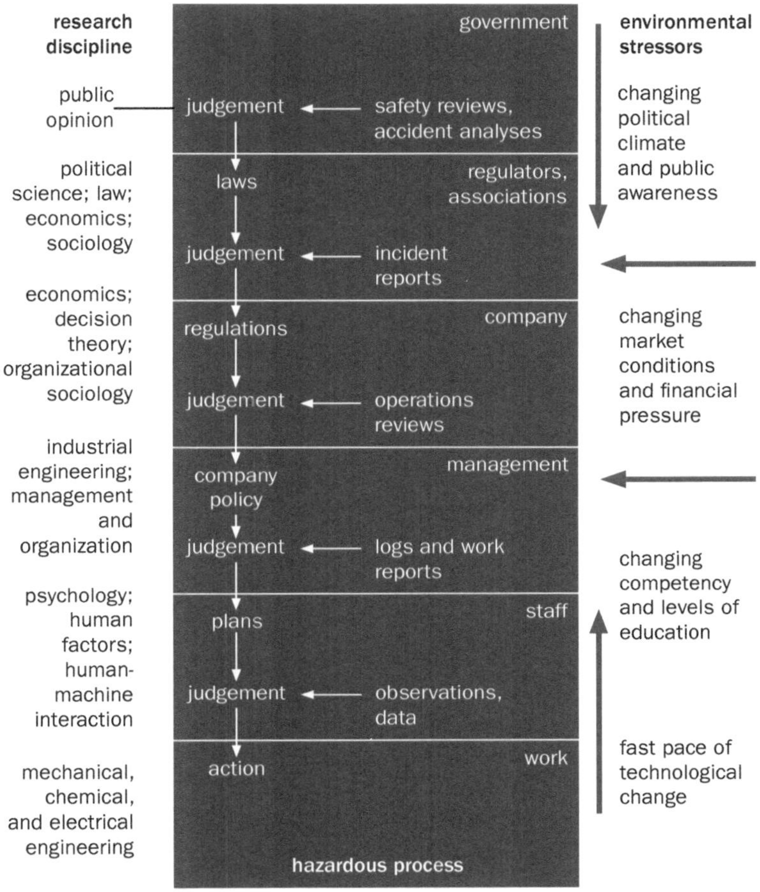

Figure 8.7 The socio-technical system for risk management

Source: After Rasmussen (1997)

'Safety science, it's not rocket science, it's much harder.' This was the title of Hudson's public lecture (Hudson 2010b), and Figure 8.8 shows why safety in high-tech-high-hazard industries is such a complex topic. The centre of the figure is an actor analysis of the parties involved in the safety of processes and production. Laws, regulations, company policies and safety plans, which ultimately resulted in safety procedures and analysis, were at all levels part of negotiations. These controls, along with legal, economic and production arguments, influenced safety. The left side of Figure 8.7 shows the different disciplines and scientific domains dealing with these different aspects. These different disciplines communicated with each other on only a limited scale and were not engaged in joint research. Researchers published in their own domain-specific scientific journals. This was the result of the above-mentioned differences in research focus – on the one hand, between sociological and organisational management research, and on the other hand, between technical and engineering research. Rasmussen and Hale and co-authors petitioned for cross- or interdisciplinary studies to model safety management according to functional abstractions rather than structural analysis (Rasmussen 1994; Hale et al. 1997). These studies had to focus on behaviour-shaping mechanisms of work system restrictions and boundaries of acceptable performance of processes, not on human failures. On the right-hand side of Figure 8.7 are factors determining the dynamics of this socio-technical system. Rapid technological changes and changing market conditions are factors that directly influence companies' decisions. Management structures adapt more slowly to changing conditions, something even truer for legislation. As an effect of market developments, banks and investors could control companies and organisations without having any affinity with the sector involved (Rasmussen 1997; Le Coze 2015).

Models and theories

Disaster incubation theory

Sometimes knowledge faded away, and sometimes it took years or decades for ideas to be accepted. One example is the task dynamics theory of Winsemius (Chapter 4). This theory was developed in the early 1950s but not referred to in publications until the 1970s. Much the same applied to the book *Man-made Disasters* by Turner, dating from 1978. The preface to the second edition (Turner and Pidgeon 1997) showed that – apart from a select group of academics – the publication went unnoticed at the time of its appearance.

> *When Barry Turner published Man-made Disasters in 1978, he hit dry ground in Europe. In those years, there was no disaster sociology, let alone a body of knowledge encompassing the more comprehensive world of crises and discontinuities. Man-made Disasters was received as a curiosity. There was no doubt whatsoever about the intriguing dimensions of the topic at large, and there was wide acclaim for the thorough examination of such well-chosen cases as the mining disaster in the Welsh village of Aberfan in October 1966. Nobody questioned the quality of the research, the more so since the author's main argument had already found its way into the prestigious American-based Administrative Science Quarterly. However, Man-made Disasters was published at a time when European social science was only beginning to recover from the turbulence of the 1960s and the early 1970s, a period marked by an emphasis on conflict, turmoil and corresponding paradigmatic instability. Turner's work therefore seemed to be at odds with the strenuous efforts of mainstream social science to study safe and sound subjects, quantifiable trends and continuities.*

Perhaps not surprisingly, a reissue of *Man-made Disasters* by the sociologist Turner and the psychologist Pidgeon appeared during the same period as the metaphors mentioned above. In the literature, this theory was known as the 'disaster incubation theory', since the concept of the incubation period of a major accident was a central concept of the book. This theory was based on a meta-analysis of reports from the British government on major accidents in various industrial sectors, including transport and recreation, and the medical and public sectors.

Various process disturbances made a production system vulnerable prior to a major accident. Initially, hidden failures and poorly understood events continued to occur and did not correspond to the existing beliefs about hazards and what the organisation regarded as normal. These disturbances and failures were not random but rather the result of the system of which they were part, and revealed themselves as poorly structured surprises (Turner 1989; Vuuren 1998). According to the authors, major accidents were a by-product of management and technical systems functioning normally. Some major accidents were the result of poor management or unprofessional behaviour, but most major accidents (70–80%) had a social, administrative or managerial origin; the remaining causes were technical. The collective failure of the knowledge of the organisation and the misconceptions of risks were caused by a lack of information and underestimation of risks. These were the ingredients of the incubation period of a major accident. The accident was developing, and management had lost contact with the operational reality. The earlier-mentioned term 'sloppy management' was introduced in the literature (Turner 1994; Turner and Pidgeon 1997).

Resilience engineering workshop

Another approach to safety came from the first workshop on resilience engineering, which took place in Sweden in 2004 (Hollnagel et al. 2006). While traditional safety science research was focused on the hazards and failures of people, installations and systems, attention was now aimed at restoration options: what caused operations most of the time to have good rather than bad endings. It was assumed that the mechanisms of failure were the same as those for success. These mechanisms had their origins in the variability of performances of tasks, processes, and so on. People never acted consistently, and variation could be a source of both positive and adverse outcomes. Later, these two opposite approaches were dubbed Safety I and Safety II. Safety I was the classical approach: the focus was on failures, process disturbances or malfunctioning process components. These included the search for deviations from procedures, and weak barriers, called 'work-as-imagined'. Safety II was a state where a system managed to recover under varying conditions and functioned error-free for the vast majority of the time. In this case, 'work-as-done' was the starting point. But in fact, there was quite a lot of similarity between this resilience engineering concept and the American high reliability theory. Neither considered errors or failures but rather focused on mechanisms maintaining a stable working system. These similarities were explicitly addressed by a few authors (Hale and Heijer 2006; Hopkins 2014). The difference between these approaches was the origin of their concepts. Reliability came from the domain of social sciences, and resilience engineering from the engineering domain.

Risk perception

There were many ways to measure consequences or losses as a result of an event (Aven 2012). These included mortality risk, number of deaths or the number of injuries due to accidents, or morbidity of occupational diseases, damage to property, shortened life expectancy or lost income due to reduced life expectancy or claims from insurance companies. Usually calculations were limited to mortality and the number of deceased people, sometimes supplemented by morbidity and the number of injured people. Examples are given in Appendix 1. Sometimes the costs of the damage were indicated. Also, mortality rates were displayed and expressed in different ways: the number of deaths in comparison to the general (national or regional) population; or to a defined-risk population; or per ton emitted (the amount of) produced hazardous substances; or millions of euros' or dollars' worth of produced products. These different ways to express risk created confusion amongst non-experts.

There was general criticism of the tendency to assume the superiority of scientific knowledge, which, among others, was initiated by *Science & Technology Studies* (Allen et al. 1992; Hermansson 2005; Lidskog and Sundqvist 2012). This was partly based on presented-risk comparisons, placing the lower risks of industrial activities next to the higher risks of everyday life. Citizens were not very convinced by these comparisons (see Chapter 6).

Risk and safety management

Both risk and safety management dealt with managing and controlling accident processes, which otherwise would lead to major accidents (Harms-Ringdahl 2004; Grote 2012; Li and Guldenmund 2018). In that respect, the two concepts were not that different. The literature gave a few definitions. Some are listed below:

> *Risk management is simply good common sense in coping with possible and actual daily mishaps and occasional major disasters that lead to financial losses and unfulfilled plans for individuals and organisations – indeed for our society as a whole.*
>
> (Kloman 1992; Smallman 1996)

> *Risk management is a systematic, statistically based and holistic process that builds a formal risk assessment and management, and addresses the set of four sources of failure within a hierarchical multi-objective framework: 1) hardware failure, 2) software failure, 3) organisational failure, 4) human failure.*
>
> (Haimes 1991; Smallman 1996)

Safety management may be defined as the aspect of the overall management function that determines and implements the safety policy. This will involve a whole range of activities, initiatives, programmes, etc., focused on technical, human and organisational aspects and referring to all the individual activities within the organisation, which tend to be formalized as safety management systems (SMS).

(Papadakis and Amendola 1997; Harms-Ringdahl 2004)

Safety management is an arrangement made by the organisation for the management of safety in order to promote a strong safety culture and achieve good safety performance.

(INSAG 1999; Grote 2012)

Risk and safety management in these definitions were a means to reach an endpoint of coping with mishaps and major accidents. A strong safety culture and good safety performance comprised a description of components of risk and safety management, including the four sources from the second definition and components of a safety policy in the second to last definition. A difference in these definitions was the way companies manage uncertainties. Uncertainties could be minimised, as in the classical safety approach, or organisations could learn to deal with these uncertainties. Managing the safety of primary processes was central in high-tech-high-hazard companies. In occupational safety, not primary processes but rather tasks of operators, employees, pilots and engineers were the focus, which were not necessarily related to a primary process but often were (Grote 2012). Current models of risk and safety management are based on a rational image of an organisation and not on a contextual picture of a complex socio-technical system. That required an understanding of the primary processes and the main task of the organisation (Reiman and Oedewald 2007). Safety management had to control hazards and risks arising from the technology. It was not realistic to expect the same system to be able to simultaneously manage product quality, process safety, the health of workers, environmental aspects and production.

The European Commission's Joint Research Centre had started two projects to elaborate these structures for the process industry: ARAMIS (Accident Risk Assessment Methodology for Industries) and its follow-up I-Risk, and the development of an integrated technical and management risk methodology for chemical plants (Kirchsteiger 2002; Papazoglou et al. 2003; Delvosalle et al. 2006; Dianous and de Fiévez 2006; Guldenmund et al. 2006; Bellamy et al. 2008b). Conformance among member states was a major goal of these projects. Risk management, risk assessment and the use of assessment results in EU countries varied greatly. Conformance would improve the transparency of decisions and facilitate risk communication. The projects used the bowtie metaphor and described generic scenarios and associated central events for process installations. Frequencies of central events were determined, based on the failure frequencies of parts of installations, on barriers and on the characterization of the material flows. To that point, the projects had shown a strong resemblance to a QRA approach. The added value of the projects was the establishment of the risk levels per installation, which were influenced by targeted activities from risk/safety management, such as ensuring barrier quality, and the overall effectiveness of safety management systems (Bellamy and Brouwer 1999; Hale et al. 2005). This impact of management activities was lacking in the QRA approach. From the literature it was not clear whether these projects had any practical applications.

Another issue was the role and position of the safety department of a company. Their task is to coordinate risk and safety management and to measure and evaluate its effectiveness. However, in many companies the safety department is positioned at the edge of the organisation. Research into the petrochemical and steel industries showed the fragmentation of these safety organisations, the poor communication about major accident processes between various departments and the presence of large groups of subcontractors active in the company. The import registration systems for data collected were restricted by the classification system in use, which often allowed storage of only technical and human information, and no information related to organisational shortcomings. Therefore, the analysis of accidents was superficial, and organisations had a perceptible resistance to unthinkable process conditions (Woods 1990b; Brehmer 1993; Delvosalle et al. 2006; Bell and Healey 2006; Guillaume 2011). This situation strongly resembled the construction industry, where management took no or only a limited responsibility for safety, bringing to mind the saying 'crap flows downhill; uphill it smells of roses' (Mostia 2010; Swuste et al. 2012b). But overall, what was missing thus far in all approaches was the socio-political dimension of organisations (see also Chapter 7). Power influences all layers of organisations, affecting the quality of decision making (Pidgeon and O'Leary 2000; Perrow 1994; Antonsen 2009). The only metaphor coming close is 'drift into danger', where aggressive market conditions are part of the concept.

North America and Australia

Determinants of major accident processes

In North America and Australia there was ample attention to determinants of major accident processes as well. 'Sloppy management', a term introduced by Barry Turner, was also used by Andrew Hopkins in his analysis of major accidents. His contribution is discussed in a later section on the contribution of Diana Vaughan. She used another term, 'normalization of abnormalities', in her analysis of the 1986 *Challenger* accident, pointing at the management practices of NASA at that time. There was also ample attention to safety culture and organisational culture. For instance, in reports of major incidents and accidents in the 2000s, including the Davis-Besse nuclear power plant, in Oak Harbour, Ohio (2002) and the BP refinery in Texas City, Texas (2005), poor safety culture of the companies involved was seen as one of the contributing factors to major accidents. Two other major contributions were the ethnographic approach towards organisational culture by Edgar Schein (Schein 1992), mentioned in the previous section, and the possibilities and limitations of inherent safe design (CCPS 1992; Khan and Abassi 1999; Joseph 2003).

Design, Layers of Protection Analysis (LOPA)

Loss prevention had addressed design by integrating inherently safe(r) design (ISD) with a 'Layers of Protection Analysis' (LOPA) (Figure 8.8). The LOPA concept stemmed from the military and was first introduced into the nuclear industry during the late 1950s.

During the late 1980s, it entered the process industry. In practice, a chemical or nuclear process was usually protected by multiple layers of independent safety barriers to control the mechanical integrity of the system(s) and to restrict or prevent the consequences of an emission or loss of containment (Pasman 2000; Summers 2003; Dowell 2011). Failures in one layer of barriers were taken care of by a following layer. Ideally, all safety barriers remained intact, preventing major accident scenarios from developing. LOPA ranged further than the plant itself, extending to community response, to media response and to the entire life cycle of the plant (CCPS 2001; US-NRC 2016). ISD and LOPA were originally conceived

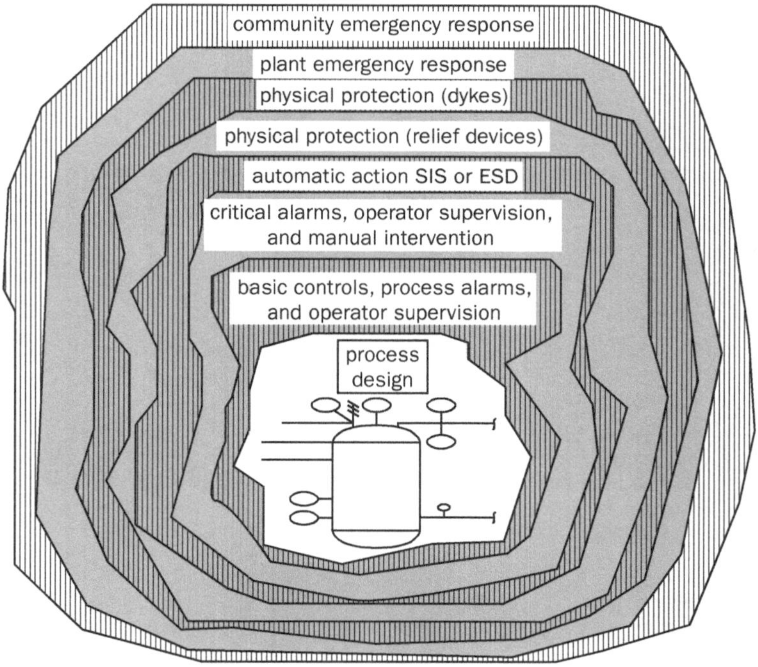

Figure 8.8 Layers of protection (LOPA)

Source: Crowl and Louvar (2002)

as essentially qualitative design philosophies. From an engineering point of view, a logical further development of LOPA was to quantify the reliability of successive protective layers. The goal was to make that process sufficiently safe.

Design, process intensification

It was not surprising that from approximately 1990 to 2000 the relative importance of loss prevention decreased. This was the consequence of changing and more challenging economic factors, such as increased international competition due to globalisation, downsizing of personnel, and general cost-cutting in all aspects of production and management. This decrease occurred while a powerful new development of process intensification (PI) was, and still is, occurring within chemical engineering and management. PI is a chemical process design approach leading to substantially smaller, cleaner, safer and more energy-efficient process technology (Figure 8.9). It encompasses safety and environment and, in this respect, actively uses the principles of inherently safer design by stressing, in particular, smaller plants. Safety is not the prime target of PI – reduction of investment costs is. Process intensification relates directly to profit. For all these reasons, process intensification is presently a 'hot topic' within chemical engineering, like loss prevention before (Etchells 2005; Becht et al. 2009; Stankiewicz and Moulijn 2000, 2004;

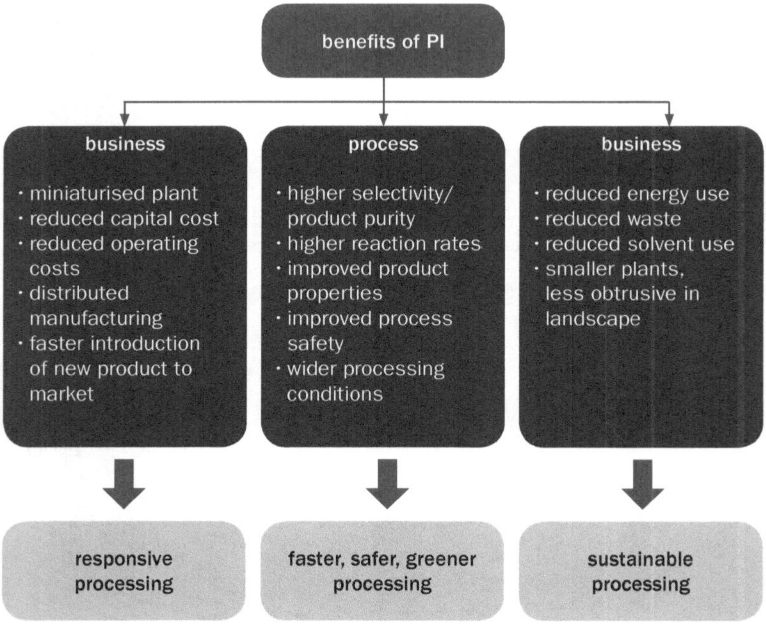

Figure 8.9 Benefits of process intensification

Source: After Boodhoo and Harvey (2012)

Grossmann and Westerberg 2000; Gerven and Stankiewicz 2009; Boodhoo and Harvey 2012; Gómez-Castro and Segovia-Hernández 2019).

Theories

Normal accident theory

The sociologist Perrow is the father of normal accident theory (see Chapter 6) (Perrow 1984). Just before the turn of the twenty-first century, a reprint of his book was published (Perrow 1999). This theory is a technologically deterministic approach to major accidents that occur when control of the technology escapes the control of the organisation (Le Coze 2015b). The theory was based on a meta-analysis of a large number of accident reports from industrial, military, transport and research sectors. The name of the theory referred to major accidents occurring in 'normal' organisations controlled by 'normal' people, similar to Turner's disaster incubation theory. Because of the complex, interactive processes and the tight coupling of process steps, these accidents became inevitable and were no longer foreseen by designers or understood by operators, workers, drivers, pilots or managers. There were some potential conflicts in decision-making and control of such systems. The complexity of the technology required employees to learn the process by making mistakes. With a tight coupling of the process steps, there was no time to correct errors, which was at odds with the trial and error approach used by operators to understand the technology. Complexity required decentralised control and decision making in order to respond appropriately to unexpected events. A tight coupling between process steps, however, required centralised control and decision making; these two conflicting demands on control would cause problems (Perrow 1994; Rijpma 1997).

The theory offered two possible routes to avoid normal accidents. In both cases, there was redundancy. The degree of coupling could be transformed from tight to loose, as could the complexity of the technology, by making it less complex. In many cases, these were only theoretical options. These options were not possible for every high-tech-high-hazard process, and it could lead to transformations that were prohibitively costly. Normal accident theory was a response to the analysis of the President's Commission of the incident at the Three Mile Island nuclear power plant, where human failure was identified as the main cause (Kemeney 1979). At the time of the process disturbance of the reactor's cooling system, the situation was far too complex for operators to understand, especially as the indicators provided wrong information. During the first half of the 1980s, attention to safety management was still quite rudimentary, as was attention to latent factors and sloppy management decisions later on. The normal accident theory did not comment on the quality of management. Perrow's publication could also be interpreted as a political statement against high-risk technologies (Rijpma 2003). The emerging focus on safety management gradually questioned the sustainability of the normal accident theory and causes of major accidents were sought in the organisation of a company's safety (Orton and Weick 1990; Hopkins 1999b, c, 2001; Shrivastava et al. 2009).

High reliability theory

Strictly speaking both high reliability theory and disaster incubation theory were not accident theories but organisational theories of high-tech-high-hazard organisations. Unlike other metaphors and theories, high reliability theory was based upon extensive observations and fieldwork, known as an 'ethnographic' approach (see Chapter 6). Although research had already started in the second half of the 1970s, publications on studies by the US Federal Aviation Administration (air traffic control), Pacific Gas and Electric, the Diablo Canyon nuclear plant in California and the US Navy's aircraft carriers appeared in the period covered by this chapter. In later years, research was carried out with fire services, health care organisations, submarines, petrochemical companies, aviation, power grids and banks (Rochlin et al. 1987; Roberts 1989, 1990; Halpern 1989; Roberts and Rousseau 1989; Rochlin 1989; Weick 1989a, 1989b, 1991, 1995; La Porte and Consolini 1991; Weick and Roberts 1993; La Porte and Thomas 1995). The research was conducted by an interdisciplinary team: Gene Rochlin is a physicist, Karlene Roberts a psychologist, Karl Weick a social psychologist and Todd La Porte a political scientist. In these organisations, complex, inherently dangerous and technically sophisticated tasks were performed with a very tight man–machine coupling under severe time pressure. The work took place at the edge of design envelopes and of human capabilities. Existing organisational theories could not explain the fact that these organisations operated virtually accident free. The study started by considering Perrow's theory predicting major accidents under these conditions. The organisations had no traditional fixation on risk and minimisation of these risks (Rochlin 1999; Swuste and Jongen 2011). There was redundancy in many respects: organisationally, technically and in terms of decision making. If one team threatened to fail a safety-critical task, another team would take over this task, causing organisational redundancy. In the case of peak operations, more teams or more experts were called in: for example, as extra eyes on the radar in air traffic control during rush hour. Technical redundancy was a backup for materials and equipment. During critical decisions, the organisation would gather a large amount of information, and critical decisions were subject to continuous cross-checking. With ample financial resources, the organisation compensated for human limitations. High reliability organisations (HROs) were operating in unstable environments: the Diablo Canyon nuclear power plant was built on a fault line, the aircraft carriers had to deal with possible enemy attacks, and both air traffic controllers and carriers had to deal with emergencies and changing weather conditions.

The main threat to HROs was the classical engineering approach, which attached greater importance to knowledge that was based on quantifiable, measurable, objective and formal factors and less to knowledge based on experience (Weick and Sutcliffe 2001). HROs had unique knowledge and behavioural elements. The first was an unambiguous and almost complete knowledge of the technical and organisational system and consequences of failure for the whole system, due to persistent training in hazardous tasks. The complexity of the technology and the organisation was understood by everyone. This enabled an almost error-free performance from both staff and installations. One example was the spotter – the junior officer on an aircraft carrier. This person could abort the landing of an aircraft when he considered the risk of a crash too high. Another example was the operators of the nuclear power plant. Looking at the control panels, they saw the total installation and not just numbers on screens.

Second, failure regimes – that is, small deviations from normally functioning machines and operations – were recognised. Failure reports were rewarded, and there was no blame laid on an operator who made a mistake once the bad news went upwards. If necessary, for example, the standard operational procedures of aircraft carriers were adapted.

Third, there were unambiguous regimes for the specification of system failures. If problems arose, small dedicated networks were formed that were self-designed and not formalised. When the problem was solved, the network dissolved. The organisation could easily switch between centralised and decentralised decision making. Decisionmaking was carried out at the lowest level where the problems arose. Following this theory, a few discussions had started in literature. First, the validity of the normal accident theory was questioned, after publication of the high reliability theory (Sagan 1993; Vaughan 1996; Rijpma 1997, 2003; Shrivastava et al. 2009). More recently, another discussion appeared in *Safety Science* on the distinction between high reliability and resilience engineering (Garbowski and Roberts 2016; Haavik et al. 2016; Le Coze 2016; Pettersen and Schulman 2016).

In the normal accident theory, actions of management of the involved companies and organisations did not play any role. In other words, these actions had no influence on the occurrence of major accidents. These major accidents were, after all, inevitable, as the complexity of the technology was high and its coupling tight. The high reliability theory started from the opposite direction. It showed that companies and organisations with sufficient redundancy could prevent major accidents by bypassing the dilemma in the normal accident theory of decentralised versus centralised decision-making. In high reliability theory, this was overcome by self-designed, transient networks operating locally at the time of danger. High reliability organisations resembled Mintzberg's 'adhocracy' (see Chapter 5). Because not all high-tech-high-hazard companies were high reliability organisations, in the literature the proposal was put forward to consider these theories as complementary.

Case studies

In publications by Sagan and Vaughan, both the normal accident theory and the high reliability theory were used (Sagan 1993; Vaughan 1996). Sagan had studied near misses in the US nuclear weapons programme during the Cold War, which could have easily escalated into large-scale nuclear catastrophes. The research was focused on the 1962 Cuban missile crisis and the 1968 accident near Thule, Greenland, where a B-52 bomber carrying a hydrogen bomb crashed at Baffin Bay. In conclusion, the study showed that a complex system, like that of nuclear weapons, made it impossible to anticipate every emergency, at both higher and lower levels in the organisational hierarchy. Complexity undermined the reliability of decisions and effective corrections of initial mistakes. According to the author, normal accident theory provided the best explanation for the incidents investigated, despite the presence of high reliability strategies.

Vaughan's book described the background of the major accident of the *Challenger* space shuttle in 1986. A little more than a minute after launch, the space shuttle disintegrated (Figure 8.10). The technical cause was a defect in a rubber O-ring intended to seal the outer lower part from the rest of the thruster. This O-ring was not resistant to the low temperature of 2° C at the time of the launch.

In the 1960s, NASA was a national icon and also a powerful organisation. In the 1980s, during the Reagan administration, this position was weakened by budget cuts, and production goals dominated space flights, which had to become operational after fewer tests. This resulted in rule violations. Critical information did not reach the highest levels of the hierarchy and there was a so-called normalization of abnormalities. Although NASA was originally characterised as a high reliability organisation, in the 1980s this was no longer the case. According to the

Figure 8.10 Iconic image of disintegration of space shuttle *Challenger*

Source: https://www.britannica.com/event/Challenger-disaster/images-videos

author, the major accident was a normal accident, a social organised risk, fuelled by an environment of constraints, competition, major production pressures and financial cuts.

A third book makes a similar comparison. Snook (2000) investigated a major accident in 1994, wherein two US Black Hawk helicopters were shot down by US F-15 fighter jets in northern Iraq. In the analysis, the concept of 'practical drift' was introduced, a concept that resembled the normalisation of deviations by Vaughan and the 'drift into danger' of Rasmussen. This practical drift came into being when established rules did not reflect situations in which pilots had to operate. Behaviour changed and adapted in those situations and the original rules were no longer applicable. As a result, feedback and coordination of these changes were not communicated between an AWACS aircraft, the pilots of the jets and the helicopters, resulting in the helicopters being misperceived as enemies, with disastrous effects. In general, tighter rules created a higher probability of practical drift (see also Amalberti et al. 2004).

The disaster incubation theory was used successfully in the analysis of major accidents in three of Hopkins' books. The first major accident was the explosion at the BHP coal mine in Moura, Queensland, Australia, in 1994, where 11 miners lost their lives (Hopkins 1999a, 2000b). The second major accident was the explosion at an Esso gas plant in Longford, Victoria, Australia, where two operators died and eight were injured (Hopkins 2000b). Management had seen only a few occupational accidents in their plant and assumed that process safety was sufficiently guaranteed. Aside from the victims, Longford's aftermath was quite serious for the Melbourne region, which was deprived of its gas supply for two weeks. The third book was on the BP major accident in Texas in 2005 (Hopkins 2008). All major accidents were the subject of extensive government investigations and legal proceedings, information that formed the basis for Hopkins' analyses. These investigations showed, among other things, that apart from the effects of sloppy management, government inspectors at the Moura mine were 'owned' by the industry, and at the Longford location, inspectors were no match for the experts from Esso. The operators were blamed for the explosion. Another case study by Hopkins, of the Deepwater Horizon major accident in the Gulf of Mexico in 2010, will be discussed at the end of this chapter.

Risk perception

As shown above, risk had many definitions and could be calculated in various ways. Every time there is a change in the definition of risk it results, obviously, in different numbers (Slovic 1999). Technical experts prefer to express risk as a number, giving the concept a scientific and objective appearance.

Subjective risk perceptions of non-experts are undervalued, and non-experts generally distrust technical experts, especially those living in the neighbourhood of hazardous industries. The technical experts were deaf to distrust, so they always used the same strategy. First, there would be a scramble to get the numbers right. Then communication about the numbers started, which was followed by an explanation about the meaning of those numbers. To clarify the meaning of numbers, comparisons were made with normal daily risks that were higher: smoking was a popular choice, as were accepted risks from the past, like air travel. This comparison was criticised as much in North America and Australia as it was in Europe (Rowe 1988; Sharlin 1989; Shrader 1990; Slovic 1993, 1998; Slovic et al. 2004; Simpson 1996). When that strategy failed in convincing non-experts, the experts tried a respectful approach towards the public, as though they were partners, to get them involved in the discussions and weighing of risks as much as possible (Fischhoff 1995; Ter Mors and Groeneweg 2016). What became clear from the discussions was the values-driven content of risks. Hazards were real, but risks were constructs within social and historical contexts that determined their significance. There seemed to be no universal approach to risk.

The Netherlands

Determinants of major accident processes

The Netherlands also worked on various determinants of major accident processes, one of them being latent factors, which were further specified in the Tripod model (Groeneweg et al. 2003). The Tripod model was originally developed for occupational accidents (see Chapter 7), and defined a set of 11 basic risk factors. The majority of these factors had an organisational origin. These factors could be seen as a further detailing of Barry Turner's incubation period and James Reason's resident pathogens. Later, the management factors or management delivery systems of the bowtie metaphor developed the concept of latent factors even further.

Sloppy management was investigated as well, but the terminology was different. In Dutch safety literature it was called a 'lack of responsible management' by the Dutch Safety Board (OVV 2013; Swuste and Reniers 2017). Lack of responsibility was a kind of formal notification of sloppy management which made companies vulnerable to poorly structured surprises, like process disturbances or scenarios induced by automation of production (Vuuren 1998; Knegtering and Pasman 2009). The Dutch also looked into the difficulties of governmental bodies to keep up with the fast-changing technological developments within companies, and the hazards and risks introduced for these companies and local residents. This is fundamentally problematic because the primary task of the government is to guarantee the safety of its citizens (Vollenhoven 2008; Dik et al. 2008). Perhaps, as a result of being behind the curve, government generally overreacts after major accidents, leading to an over-regulation of the safety domain. The situation was not helped by the fact that politicians were not very interested in an independent review of the quality and effectiveness of safety laws, inspections and other legal duties imposed on industry (Hale, Goosens et al. 2002; Gundlach 2002; Mertens 2011).

Metaphors

Bowtie

The bowtie is a Dutch metaphor for the accident process, combining an engineering with a socio-technical approach. This metaphor was introduced in Chapter 7 for occupational accidents. A bowtie presentation was also applicable for major accidents in the high-tech-high-hazard sectors. Barriers are important elements in this metaphor, effective in technical processes for the prevention of explosions and fires, as in the Swiss cheese metaphor. The organisational factors of major accidents of the bowtie metaphor are not expressed as holes in barriers, as in the Swiss cheese metaphor, but in 'management delivery systems'. These are actions taken by management to ensure and monitor barrier quality (Guldenmund et al. 2006). Unlike the Swiss cheese metaphor, the bowtie is based on multiple scenarios leading to adverse consequences. The metaphor also introduces the central event', a state in which a hazard has gone out of control and might manifest in the form of an incident or even a major accident.

An adopted version of the bowtie metaphor is developed for domino effects. The main determinants of accident processes leading to domino effects are known, coming mainly from case descriptions of past major domino accidents. One way to visualize these accident processes is a 'double bowtie' (Swuste et al. 2020c) (Figure 8.11).

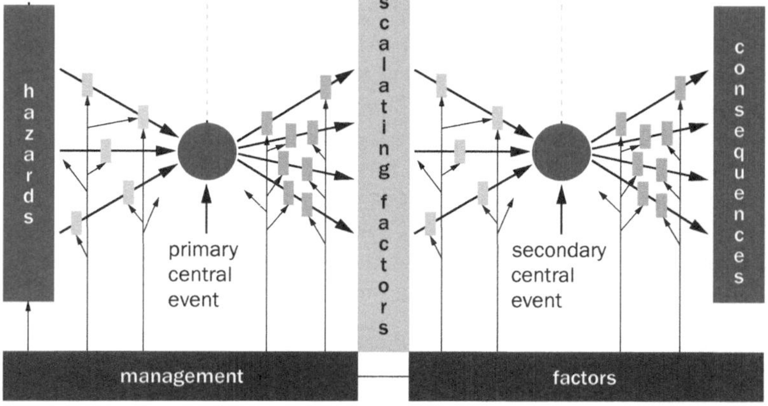

Figure 8.11 A domino effect double bowtie

Source: Swuste et al. (2020)

This bowtie illustrates on the left-hand side the onset of an accident process, starting from a hazard. Several accident scenarios are shown at the top, from left to right (the grey arrows). These scenarios may lead to the central point of the domino effect: the primary central event. This is a situation where the hazard has become uncontrollable, leading to escalating factors, the tall grey rectangle in the middle of the figure.

Cozzani and co-authors (2006, 2007) defined the propagation of these dominoes, their escalating factors and physical effects. They provided a follow-up trajectory with the domino scenarios and a secondary central event (see Table 8.2). This secondary central event may lead to consequences (right-hand side of the figure) which are greater than the consequences of the primary accident process. The figure shows the accident process of a single domino. In principle, the primary accident process may result in multiple secondary central events. The strength of the model concerns the influencing parameters. These parameters can prevent primary and secondary central events (the circles in the figure), or limit the consequences (the small grey rectangles). Two types of influencing parameters can be identified. The first are the safety barriers, represented by the black rectangles in the figure. These are physical or technical entities that interrupt the accident scenarios. The second are management factors, the rectangles at the bottom of the figure, influencing the quality of barriers, scenarios and hazards through the vertical arrows. The lines represent non-physical or organisational and human aspects. The line leading directly from the management factors to the hazard represents inherent safe design options. To manage domino effects adequately, both primary and secondary domino scenarios must be controlled.

Quantification of risk

In the period after 1990, risk and risk-based approaches for safety were accepted and established in the Netherlands. The high population density of this country, combined with a high level of industrialisation, favoured the quantification of risk as a basis for better controllability and acceptability for future industrial expansion (Ale 2002, 2003). The risk-based approach was seen as a rational tool for decision-making about investments for additional control measures. At that time, the textbook series of 'coloured books' formed a complete methodology for carrying out quantitative risk assessment (QRA) in the area of hazardous substances and received much international recognition.

The fourth and final part of the textbook series of coloured books, the Purple Book (CPR 18 1999), was published at the end of the 1990s. The Purple Book included calculation rules and criteria for performing a QRA. Criteria were given for deciding which installations of a certain industrial plant should be part of a QRA and which could be excluded. This final publication in particular was strongly criticised for its limited assessment of hazards. The Purple Book presented the agreements on selecting the installations to be analysed, determination of the relevant occurrences of LOC (loss of containment), relevant subsequent events, corresponding effects and damage calculations and, finally, presentation of the results. The Purple Book reflected the Dutch approach and formed a basis for Dutch regulation. The book described the criteria for 'containments' to be included in a QRA, with more than a certain amount of a specific hazardous substance. Subsequently, for these containments the probability of failure was determined, based on historical data. Criticism focused on the need for a more thorough but also time-consuming method of a systematic identification of hazards. This included describing and analysing relevant detailed scenarios using hazard and operability studies and other safety studies. These methods were available at that time (Lawley 1974). A QRA based on a previously carried out hazard identification and Hazop would give a better view of the failure probabilities, as well as of conditions and process data specific to the site in question, of relevant human behaviour, and of where and how safety could be improved.

From the first drafts of the coloured books, there was much discussion on the uncertainties of the models used. Uncertainty in the parameters of the effect models gave variations in the results of a factor of two to six. The damage models also contained large uncertainties. The Green Book mentions model uncertainties, inter-individual variation and differences between buildings, and parameter uncertainties like population data or toxicity data, and the influence of escape behaviour on exposure time. These uncertainties resulted in a large variation in the final calculated risk. When comparing different research teams, the calculated risk varied by orders of magnitude (Pasman 2011), with large differences for the risk contour around an installation (AGS 2010). Nevertheless, quantification of risk became an important part of nearly all safety decision-making processes. Methods for quantifying risk are now widely applied and accepted; however, risk perception and risk in the political decision-making process were and still are being debated (Rip 1986).

Acceptable risk levels

Also, in the field of delayed health effects, in particular cancer caused by exposure to carcinogens, risk was a basic concept used in setting standards for exposure to toxic chemicals. These discussions on acceptable risk levels for toxic chemicals played a pioneering role in the safety domain. A standard for a maximum acceptable risk for exposure to genotoxic substances was set by the Dutch Board for Occupational Health and Safety in 1992. The board referred to the ministerial memorandum 'Dealing with Risks of Radiation: Ionizing Radiation Limits for Workplaces and the Environment' from 1990 (Ministry of Housing 1990). On the board, delegates from employers and employees had different opinions about the maximum difference in risk level between the general population and labour force employers and employees:

> *The delegates of the employees believe that the risk for workers in the field should be equal to that of the general public, i.e.* 10^{-6} *per year. (. . .) The employer delegates (. . .) are of the opinion that the standard risk limit of*

> *incidents with fatal results in the so-called 'safe industry' should be held, namely 10^{-4} per year (. . .).*

The board proposed striving for a risk level for each substance of 10^{-6} per year, the 'target level', and in any case not accepting a higher risk of a maximum 10^{-4} per year, the 'banning level'.

Risk perception

The Dutch Health Council warned the government: 'Risk has many dimensions and cannot be captured by a single number [. . .]. Risk characteristics that are quantified too much into one single risk expression are counterproductive and raise many questions' (Gezondheidsraad 1995, 1996). The technical framing of risk meant that citizens simply had to choose to trust or distrust experts. For experts, risk was usually a technical assessment of the probability of mortality. For non-experts or citizens, other characteristics of hazards and risk were relevant. In addition to the uneven distribution of risks, benefits and the degree of control, other arguments also played a role. Examples included the potential for catastrophe, the threat to future generations or the uncertainty of the calculation. In the period under study, the statistical approach to risk was not convincing for the public. Proponents of the formal risk approach saw citizens' intuitive reactions as irrational. Their perception of risk was considered to be a product of 'feelings'.

Opposition to a formal risk approach was evident during the occupation of the Brent Spar oil platform by Greenpeace in 1995. The platform, which provided oil storage for shuttle tankers, was not needed anymore, after submarine pipelines were operational. The owners, Shell, and Esso, planned to decommission and sink the platform. Greenpeace appealed that decision, arguing that heavy metals from the storage platform were harmful to the maritime environment. During the resulting discussion, Shell came up with an objective risk assessment, which in hindsight turned out to be correct. Although Greenpeace had to admit that the objective risk analysis was correct, it had already launched a consumer boycott, which proved to be effective in Germany, the Scandinavian countries and the Netherlands. It was a David versus Goliath battle, in which Greenpeace argued that the sea was not a dumping ground (Löfstedt and Renn 1997; Ravetz 2001). This argument was persuasive and, ultimately, the oil platform was not sunk. This was an example of a small non-governmental organisation that could influence public attitudes very effectively by addressing corporate responsibility. A similar case in the same year was the public outcry against the execution of Ogoni chiefs in the Niger delta in Nigeria who protested non-violently against extreme environmental damage from decades of indiscriminate petroleum waste dumping.

Risk and safety management

The socio-technical approach made its entry into the safety domain; the organisation, it was now believed, should be able to guarantee safe operations by both predicting and controlling technical, organisational and human aspects of major accidents. This required a robust design based on defence-in-depth and an active and effective risk or safety management system. Defence-in-depth means that the organisation monitored the quality of the material and immaterial barriers that prevented hazards developing through scenarios. This was the basis of risk and safety management (Hale and Hovden 1998; Li and Guldenmund 2018).

Risk and safety management are two concepts that are both used in the safety literature for high-tech-high-hazard companies. Different sectors and different professional domains preferred either one term or the other. Risk management focused on the quantification of scenarios and their consequences. Often, financial consequences of a major accident were a component of the management approach, but it was questionable whether managers were interested in a rational consideration of cost. In their view, it was often a good decision to introduce safety measures, especially when they personally had witnessed a major accident (Hale 2014).

The literature on risk and safety management is divided into three groups: the developed models, the functioning of these systems based on fieldwork at companies and, finally, their quality based on retrospective studies and audits (Li and Guldenmund 2018). The model of risk and safety management, including its overall structure and its elements, was known and there was not much discussion about it. The first element was the primary and supporting processes of the company as the starting point for all the possible scenarios leading to major accidents and to the identification of barriers to managing (controlling) these scenarios. The second element was what was called the 'life cycle approach'. This approach described the design of system elements, their purchase, maintenance, use, modifications and dismantling. The third element was the problem-solving approach, which addressed risks during normal and disturbed processes and, additionally, had plans and procedures for online risk management and a policy to regularly test the quality of the management and, if necessary, to adjust (improve) it. Finally, staffing and resources are needed to carry out these aspects adequately. These elements of a management system were developed from the results of research at large companies, or 'professional bureaucracies'. But major accidents also occur in smaller companies, as was the case in the Netherlands in 2000 when an explosion at a fireworks storage facility in Enschede killed 23 people and wounded 950, and when a fire in a cafe in Volendam in the same year killed 14 people and injured 241. It is still unclear how risk and safety management should be organised for smaller companies (Hale 2003).

These accidents and fieldwork in high-tech-high-hazard corporations both highlighted sloppy management. The fieldwork showed that this could happen in an interactive sense-making process. The American author Weick described this as the way the organisation and its members gave meaning to the dynamic reality as it occurred (Weick 1993, 1995). The momentum was fuelled by internal and external factors, as shown in the 'drift into danger' metaphor by Rasmussen (Figures 8.6 and 8.7).

Deepwater Horizon

On 20 April 2010, an explosion occurred on the giant platform Deepwater Horizon in the Gulf of Mexico. Two days later the platform sank off the coast of Louisiana (Figure 8.12). Of the 126 employees, 11 lost their lives.

The so-called cementing of the head of the well – an action intended to prevent a blowout – had gone wrong. The well was more than 1,500 metres below sea level. For ten weeks, oil continued to flow from the well – 200 million gallons, equivalent to 20 times the pollution caused by the *Exxon Valdez* oil spill in 1989. The oil threatened the ecosystem of the Gulf of Mexico and reached the coasts of Louisiana, Mississippi, Alabama and Florida, as well as the Mississippi River (Steffy 2011; Hopkins 2012).

BP was blamed, but it was not actually the owner of the platform – Transocean was, and Halliburton was responsible for cementing the well. Deepwater Horizon operated the Macondo oilfield. Ironically, in the novel *One Hundred Years of Solitude* by Colombian Nobel Prize winner Gabriel García Márquez, Macondo was depicted as a sleepy town. The town came into bloom and deteriorated after a four-year rainy period.

But in many ways, the accident was no exception. Incidents and major accidents, blowouts and oil emissions occurred almost every year in the Gulf of

Figure 8.12 Deepwater Horizon

Source: Hopkins (2012)

Mexico and often several times a year (see Appendix 1). What made BP stand out was that the company had experienced several incidents and major accidents within a short time period: at the refinery in Grangemouth, Scotland (2000), at the Texas City refinery (2005) and the oil leaks in the Prudhoe Bay pipeline in Alaska (2006). BP was a company where process safety was guided by occupational safety. Low numbers of personal accidents were seen as indicators of safe production, which was an obvious misconception.

BP's head office mainly focused on financial risks and not on safety risks. The head office had a policy of minimal effort, resulting in understaffed teams and activities. Engineers' bonuses were directly linked to output, creating a time pressure on drilling operations. For drilling activities at extreme depths, a number of barriers existed, in accordance with the defence-in-depth concept, to prevent blowout scenarios or, in the case of a blowout, to mitigate the effects (Hopkins 2012) (Figure 8.13).

This concept could be effective only if barriers functioned independently. At Deepwater Horizon, barriers were made to be interdependent – not so much technically, but related to communication about the success of the first barrier.

When it was reported that the cementing had been completed successfully, an evaluation of the cementing was not performed. The notion of defence-in-depth was not sufficiently clear to operators; one working barrier was sufficient for them.

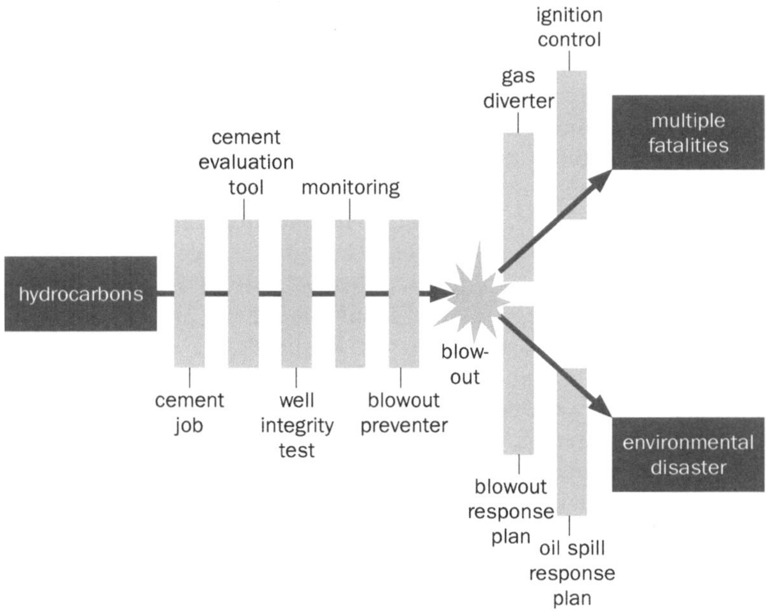

Figure 8.13 Failing barriers, Deepwater Horizon

Source: After Hopkins (2012)

Thus, these two other barriers – measurement and testing of the integrity of the well – were not adequately installed. The last barrier, the blowout preventer, was a big installation, placed on top of the well. During an emergency, the preventer could pinch off the drill pipe, locking the well. But the preventer was not designed to stop an already developing blowout and it failed to stop the outflow of oil. The barriers that were supposed to operate after the blowout were not very effective because the designers had relied on pre-blowout barriers. Just like a house of cards, all barriers to prevent a blowout tumbled down (Hopkins 2012).

The example of the Deepwater Horizon in the Macondo oilfield marks the end of the period discussed in this book. Knowledge development of accident processes in high-tech-high-hazard sectors was mainly guided by major accidents, and this justified the selection of these accidents in this chapter and in Chapter 6. Many major accidents showed a number of similarities, such as time pressure, a primarily business focus, understaffing and a lack of oversight on safety-critical activities. These similarities were also called 'sloppy management', where organisational factors and latent failures played a dominant role in accident processes. Additionally, the analysis of Deepwater Horizon showed in detail the vulnerabilities of barrier systems.

Contributions to the knowledge development of accident processes were present in all three geographical areas discussed. Like the 1980s, the 1990s were 'golden years' for safety science, and the knowledge development was even more internationally oriented than in periods before. Concepts of risk, safety and organisational culture, domino effects and design could not be identified with a specific country, or group of countries. Also, the distress about major accidents was an important driving force behind this development. The socio-technical approach towards major accidents became dominant. The American high reliability theory is an example, followed by the Resilience Engineering workshop more than 15 years later in western Europe and Nordic countries. This socio-technical approach combined an engineering view of accident causation with a sociological and psychological view. This became visible in safety metaphors, models and theories such as Swiss cheese, drift to danger and the reissue of the disaster incubation theory, which all originated in western Europe and Nordic countries. In the Netherlands, the basic risk factors from the Tripod model were one of the first attempts to specify the latent factors which were dominant in major accident processes. The bow-tie metaphor went further, by postulating 'management delivery systems'. But complexity was becoming an increasingly important theme. Companies could be safe on paper, but practice could deviate considerably from paper. In contrast to occupational safety, the focus on human behaviour and wrong actions disappeared into the background. The following chapter provides a concluding summary and takes stock.

9 Epilogue

More than 150 years of knowledge development within the safety domain has been reviewed in eight chapters. Periods with a strong dynamic development and periods with a certain standstill alternate. This chapter provides a summary and takes stock.

DOI: 10.4324/9781003001379-9

This book started with questions about knowledge development in the safety domain and about the influence of general management schools on developments in safety management. Before the Second World War, the United Kingdom and the United States were front-runners in safety, and the Netherlands is presented as a case study showing how these Anglo-Saxon ideas on safety were translated into the Dutch context. After the Second World War, safety theories, models and metaphors increasingly came from other countries too, notably Nordic countries, Canada and Australia.

An overview of major accidents is provided in Appendix 1, and the developments in knowledge, in both occupational safety and high-tech-high-hazard safety, are summarized in Appendix 2, including management trends and trends in safety management. Finally, Appendix 3 provides information about occupational safety from the 1800s to date. This shows the relationship between knowledge development and these accidents from the Second World War onwards. The American dominance in publications on general management trends and safety management is striking: managing production and safety had its origins in the United States.

Acts of God

In the nineteenth century the machine had an almost holy status. Labour productivity increased spectacularly due to mechanised production. But inside factories and workshops there was a sinister downside: machines were dangerous. Although there is some anecdotal evidence that some people in the nineteenth century considered external causes for accidents, the common belief was that accidents were acts of God; they were unavoidable, went with the work and had a streak of God's punishment for bad behaviour. It did not help that workers were not seen as a separate group that deserved special attention for safety reasons. They were just *the poor*, as is so vividly described in the novels of Charles Dickens, such as *Hard Times* (Figure 9.1) (Dickens 1854).

Dickens visited factories in Manchester in 1839, where production workers were in fact known as 'the hands' and thereby extensions of the machines. These conditions, what Friedrich Engels called 'economic barbarism', were also reported in the Netherlands, some 40 years later, and in the United States, some 50 years later (Freeman 2018). In these two countries the Industrial Revolution started later than in the United Kingdom.

The individual and environmental hypotheses, pre-war period

Different explanations for accidents emerged at the end of the nineteenth and in the first part of the twentieth century. In literature these explanations were called the 'individual' and the 'environmental' hypotheses (Hale and Glendon 1987). The individual hypothesis focused on human failures and explained accidents by pointing at less competent workers, who were more liable to have accidents, regardless of the nature of the hazard. The environmental hypothesis was concerned with external factors, such as machine-related hazards, workplaces

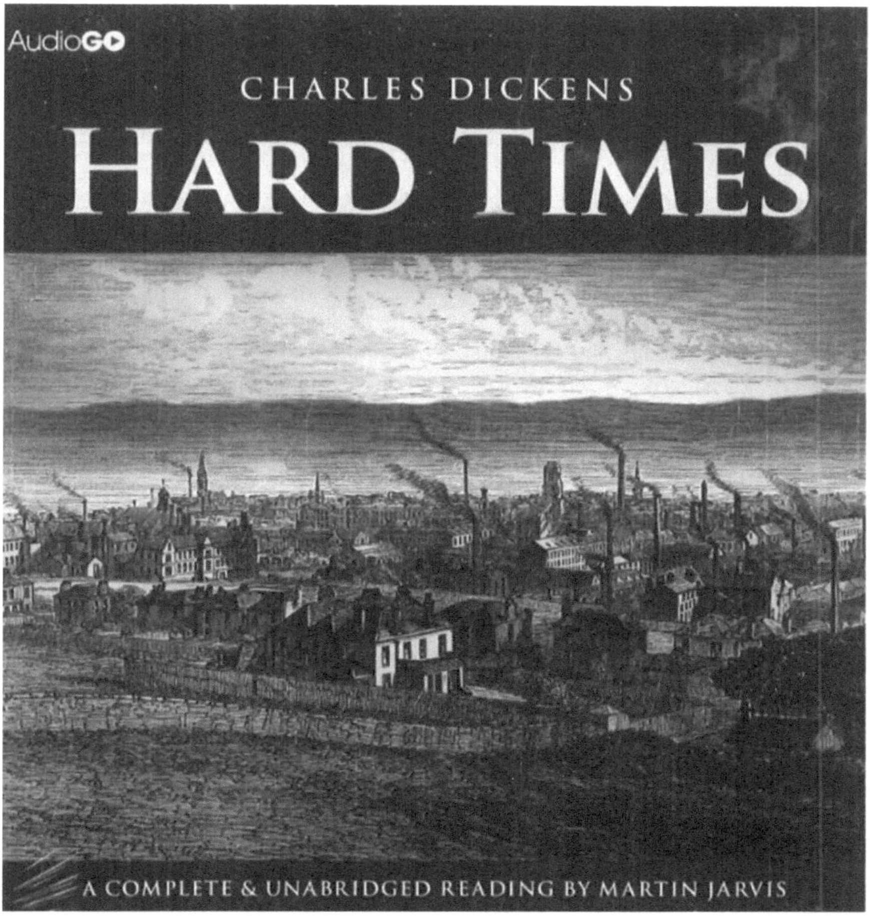

Figure 9.1 Hard Times

Source: Dickens (1854)

crammed with dangerous machines, increased pressure of work and long working hours.

Safety technique originated from a technical approach to safety by fencing off moving parts of machines and safeguarding workplaces at heights. Safety technique was an example of the environmental hypothesis. The first reports came from the United Kingdom in 1832, from Charles Thackrah, a physician-hygienist. In 1876, the Dutch physician-hygienist Samuel Coronel published a similar report. In 1837, a British court case of an injured worker introduced what nowadays is called 'duty of care of the employer'. Much of the British health and safety legislation was based upon the control of these environmental factors. The 1844

Factory Act was the first to order that factory inspectors should safeguard the machines. The first overview of safety technique applications in various industries was published in 1899. This was a practical guide to the laws on safeguarding, safe working and safe construction of machines in industry by the British engineer and factory inspector John Calder. Safety now became the domain of the technician. The safety technique remained an important safety approach until well after the Second World War.

In 1895, the Netherlands followed the British example by adopting De Veiligheidswet (The Safety Law), which was similar to the British Factory Act. This was three years after the 1892 National Exhibition to promote safety and health in factories and workplaces was organized in Amsterdam. The exhibition became a permanent event, giving birth to the Safety Museum in 1893, which was one of the first safety museums in the world. Safety technique was further promoted by the engineer Frederik Westerouen van Meeteren in his 1893 reference book on occupational safety and health and his professional periodical *De Veiligheid* (The Safety Journal), which was issued between 1893 and 1904. A second Dutch safety promoter was the physician-hygienist Louis Heijermans. He published books on hygiene for workers and taught a course on social and technical hygiene at the forerunner of Delft University of Technology. His magnum opus from 1908 was his reference book on occupational diseases, including occupational safety, and was based on his frequent visits to, and observations on, shop floors in various factories. He also categorized accidents according to the hazards involved. These categories were a first attempt to define accident scenarios, and they closely resemble today's categories.

In the nineteenth century and in the first decade of the twentieth century, the United Kingdom and the Netherlands were international leaders in knowledge development in occupational safety. These developments took place against the backdrop of the emancipation of the working class and the rise of the unions. Workers were lifted from the undefined group of the poor to a valuable source of labour that required attention from both a public and an occupational health and safety perspective. Workers had to be protected from machines; that much was apparent in the British survey of the Departmental Committee on Accidents in Factories and Workshops. In 1911, the committee sounded the alarm bell over the stark rise in fatal accidents due to 'industrial fatigue' and a lack of safeguarding of machines. Two decades later, Horace Vernon wrote his magnum opus on the state of safety in the 1930s. Vernon was an academic with degrees not only in medicine but also in chemistry, biology and physiology, and was a proponent of the environmental hypothesis. His major publication on occupational safety was of excellent quality, and in it he produced a categorization of causes of accidents that resembled the one from Heijermans in 1908 to a great extent.

Although it is generally accepted that the individual hypothesis was quite central at the beginning of the twentieth century, there were not many critical references to it in the literature. Criticism came from Heijermans, who argued against the common opinion that workers themselves were responsible for their accidents, that it was primarily their indifference and carelessness that caused accidents at work. Heijermans showed that indifference was the consequence of daily confrontations with danger, the monotonous tasks they had to perform and the very long working hours.

At the end of the first decade of the twentieth century until the 1970s, the United States took over the leading role in knowledge development in occupational safety and the management of safety. The classical management school was dominant at that time and placed top company managers at the centre of decision-making, which was a revolutionary concept at the time. Up till then, managers had no interactions with the shop floor. This was the domain of the supervisor or foreman, who determined the course of production. In contrast to those of the United Kingdom and the Netherlands, American publications emphasised the management of safety, stimulated by Taylor's 1911 publication, *The Principles of Scientific Management*, as a representative of the classical management school. The technique and the rationalisation of production were central to this approach. In line with scientific management, management of safety in companies was concerned with standardisation of safety procedures and appropriate selection and training of employees to prevent hiring reckless and inattentive employees. Safety was directly linked to the efficiency of production. These were main elements following the individual hypothesis and linked classical management to the need to manage occupational safety. Also, before Taylor's publication, the American company US Steel had started the Safety First movement in 1906 for production-related reasons. A few years later, it scaled up to a national movement with its appealing safety metaphor, the 'road to happiness'. Safety was the number one priority of a company and a necessary condition for efficient production. Workers had to be educated in the *safety idea*: they had to become safety conscious. Managing safety was achieved through the selection and training of employees.

The individual hypothesis received scientific support in the second decade of the twentieth century from the United Kingdom. The accident proneness theory was born, and its influence continued to grow with the emergence of the behavioural management school. This school stemmed from the emerging fields of industrial psychology and behaviourism, the latter being a movement in psychology with an emphasis on visible behaviour. Shortly after the Second World War, the accident proneness theory was criticised by several authors in the scientific press, who termed it a 'shotgun approach', in which a large number of variables were measured without any theoretical basis for their inclusion in the study. Under these conditions it could be expected that significant results would occasionally be obtained simply on the basis of chance alone. Nevertheless, even nowadays the popularity of theory amongst safety professionals has not diminished.

In the United States, the individual hypothesis was prominent but not dominant. The environmental hypothesis gained support from the Pittsburgh Survey, conducted by the American sociologist Crystal Eastman. In 1910 Eastman came to similar conclusions to Heijermans. Like Heijermans, she conducted her extensive fieldwork at the shop-floor level. Her results, gleaned at the Pittsburgh US Steel plant, supported the environmental hypothesis. Eastman pointed to the extraordinary number of industrial fatalities in the Pittsburgh plant as a huge social waste.

She noticed that the decisions by foremen to send inexperienced workers and child laborers to dangerous workplaces were one of the main causes of the fatal accidents. Like Heijermans, she also categorised accidents by specific hazards, thereby generating a first impression of accident scenarios particular to steelworks.

In the second decade of the twentieth century, many publications on safety technique appeared in the United States and were tailored to the conditions of different branches of industry. These publications presented a symbiosis of the environmental and individual hypotheses. This symbiosis was also reflected in the '3-E' slogan of the American National Safety Council in 1914. The slogan referred to engineering (safety technique), education (training workers in 'the safety idea') and enforcement (rules, procedures and supervision). Financial issues related to accidents and lack of safety were always a major item. In the United States, safety was the domain of large companies and insurance companies. In the United Kingdom, with its extensive social legislation, occupational safety was initiated not by market parties but by governmental committees. These committees conducted surveys and research into the causes of accidents and the conditions in companies and workshops. In the Netherlands, no systematic research in the field of occupational safety was conducted until the Second World War, with the exception of the Eighth Parliamentary Inquiry of 1887 and two doctoral theses on occupational safety. The Eighth Parliamentary Inquiry created the Dutch Factory Inspectorate, and this was the introduction of social legislation. The two doctoral theses, published just before and during the war period, reported the results of tests on accident-prone workers, with limited success. Unlike in the United States, in the United Kingdom and in the Netherlands no special attention was given to the management of safety.

Hazards and unsafe acts, pre-war period

Hazards and unsafe acts were the main components of the environmental and individual hypotheses. But these two concepts still deserve special attention, due to publications by two Americans, the safety engineer Lewis DeBlois and the technician Herbert Heinrich.

DeBlois introduced the concept of accident probability and, following Eastman and Heijermans, the concept of hazard as the starting point of an accident process. Hazard, in his view, equalled energy, be it potential, kinetic, electrical or any other form of energy. An accident was considered to be a sequence of events, initiated by process disturbances. This view was revolutionary at the time. He also introduced the term 'exposure to hazard(s)'. Prevention was a logical consequence of his ideas: reducing hazards or reducing exposure. DeBlois formulated the main concepts of safety science in his publication. Unfortunately, the influence of these ideas remained limited, perhaps due to his few publications.

Heinrich, on the other hand, who worked at an insurance company, published 16 articles and five books on safety from 1927 onward. Although a technician, he believed that psychology in accident prevention was fundamentally important. With his focus on unsafe acts of workers, Heinrich was a major influence on the popularity of the individual hypothesis. Heinrich was able to reduce complex safety concepts to simple numerical ratios on the costs, causes and mechanisms of accidents. His metaphor of the iceberg, or 'safety pyramid', and his domino metaphor became famous. This domino metaphor, from 1941. presented accidents as a process or a sequence of events, just as DeBlois had proposed 15 years earlier. For Heinrich, the unsafe act was at the centre of a series of dominoes and the focus of preventing accidents at work.

Occupational safety, post-war period

Management and safety management, post-war period

Like most wars and crises, the Second World War brought major changes. In post-war Japan, the modern management school took off. Modern management regarded a company as an open system in which several actors determined the quality of production and products. The Americans William Deming, a statistician, and Joseph Juran, a businessman, were early pioneers of this approach, the introduction of quality control being their main achievement, with a major role being played by employees and customers. Total quality management, or TQM, and corporate social responsibility as expressions of the modern management school entered the Western world much later, in the 1980s. Also prompted by Anglo-Saxon management concepts, the emerging quality management school introduced an accountability of companies towards customers and the government. This required a need for a greater transparency in decision-making and initiated an extensive bureaucratisation. Quality management had a major impact on safety management, and the 'Deming cycle' was introduced in the safety domain.

As early as 1950, Heinrich published the first study on the structure of safety management. He presented his 'safety ladder' metaphor, with five steps of a stairway representing the management activities for safe and efficient production. But this contribution was not recognised in formal scientific literature. Serious attention to this topic had to wait until after the Piper Alpha major accident, in the North Sea, in 1988. Meanwhile, in the United States, safety was expanded with damage control in the 1960s. The focus on property damage had a financial incentive. It was claimed that the costs of property damage caused by accidents were generally higher than the costs of injuries to workers. Until the 1970s, in the United States seven textbooks on occupational safety and safety management were published, none in the United Kingdom and only one in the Netherlands.

This changed in 1971 when the term 'safety management' was introduced by the American engineer Dan Petersen, who also created his first prototype of a safety audit. These audits allowed companies to gather information on the quality of their safety management system and causes of occupational accidents. The importance of safety was also stressed by Maslow in 1973, in his hierarchy of human needs. Another interesting development was the concept of safety climate, introduced by the Israeli organisational psychologist Dov Zohar in 1980. Later, in the 1990s, this would inspire a substantial body of research in North America,

Australia, western Europe, Nordic countries and the Netherlands. Safety climate and safety culture were seen as the motor of a safety management system and of successful safety improvements, in both occupational safety and in the high-tech-high-hazard sectors. In 1998, the psychologist Andrew Hale and the political scientist Jan Hovden named the focus on safety management combined with safety climate and culture the 'third era' in the development of safety.

Though the United States was leading the way, western Europe and Nordic countries increased their research on safety and started their critique of contemporary safety methods. The British Robens Report of 1972 showed the inefficiency of safety legislation, which was far too fragmented. It was the first time that self-regulation of safety for companies had been proposed. Later, Norwegian research stressed the conditions for self-regulation, which should involve a high degree of consensus on basic safety values between public safety and health executives and company managers, and between employers and employees and their organisations. But in most manufacturing and high-tech-high-hazard companies, those who had a safety management system in place were still the odd ones out. Similar comments came from the United States in the 1970s, where researchers pointed at the low quality of research on safety training and the doubtful effects of safety legislation, inspections, safety statistics and governmental safety standards, due to a lack of evaluation studies.

Ergonomics, post-war period

The rise of ergonomics dates back to the Second World War, when military equipment and machinery became increasingly complex and created serious control problems with increasing demands on operators' skills and cognition. Human factors, human reliability analysis and ergonomics became academic domains in the early 1960s. This new domain of man–machine interactions was an important influence in the safety domain. Ergonomics raised the question of whether workers should adapt to machines or vice versa (Waterson and Eason 2009; Waterson 2011). Many workplaces suffered from poor human factor conditions. A forerunner was the Dutch physician and psychologist Willem Winsemius with his investigation into industrial accidents at the Dutch steelworks Hoogovens. Winsemius proposed the task dynamics theory. This theory addressed, for the first time, the man–machine interaction during process disturbances. Accidents could be prevented by eliminating or reducing these process disruptions and improving the ergonomics of the workplace. A similar conclusion came from a Belgian study of miners in the late 1960s and from a British prospective study of four production companies in the early 1970s. The British study also revealed the impact of the general atmosphere in a company: the *them versus us* polarity, the sharp difference in perceptions of safety and risk between shop floor workers and management. Management could find safety an important issue for the company, but this statement remained simply a message on paper without any consequences for business operations.

'Human factors' was the term used in the United States, with a strong focus on quantification of human errors. But it was extremely difficult to relate human failure frequencies to system failure modes. This strong engineering orientation towards human failures denied the intrinsic variations in human behaviour. British developments in ergonomics followed a different approach. In Britain there was general criticism of companies that organised their production according to Taylor's principles of scientific management. This resulted in a focus on the well-being and the health of workers and the response of workers to work stress. Ergonomists developed the ergonomics of information, how information flows in a company could be disturbed and what information was available to the workers just before an accident occurred. This led to the information-accident model of Hale in 1970. Other models for occupational accidents, coming from the United States, the United Kingdom and Nordic countries at the beginning of the 1980s adopted a system approach. An accident was an outcome of a process disturbance, or the combination of individual, job, environment and materials acting in a dynamic system. Injuries occurred when risk factors came into contact with individuals.

The environmental and individual hypotheses, post-war period

Just after the Second World War, technical factors of accidents received increasing attention in publications by physicians, like the American John Gordon. The symbiosis between the environmental and individual hypotheses, present in the early years of the twentieth century, disappeared. Gordon introduced the epidemiological triangle. Accidents could be prevented by breaking the relationship of barriers between the danger, the environment and the victim, the corner points of the triangle. Based on this idea, the American physician-engineer William Haddon introduced in 1963 the hazard-barrier-target model. According to this model, a barrier reduces or stops the energy flow of the hazard. From this model, Haddon logically derived his ten prevention strategies, published in 1973. The input of these physicians was also important in other ways. They were not impressed by the quality of safety research at the time, and they disputed the dominance of safety professionals in companies. These professionals gave attention only to unsafe acts of workers, which meant primarily finding someone to blame for the accident. Accidents at work were seen only as a product of habits and behaviour, which was inconsistent with the modern world. Prevention was limited to training workers the right (read safe) way. Unsafe acts in the 1970s were more clearly explained as symptoms of faulty management, rather than causes of accidents. The term *accident-prone conditions* emerged in the United States. This was in contrast to 'accident-prone workers' or 'unsafe acts', which were rejected as outdated.

Meanwhile, it became more and more clear that occupational accident processes and accident scenarios were complex phenomena and therefore not easily predictable, as was noted earlier by Haddon. Also, solutions, and what later would be called 'best practices', were difficult to implement due to what in Australia was called the 'knowledge-application gap'. The inability to control safety and health risks adequately was due not to a lack of knowledge but to a lack of applying existing knowledge. The physician Kazutaka Kogi from the International Labour Office was one of the first in the late 1980s to collect and disseminate simple solutions for workplace safety and health problems in small and medium-sized industries. But the exchange of solutions was not a simple matter. In the Netherlands, research was carried out on the topic of transferability of solutions from one company or location to another. This was successful only when sections of production processes with successful solutions for safety problems were comparable with similar sections of other companies where safety problems occurred.

At the beginning of the 1970s, the concept of risk entered the occupational safety domain. In manufacturing companies, risk categories with relatively high probabilities of occurrence were present, but with relatively minor effects. In 1971, the American mathematician William Fine presented a simple, semi-quantitative method for risk assessment that was suitable for these risk categories, namely the risk ranking method. Other developments in the 1970s were based on a similar criticism of the principles of Taylorism to that in the United Kingdom. There was a call for workers' well-being, like job enrichment, matching tasks to humans and giving workers greater autonomy over their work.

Despite the comments on unsafe acts and accident-prone workers, the popularity of Heinrich's ideas on accident processes faded little, if at all. The human error model was still popular amongst safety practitioners. And, for instance in the Netherlands, the Lateiner method taught the relatively simple ratios of Heinrich on the causes and costs of accidents at vocational safety courses between 1960 and the late 1980s. The American consultant Alfred Lateiner worked with Heinrich. Lateiner's influence is still noticeable in the Netherlands in the present-day exam requirements of the Contractors Safety, Health and Environmental Checklist and the contemporary education of Dutch safety professionals.

A safety theory from 1982 that also stressed the behavioural aspects of accidents was the risk homeostasis theory, from the Canadian psychologist Gerald Wilde. This theory dealt with risk compensation, suggesting that people adjusted their behaviour according to the perceived level of risk. They became more careful where they sensed greater risk and less careful if they felt more protected. This theory, from the traffic safety domain, was applied in occupational safety but proved to be less successful there. More successful was the growing research on safety climate and culture in the 1990s. Behaviour as a primary starting point for the prevention of accidents at work made a comeback with the emergence of climate and culture surveys in companies. In the literature it is suggested that this renewed interest in behavioural interventions, which started in America, amounted to a resuscitation of the former accident proneness theory. Behavioural interventions were thought to improve organisations' safety performance through behaviour-based safety (BBS) programmes.

High-tech-high-hazard safety, post-war period

In the post-war safety of high-tech-high-hazard sectors there is not a leading country anymore. All three geographical areas discussed contribute to different topics within the safety domain. Attitudes toward technology became increasingly optimistic in the 1950s and early 1960s, after the introduction of supersonic flights and the Apollo space programme. Another example was the reaction to a blowout during a gas drilling in Schoonebeek, the Netherlands in 1965 (Figure 9.2).

> *The gas was howling around your ears. A fountain of mud and water spouted metres high into the sky with great power. The island seemed to be heaving and the derrick of thick steel flew almost 50 metres high, briefly bent forward and disappeared gracefully into the gas crater after a few seconds. Swallowed by the troubled earth. It was never seen again, said an eyewitness.*
>
> (Westerman 2017)

The reactions to the earthquakes in the Netherlands because of gas drillings 45 years later were quite different. In the early days there was no distrust in technology, or in the responsibility taken by companies, and no societal pressure to reduce these drillings. Technology optimism was visible in the rise of new technologies and the nuclear sector, in the scaling up of the process industry and in various popular publications depicting the mysteries of the atom. This lax approach inevitably led to major accidents, which, initially, hardly received any international public attention. Even major accidents like the 1957 British Windscale reactor fire, the 1966 French Feyzin LPG explosions and the 1966 British Aberfan mine accident attracted local, and national media attention. This

Figure 9.2 Blowout at Schoonebeek

Source: Nederlands Dagblad (1965)

changed dramatically in the 1970s, when technology optimism shifted to the awareness that technology could fail, with devastating consequences. Major accidents were widely reported in the international press. Only since then have accidents become the starting point for new theories, models and metaphors. Lasting until the end of the twentieth century, these were definitely the *golden years of safety science.*

Organisational factors

Many investigations of major accidents, including Aberfan, Clapham Junction, Piper Alpha, Challenger, Deepwater Horizon and many more, pointed at flaws in the organisation and in their risk and safety management. The term 'sloppy management' was introduced in the safety literature, and it became clear that not only technology but also management could hopelessly fail. Two contributions provided an understanding of organisational functioning: firstly, the publication of the Canadian management specialist Henry Mintzberg on the structure of organisations; and secondly, the search for latent factors in organisations.

Mintzberg's theory described the division of labour and the coordination of tasks, factors that primarily determined the structure of an organisation and its decision-making processes. The influence of a safety department and the impact of its arguments was directly related to the particular structure of the organisation and the position the department had vis-à-vis other departments.

From a safety point of view, the British chemical engineer Trevor Kletz, the American sociologist Charles Perrow and the British sociologist Barry Turner argued from the late 1970s onwards that major accident processes often started from less noticeable events. It was Turner who postulated the incubation theory, after his research on the major mine accident at Aberfan. This theory convincingly showed how organisations became blind to major hazards and risks. Perrow's 'normal accident' theory from 1984 was a reaction to the major nuclear near-accident at Three Mile Island. It was the first time that a classification of hazardous industrial clusters had been presented, based on just two determinants of technology: the connection between various process steps and the complexity of production processes. These less noticeable events paved the way for latent factors in an organisation and the associated incubation period of major accidents. Not long after Perrow's publication, the Berkeley group came forward with its high reliability theory (HRT). In fact, both HRT and the incubation theory were, strictly speaking, not safety theories but organisational theories. The incubation theory stressed the organisational failure mechanisms, while HRT outlined the conditions an organisation had to meet to safely manage a complex technology. While Perrow placed the emphasis on technological redundancy, for HRT the redundancy of the organisation was the focus to prevent latent factors from becoming problematic. In Mintzberg's terms, an HRT organisation should function permanently like an adhocracy.

After analysing the 1986 Chernobyl major accident, the British psychologist James Reason used the metaphor of resident pathogens for latent factors, with role and position later being visualised as holes in barriers in his well-known Swiss cheese metaphor. The origin of these holes lay in the decision-making processes of the 'blunt-end managers' and the impact on front-line, 'sharp-end' operators. For the first time, the Tripod model made the concept of latent factors operational with its basic risk factors. And 30 years after the publication of Turner's book *Man-made Disasters*, the operational latent factors were part of the management delivery systems of the bowtie metaphor. These delivery systems were necessary management processes to ensure the presence and quality of barriers. Latent factors are now investigated empirically, and they have gained an important place in the current understanding of complex accident processes.

Another view on organisational factors in high-tech-high-hazard sectors was based on the concept of safety culture and found its origin in anthropology. This discipline's approach, used to study cultures of extended families and tribes, was applied to describe the culture of organisations. In addition to this mostly academic view on culture, psychologists also developed a more analytical approach, known as the 'measurement of safety climate'. A normative approach, based on practical experience from companies, also appeared at the time, with names like 'Hearts and Minds' and the 'Safety Culture Ladder'. But until now, none of these approaches to safety culture and climate were able to establish a clear relationship with the safety level of a company.

The combination of technology, behaviour and organisation

There was still a focus on the actions of front-line operators. Research in the 1980s led to a new theory from an electrical engineer from Denmark, Jens Rasmussen. Operator behaviour during various process conditions was captured in three different categories: skill-, rule- and knowledge-based. This taxonomy of behaviour was a first step towards making human functioning part of design options for human–machine interfaces. The taxonomy initiated the domain of cognitive engineering, the design of man–machine interfaces to support operators in their complex tasks of dealing with unexpected process disturbances and deviations. Another of Rasmussen's contributions was the drift into danger. An actor analysis of his model showed the complexity of safety in workplaces and how dominant external factors pushed a process to the boundaries of functionally accepted performance.

Safety and risk management

Under the influence of military efforts, operational research emerged. Operational research was a statistical approach to military operations. Later this name was changed to 'operations research'. After the Second World War, this approach was applied as support for management decisions. Quantitative management was born and was fuelled by a second development, reliability engineering. Reliability engineering introduced systems thinking in the safety domain and facilitated the calculation of failure probabilities of components, of technical installations and also of human failures, as in the technique for human error rate prediction (THERP). The calculation of safety started with the introduction of the risk concept.

Risk and safety management are two management concepts that are not very different. Both dealt with managing and controlling accident scenarios of primary processes, which otherwise could lead to major accidents. Models of risk and safety management were based on a rational image of an organisation and not on a contextual picture of a complex socio-technical system. In literature, doubts were expressed about the formal rationality of organisations. Actions and decisions of management seemed rational only in retrospect. In a contribution to a World Bank conference in 1988, the American sociologist Ron Westrum put forward a classification of how organisations deal with safety issues, ranging from pathological to calculative and to generative.

Risk and risk perception

The risk concept was introduced in the high-tech-high-hazard safety domain in the 1950s and 1960s. The nuclear sector and the process and chemical industries were leading the way. Quantitative risk assessment (QRA) was introduced in the nuclear sector under the name 'probabilistic risk assessment' (PRA), with the famous WASH 740 and WASH 1400 reports of the American Office of Nuclear Reactor Regulation as landmark studies. And in the Netherlands, the series of 'coloured books' played an important role in the development and popularity of QRA outside the nuclear domain. Process engineers and safety professionals organised themselves in a movement called 'loss prevention', focusing on both a qualitative design philosophy and the quantification of risks. Risks were reduced to an acceptable level and inherently safe design was introduced by Kletz together with the American concept of 'layers of protection' (LOPA). In addition, chemical engineer Frank Lees influenced this young field with his now-standard work *Loss Prevention in the Process Industries*, from 1980. The term 'loss prevention' showed that, next to safety, financial consequences of major accidents were also a topic for this movement. The relative importance of loss prevention decreased in the 1990s as a result of the increased influence of market forces and the resulting pressure for cost reduction. At the same time, a new movement emerged called 'process intensification'. Process intensification primarily promoted smaller production units, using the principles of inherently safe design.

The increased media attention given to process incidents initiated social discussions about these major industrial accidents. The introduction of risk led to questions about the acceptability of risk. Occupational physicians, toxicologists and occupational hygienists developed criteria for acceptable risks of hazardous chemicals at work: the maximum acceptable concentrations (MAC values). The quantitative risk approaches seemed useful in the social debate on acceptable industrial risks. Nevertheless, there was criticism from the social sciences. The calculations did not consider various factors determining the risk perception of citizens and residents living near hazardous industries. After all, risk perception was defined by its context, which was divergent in different geographical areas and in different time periods. However, some aspects of risk perception remained remarkably constant. Despite the many attempts by experts to reconcile perception and behaviour with probabilistic risk calculations, citizens continued to consider other aspects and risk factors important. Risk communication and risk perception became two new areas of research.

Again theories, models and metaphors

Technicians, physicians, psychologists and sociologists developed knowledge about accident processes, and this knowledge went through a few transitions, starting with the role of environmental factors as causes of accidents, followed later by the behaviour of workers and then, specifically, unsafe actions. After the Second World War, these two approaches merged. Environmental factors and unsafe actions were integrated into what was called 'man–machine interaction' and is now known as 'socio-technical systems'. Several theories, models and metaphors were developed for retrospective analysis. Unfortunately, prospective analyses are still impossible.

Amongst researchers there is still no agreement on a central theory, model or metaphor for accident processes. This is an indication of the relatively young age and pseudo-scientific status of the discipline. Safety science became an academic discipline only in the 1970s and 1980s (Hale 2006). Researchers from various disciplines populate the domain. Social scientists usually approach organisational determinants of the accident process. Psychologists are focused on behaviour. And researchers from other sciences, like physicians, begin with hazards, scenarios and barriers as starting points of accident processes. A separate group consists of the technicians who calculate the chances of damage and injuries using a risk approach. These calculations are often limited to a small number of scenarios and do not provide information about determinants of possible accident processes (Pasman 2015).

This diversity of disciplines has advantages if the research is interdisciplinary. Various authors, such as Rasmussen and Hale, argue for such an approach, but conclude that the research is still too *multi*disciplinary and not sufficiently *inter*disciplinary. Every discipline publishes in journals of the author's own field, which researchers from other disciplines hardly ever read.

The interbellum period and the 1970s–1990s are the most productive periods of safety-related knowledge development. In the twenty-first century, few new theories, models or metaphors have been developed. It is mainly old wine in new bottles such as, for example, the concepts of Safety I and Safety II, which are comparable to the concepts of high reliability theory. Another worrying development is the lack of academic education and research in the safety domain. This is especially true for safety in the high-tech-high-hazard sectors. Knowledge about major accident processes has not increased, except for the observation that conditions that lead to a major accident are very complex and difficult to reproduce (Rademaker et al. 2014). The Netherlands and its neighbouring countries have relatively many high-tech-high-hazard industries. That should justify an investment in university research and education (AGS 2009; Jongen and Swuste 2017).

The fruits of progress

Did all these knowledge developments in the safety domain lead to a safer world, to a demonstrable effect? When looking at the last 150 years, the answer is yes. But looking at the period between 1990 and 2010, the image is more complex. If we focus on the data that are available at national level on the number of industrial accidents and major accidents, the answer to the above question varies between 'no' and 'maybe'. The answer simply depends on the interpretation of the numbers. The number of occupational accidents per year is an estimate and it remains constant or shows a slight increase. For major accidents this is more complicated. Because this type of accident occurs much less frequently, a much longer period than one year must be considered. The tables in Chapter 8 and Appendix 3 also justify a 'maybe' answer. Although conclusions are often drawn on the basis of absolute numbers, these numbers reveal very little. Those who are used to handling registration data first ask a few fundamental questions, such as: how reliable are the numbers registered and how large is the group of companies or activities where the accidents occur? The latter is unknown and therefore no definite answer to this question is available. At the level of a company or organisation, or of society as a whole, things can be different. There can be a decrease in absolute numbers, but here too the variation in absolute numbers can be determined by a variation in high-risk activities, or in process failures, or by a variation in personnel. The second approach, based on accident scenarios, seems possible only at the level of companies or organisations. There has been little experience with this approach to date, although recent research has been started to derive indicators from accident scenarios and the quality of barriers. These provide insight into the course and influence of these scenarios for both occupational safety and process safety (Swuste, Theunissen et al. 2016; Nunen et al. 2018; Schmitz, Swuste et al. 2020, 2021; Schmitz, Reniers et al. 2021; Schmitz et al. 2021a, 2021b).

Worrying between thinking and doing

Scientists think in terms of causes, while practitioners think in terms of remedies (Zwaard 2007). Scientists think critically and ask fundamental questions about the nature and content of models and the efficiency of interventions taken. As in other applied fields, the safety profession and science have their own dynamics (Boer 1967; Zwaard 2009). The developmental pathways sometimes run far apart. The system approach is accepted by academic safety experts, as is the socio-technical approach to the accident process, but the question is whether this also applies to the safety profession. Here, influencing the behaviour of employees and managers seems to be an important aspect. It is remarkable that safety professionals, who are generally technically trained, are so emphatically involved in behavioural psychology nowadays.

An inherent problem with the scientific approach is that it appears to be difficult to design unambiguous experiments to determine whether safety programmes, interventions, concepts or optimisations actually work. A purely scientific approach, using randomised controlled trials as a research design, makes modelling and quantifying the working ingredients of an intervention with 'real people' in 'real organisations' in an open organisation almost impossible. This research design, enabling various forms of bias to be addressed, runs up against a number of insurmountable practical and ethical problems in safety science research while studying the effects of interventions (Molen et al. 2007). If we take the example of 'life-saving rules' as a set of simple rules to limit the consequences of an incident, such as requiring fall protection to be worn when working at heights, it is hard to test scientifically what exactly the contribution is to the safety records (Peuscher and Groeneweg 2012). But from a practical point of view, life-saving rules offer simple and clear guidelines for safe behaviour that can be implemented and supported throughout. This dichotomy makes it difficult to publish articles on change processes in organisations in scientific journals. As a result, there appears to be a division between the real and the scientific world. Real-world knowledge is mainly described in handbooks and contains models with (some) face validity, supported by hard-to-verify but appealing case studies and anecdotes. In occupational safety and process safety, there are also organisational standards that relate, for example, to the presence of procedures, permits and registration systems, to frequencies of consultation or to establishing responsibilities. A large part of all these standards fits in with a deterministic approach where it is 'safe' if the established standards are met. These standards have their limitations. They are varied and sometimes poorly substantiated, have a limited scope and indicate what is presently accepted as important. Standards usually relate to the means used to achieve the goal, safety. There is little evidence yet as to whether these organisational standards are actually also effective.

There are, for example, quite a few companies measuring the quality of process safety against occupational safety performance. For example, if a company has few occupational accidents, the company takes this as an indication that it has everything under control in the area of process safety. Often reference is made to Heinrich's accident pyramid. Maybe for companies it is relatively easy to monitor and register occupational accidents, but the relationship between occupational safety and process safety performance has never been supported with scientific proof. One does not expect a relation, because accident processes and scenarios differ substantially between these two types of accidents.

The scientific approach tends to be reported in dedicated journals that require the application of scientific methods to relatively narrow aspects of safety. This scientific approach rarely produces inspiring stories and only sporadically produces readily applicable instruments. At the end of scientific publications, limitations and shortcomings of the study are listed and assessed but scientists often do not take the trouble to translate scientific insights into practical information that the professional in the field can apply. For many safety professionals, scientific insights are almost irrelevant (Kampen et al. 2010). There is a big difference here between the safety scientists involved in the development of the profession and safety professionals, including government officials, who are confronted with situations in which 'something needs to be done'. It is this *worrying between thinking and doing* that is so characteristic of the safety-related profession (Zwaard 1993b).

Safety as a science?

New questions arise after a description of 150 years of safety-related knowledge development, such as the question of what exactly this knowledge has produced, apart from the many books and papers. Is there an independent science of safety? In other words: is there anything distinct about safety science? Clearly, a safety language has developed within the domain. Safety experts, whether they are scientists or working within companies and organisations, speak of hazards and risks, of fail-safe installations, of inherent safety, of safety culture, of safety factors and barriers, of latent factors, of system redundancy and resilience of organisations, and so on. This suggests that safety science is a separate scientific domain. The central research theme of safety science as a science is the accident process and the determinants of this process as a first step to influence its outcomes. This applies to all safety-related themes and domains, including occupational safety and safety within the high-tech-high-hazard sectors.

Metaphors and models about the accident process were developed within safety science. A science focuses on the validation of models and metaphors. Theories are thus developed from empirical research. Nevertheless, this book shows that research into safety-related issues is taking place in many other scientific disciplines. Indeed, safety science has borrowed many concepts from other sciences. The progress in scientific thinking about safety is therefore strongly and indirectly dependent on developments in those sciences. For instance, the concept of culture comes from anthropology, the latent factors from sociology and resilience from the social sciences. Other common safety-related concepts often seem to be a matter of logical thinking and 'common sense' (Swuste et al. 2020c). However, these concepts, and concepts from other sciences, have been applied to safety-related problems and a safety-related jargon has developed. Nevertheless, given the current state of its theories, safety science as a scientific domain does not yet have a generally accepted paradigm.

Room for optimism

It is not all bad news. On the contrary, there is room for optimism. About 100 years ago, safety was still an exclusively practical matter. Only 50 years later, the mood started to change with a call for 'cooperation between science and safety' (Boer 1967; Zwaard 2009). Nowadays, safety professions and science are increasingly able to find each other and the boundaries between the separate domains are slowly dissolving.

A lot has changed since Dickens' *Hard Times*. A framework has been developed and a language has emerged of concepts and models, of words and metaphors.

That framework and that language can also be understood and applied outside of occupational safety and high-tech-high-hazard safety. For example, the concept of risk, as a 'crown jewel' of safety, is widely applicable within all kinds of safety domains. Models and instruments have been developed that do justice to their hard aspects, 'risk as a number', as well as to the soft aspects, 'risk as feeling'. The bowtie and the Swiss cheese metaphor, for example, appear to be widely applicable in various safety domains.

The time traveller again

Time to return to a visitor to the 1890 safety exhibition in Amsterdam. At the start of this book we fantasised that our time traveller would move to the present in a time machine, and that he would notice spectacular changes in our safety thinking nowadays.

It may be clear, even to our time traveller, that prevention is more important, and more efficient, than cure after an accident has materialised. But only in the final decades of the past century were innovative theories, management systems and technical provisions devised, optimised and implemented to put this wisdom into practice. Prevention and source control have now been given preference over all other measures, and the broad choice in solutions would be overwhelming to our time traveller. Fortunately, he or she is helped by innovative graphical representations that facilitate the understanding of different approaches. This includes approaches that vary from Swiss cheese, bowties, drift into danger, triangles and icebergs to dominoes, ladders and so on (Swuste, Gulijk et al. 2019). Methods like 'inherently safe design' and 'layers of protection', and various safety indexes, are developed to support effective implementation of safety. The safety professionals of today learn the approaches and methods and their utility in special training taking place at different levels. In contrast to the 1900s, there is a wealth of transferable and immediately applicable knowledge. It is a major achievement that safety and control measures are now being investigated and taught even at an academic level (Swuste and Sillem 2018). Our time traveller would notice that people from very different disciplines collaborate on safety. Nobody today is surprised when psychologists, sociologists, political scientists, safety professionals, physicians and mathematicians come together to investigate, for example, the role of fatigue in the functioning of humans in complex decision-making tasks, or in managing the complexity of modern production. To our time traveller, who might have worked in a safety team consisting of a manager, a safety engineer and a handful of blue-collar workers, it would be completely alien. Safety is no longer the exclusive domain of the technician in industry; it is a melting pot of expertise and experts.

Optimise or innovate?

Our time traveller would notice remarkable improvements in all industries: a larger and wider understanding of factors contributing to present safety, and widespread implementation of safety measures. But he or she would also recognise the growing complexity of technology and organisations and the need to develop safety science further. Continuous improvement and optimisation remain as essential as they were more than a century ago, and stagnation of safety performance must be prevented at all costs. This mantra can be heard in academic circles, and it reverberates in many industries. Some scientists propose a complete overhaul of safety thinking. According to these scientists, traditional ways of thinking no longer lead to any further improvements. They dislike the accident-related focus of safety and prefer to look at what goes well rather than what goes wrong. To the authors of this book, and perhaps to our time traveller, such a call for change is an incentive to develop new innovations for safety. It could actually serve as a wake-up call to reconsider the way safety processionals communicate about their initiatives. As with any innovation, there are some reservations, though, in the sense that progress should not lead to a misunderstanding of what has already been achieved and to rejecting the achievements of knowledge and tools developed in the past.

The need for cooperation

For one thing, there is a need for closer cooperation between business, government and scientific institutes. The way in which safety in hospitals is approached can be an inspiring example: the scientific researchers are often actively involved in the subject of their research. Academically trained surgeons help to develop safer methods for surgical procedures. With answers they are a stakeholder in their own research, and that increases the practical relevance of the research. By contrast, the number of scientists working in a refinery, for example, is very limited. A useful compromise could therefore be to involve organisations more actively in the scientific research process. Only recently did academic safety researchers have access to the practical knowledge and experience of operational experts, and funding from the petrochemical industry and aviation in, for example, the development of the safety management system Hazop, the Swiss cheese metaphor, the Hearts and Minds programme, the bowtie, barrier-based accident analysis methods and the life-saving rules. This has ensured that the practical comprehensibility and applicability of the developed concepts and metaphors is never forgotten. Anyone who opens a recent issue of, for example, *Safety Science* will find that there is still a lot to be gained in the scientific world in that regard. In addition to developing substantive knowledge, it is also important for academics to think about how to translate and disseminate this knowledge. The modern means of communication offer many hitherto underutilised opportunities to disseminate knowledge outside of peer-reviewed journals and scientific conferences.

To optimise existing approaches and develop new ones, much research is still needed, as well as intensive collaboration between scientists, professionals, administrators, governments and operational experts. Of course, with such an approach, it is of great importance that the various parties respect one another's independence, but close cooperation seems essential for further improvement of safety.

Sorcery

For our time traveller from the second half of the nineteenth century it is a foregone conclusion that in the field of safety remarkable progress has been made. Safety-related knowledge development has played an important role in this. The radical change in thinking about safety and the successes achieved are huge. When travelling back to the 1890s, our time traveller does not even try to explain this to the other visitors at the Amsterdam safety exhibition. They will probably dismiss his story as a lot of baloney.

In 1962, the British science fiction writer Arthur C. Clarke formulated his third law: *Any sufficiently advanced technology is indistinguishable from magic* (Clarke 1973). Our time traveller cannot give a bigger compliment to all those parties who have worked on improving safety in recent decades. The sorcery of safety science has saved the lives of hundreds of thousands of people and prevented the suffering of millions in the past century. Strengthened by this success, safety professionals and risk professionals must continue along this path together, so that our time traveller, if he travels to 2090, will see that a revolutionary improvement not yet imaginable to us has taken place: The safety of future generations will continue to be a joint commitment necessary to make the next leap forward.

Appendix 1

Reported 'man-made' incidents and major accidents from public literature, 1990–2010

Note: Empty cells indicate no information is available.

(Lees 1996, 2005; Rasmussen and Gronberg 1997; Khan and Abbasi., 1999; Chemical Safety Board 1988–2010; Office of Nuclear Regulations 2000–2010; HSE 2003a; Kinnersley and Roelen 2007; Clough 2009; OGP 2010; Abdolhamidzadeh et al. 2011; Mihailidou et al. 2012; Marsh 2012; Thomson 2013; Van Dort 2014; Kraaijvanger 2014; Wikipedia 2016; Zwaailichten 2016)

Date	*AVIATION Country, Location, Company (If Known)*	*Type*	*Direct Cause*	*Event*	*Deaths †; Wounded (Wou)*
1990					
	North Sea – Brent Spar Helideck	helicopter	collision	crashed	6†
Jun 10th	UK – Didcot Oxfordshire British Airways	BAC	decompression	safe landing	
Jan 4th	US – Madison FL NorthWest Airlines	Boeing	lost engine	safe landing	
Jan 25th	US – Cove Neck NY Avianca	Boeing		crashed	73†
Nov 25th	US – Denver CO	storage airport	jet fuel	fire	
Dec 3rd	US – Detroit MI NW Airlines	DC-Boeing		collision airport	8†
Nov 14th	Switzerland – Zürich Alitalia	DC		crashed	46†
1991					
Feb 5th	Greece – Mount Othrys Elleniki Polemiki	Lockheed		crashed	63†
	Norway – Noo Ekofisk veld	helicopter			3†
Feb 1st	US – Los Angeles USAir	Boeing-Fairchild		collision airport	
Mar 3rd	US – Colorado Springs CO United Airlines	Boeing		crashed	25†
Apr 5th	US – Brunswick GA Atlantic Set	Embrear		crashed	23†
Sep 11th	US – Eagle Lake TX Continental Expr	Embraer		crashed	14†
Dec 27th	Sweden – Stockholm Scandinavian Airline	MD		emergency landing	25 wou
1992					
Jan 20th	France – Barr Air Inter	Airbus		crashed	87†
Oct 4th	Netherlands–Amsterdam Bijlmer El Al	Boeing	lost engines	crashed	43† 25 wou
Dec 21st	Portugal – Faro Martinair	DC	windshear	crashed	56† 106 wou
Mar 22nd	US – New York USAir	Fokker	icing	crashed	27†
Jul 30th	US – New York NY TWA	Lockheed		fire on airport	
1993					
Jan 6th	France – Paris Lufthansa City Line	Havilland		crashed	4†
Apr 18th	Japan – Hanamaki Jap Air	DC	windshear		-

(*Continued*)

(Continued)

Date	*AVIATION Country, Location, Company (If Known)*	*Type*	*Direct Cause*	*Event*	*Deaths †; Wounded (Wou)*
Apr 4th	Netherlands – Schiphol	Cityhopper	oil pressure engine	crashed	3† 9 wou
Oct 27th	Norway – Overhalla Widerøe	Havilland		crashed	6†
Apr 6th	US – Aleutian Eilanden AK Chi E-Airl.	MD		crashed	2†
Dec 1st	US – Hibbing MA Northwest Airlines	Jetstream		crashed	18†
1994					
Apr 26th	Japan – Nagoya Chi Airlines	Airbus	pilot error	crashed	264†
Apr 4th	Netherlands – Amsterdam KLM Cityhopper	Saab		crashed	2† 9 wou
Dec 21st	UK – Coventry	Boeing		crashed	5†
Jul 2nd	US – Charlotte NC USAir	DC	violent storm	crashed	37†
Sep 8th	US – Aliquippa PA US Air	Boeing		crashed	132†
Oct 31st	US – Roselawn IN American Eagle	ATR		crashed	68†
Nov 22nd	US – St Louis MO TWA	MD-Cessna		collision airport	2†
1995					
Dec 13th	Italy– Verona Banat Air	Antonov		crashed	49†
Jun 9th	New Zealand – Tararua Ranges Ansett NwZ	Havilland		crashed	4†
Aug 21st	US – Carrollton GA Atlantic Southeast	Embraer		crashed	9†
1996					
Feb 6th	Atlantic Ocean – Puerto Plata Birgenair	Boeing		crashed	189†
Jun 13th	Japan – Fukuoka Garuda Indonesia	DC	motor	accident runway	3†
Jul 15th	Netherlands – Eindhoven	Hercules	bird swarm	crashed	34† 7 wou
Sep 25th	Netherlands – Den Helder	Dakota	mechanical failure	crashed	32†
Aug 29th	Norway – Svalbard Vnukovo Airlines	Tupolev		crashed	153†
May 11th	US – Florida Everglades ValuJet	DC		crashed	110†

Date	*AVIATION Country, Location, Company (If Known)*	*Type*	*Direct Cause*	*Event*	*Deaths †; Wounded (Wou)*
Jun 9th	US – Richmond VA Eastwind Airlines	Boeing		emergency landing	-
Jul 17th	US – East Moriches NY TWA	Boeing		crashed	230†
Dec 22th	US – Narrows VA Airborne Express	DC		crashed	
1997					
Dec 17th	Greece– Tessaloniki Aerovit	Yakovlev		crashed	70†
Dec 28th	Japan – Tokyo United Airlines		weather		-
Sep 8th	Norway – Noo Norne FPU	helicopter		crashed	12†
Jan 9th	US – Ida MI Comair	Embraer		crashed	29†
Jul 31	US – Newark NJ FedEx Express	MD		crashed	5†
Aug 6th	US – Asan Guam Stille Oceaan Kor Air	Boeing		crashed	228†
Aug 7th	US – Miami FL Fine Airlines	DC		crashed	
1998					
Sep 2nd	Canada – Halifax Swissair	MD		crashed	229†
Sep 25th	Spain – Malaga PauknAir	Bae		crashed	38†
1999					
Sep 2nd	Canada – Peggys Cove NS Swissair	DC		crashed	229†
Nov 25th	Italy – Adriatic sea	helicopter		crashed	13†
Jan 13th	Netherlands – Schinveld		mechanical failure	crashed	4†
Sep 14th	Spain – Girona Britannia Airways	Boeing	tempest	crashed	1† 42 wou
Jan 12th	UK Channel Islands – Guernsey	Fokker		crashed	
Dec 22th	UK – Great Hallingbury Kor Air Cargo	Boeing		crashed	4†
Jun 1st	US – Little Rock AR American Airlines	MD		crashed	11† 86 wou
Oct 21st	US – Nantucket MA Egypt Airlines	Boeing		crashed	
Oct 31st	US – Nantucket MA Egypt Air	Boeing		crashed	217†

(*Continued*)

(Continued)

Date	*AVIATION Country, Location, Company (If Known)*	*Type*	*Direct Cause*	*Event*	*Deaths †; Wounded (Wou)*
2000					
May 25th	France – Paris Air Liberté	MD	collision runway	crashed	1† 1 wou
Jul 25th	France – Paris Air France	Concorde		crashed	104†
Jul 4th	Greece – Tessaloniki Malév	Tupolev		crashed	-
Jul 12th	Austria – Wenen Hapaq-Lloyd	Airbus		crashed	-
Jan 31st	US – Anacapa Islands CA Alaska Airl	DC		crashed	88†
Feb 16th	US – Rancho Cordova CA Emery Worldwide A	MD	collision	crashed	3†
Mar 5th	US – Burbank CA Southwest Airlines	MD		crashed	43 wou
Jan 10th	Switzerland – Niederhasli Crossair	Saab		crashed	10†
2001					
Oct 8th	Italy – Milan SAS private	DC-Cessna		mid-air collision	118†
Nov 12th	US – New York NY Am Airlines Boeing	Airbus	turbulance	crashed	265†
2002					
Jul 1st	Germany – Überlingen Bashkirian – DHLB	Tupolev Boeing	controller system	mid-air collision	71†
Nov 6th	Luxemburg – Luxembourg	Fokker	fog	crashed	20†
Jul 16th	UK – Noo Leman field	helicopter		crashed	11†
Jul 10th	Switzerland – Werneuchen Swiss International	Saab	tempest		-
2003					
Oct 14th	Canada – Halifax	Boeing		crashed	7†

Date	*AVIATION Country, Location, Company (If Known)*	*Type*	*Direct Cause*	*Event*	*Deaths †; Wounded (Wou)*
Jan 8th	US – Charlotte Air Midwest	Beechcraft		crashed	21†
Mar 24th	US – Gulf of Mexico	helicopter		crashed	11†
Apr 8th	US – Gallup NM Giant Industries	refinery	maintenance	LOC/fire	4 wou
Aug 13th	US – Covington KY	Convair		crashed	1†
Oct 19th	US – Kirksville MO	Jetstream		crashed	13†
Dec 18th	US – Memphis TN FedEx Express	MD		crashed	-
2004					
2005					
Aug 14th	Greece – Grammatiko Helios Airways	Boeing		crashed	121†
Aug 6th	Italy – Palermo Tuninter	ATR		crashed	16†
Feb 20th	UK – Manchester British Airways	Boeing	motor	emergency landing	-
Jun 9th	US – Boston US Airways	Airbus	fire	crashed	-
Sep 21st	US – Los Angeles CA JetBlue Airways	Airbus	landing gear	emergency landing	-
Dec 8th	US – Chicago IL Southwest Airlines	Boeing	snow	crashed	1†
Dec 19th	US – Miami FL Chalk's Ocean Airways	Grumman		crashed	20†
2006					
Aug 13th	Italy– Air Algérie	Lockheed		crashed	3†
Oct 10th	Norway – Stord Atlantic Airways	Bae		crashed	4†
Dec 28th	UK – Noo Morecambe baai	helicopter		crashed	7†
2007					
Sep 9th	Denmark – Aalborg Scandinavian Airlines	Havilland	landing gear	crashed	-
Oct 27th	Denmark – Copenhagen Scandinavian Airlines	Havilland	landing gear	crashed	-

(*Continued*)

(Continued)

Date	*AVIATION Country, Location, Company (If Known)*	*Type*	*Direct Cause*	*Event*	*Deaths †; Wounded (Wou)*
Aug 9th	France – Polynesië Air Moorea	Havilland		crashed	11†
Aug 20th	Japan – Naha China Airlines	Boeing	fire	crashed	165 wou
2008					
Oct 7th	Australia – Exmouth Qantas	Airbus		crashed	70 wou
Nov 10th	Italy– Rome Rayanair	Boeing	bird swarm	emergency landing	10 wou
Aug 20th	Spain– Madrid Spanair	DC		crashed	154†
Jan 17th	UK – London British Airways	Boeing	fuel	crashed	-
Dec 20th	US – Denver Continental Airlines	Boeing	breakdown motor	emergency landing	38 wou
2009					
Jul 1st	Atlantic Ocean St Peter-Paul Air France	Airbus		crashed	228†
Mar 20th	Australia – Melbourne Emirates	Airbus	tail hit runway		-
Mar 12th	Canada – Newfoundland Atlantic Cougar Heli	heli Sikorsky	breakdown motor	crashed	17†
Feb 25th	Netherlands– Haarlemmermeer Turkish	Boeing		crashed	9† 121 wou
Oct 22nd	Netherlands– Bonaire Divi Air	Britten-Norman	motor	crashed	1†
Apr 1st	UK – Aberdeen Bond Eurocopter	helicopter	motor	crashed	16†
Jan 15th	US – New York US Airways	Airbus	bird swarm	landing Hudson	-
Feb 12th	US – Clarence Center NY Colgan Air	Bombardier		crashed	50†
Jun 30th	US – Charleston WV SW Airl	Boeing	decom-pression	emergency landing	-

Date	*TRANSPORT (Train) Country - Location (If Known)*	*Direct Cause*	*Event*	*Deaths-†; Wounded (Wou)*
1990				
Aug 5th	Canada – Kinselle Alberta		oil truck-train collision, fire	3†
Feb 2nd	Germany – Rüsselsheim		train-train collision	17† 80 wou
Mar 7th	US – Philadelphia PA		derailment	3† 150 wou
Dec 12th	US – Boston MA		train-train collision	453 wou
1991				
Oct 16th	France – Melun		train-train collision	16†
May 14th	Japan – Shigaraki Shiga			42†
May 16th	UK – Bradford-on-Tone Somerset		derailment fire	
Jul 21th	UK – Glasgow			4† 22 wou
Nov 12th	UK – Stanedge West Yorkshire		derailment tunnel fire	
Dec 7th	UK – Newport Severn Tunnel	collision		185 wou
Jul 15th	US – Dunsmuir CA	herbicides	herbicides LOC	
Jul 29th	US – Ventura CA	chemicals	chemicals LOC	
Jul 31th	US – Lugoff SC		train-train collision	8† 76 wou
Aug 28th	US – New York NY		derailment	5† 200 wou
Feb 16th	Switzerland – Saxon		crane-train collision	
1992				
Sep 25th	Denmark – Naestved	collision	emission CH_2N	
Nov 15th	Germany – Northeim		train-train collision	11† 52 wou
Oct 31th	Netherlands – Eindhoven		train-train collision	11 wou
Nov 30th	Netherlands – Hoofddorp		derailment	5† 33 wou
Apr 29th	US – Newport News VA		truck-train collision	1† 54 wou
Aug 12th	US – Newport News			12 wou
Aug 8th	Switzerland – Zurich		tram-train collision	1† 9 wou
1993				
Feb 5th	Netherlands – Den Haag		train-train collision	21 wou
Sep 22th	US – Mobile AL	bridge	derailment	47†
Nov 11th	US – Kelso WA		train-train collision	5†
1994				
Sep 29th	Germany – Bad Bramstedt		train-train collision	6† 67 wou
Jun 25th	UK – Greenock Schotland		derailment	2†
Oct 15th	UK – Cowden		train-train collision	5† 12 wou
May 16th	US – Selma AL		container-train collision	1† 100 wou
Aug 3rd	US – Batavia NY		derailment	108 wou

(*Continued*)

(Continued)

Date	*TRANSPORT (Train) Country - Location (If Known)*	*Direct Cause*	*Event*	*Deaths-†; Wounded (Wou)*
Dec 14th	US – Alray CA		train-train collision	
Mar 8th	Switzerland – Zurich		derailment	3 wou
Mar 21th	Switzerland – Däniken		crane-train collision	9†
1995				
Aug 11th	Canada – Toronto Ont		metro-train collision	3† 30 wou
Dec 25th	Spain – Jaèn	bridge	derailment	2†
Jan 31th	UK – Ais Gill Cumbria		derailment	1† 30 wou
May 26th	US – Flomaton AL		train-train collision LOC C_2H_3Cl	
Jun 16th	US – Gettysburg PA	boiler	explosion	3 wou
Oct 9th	US – Palo Verde AZ		derailment	1† 78 wou
Oct 25th	US – Fox River Grove IL		bus-train collision	7†
1996				
Apr 21st	Finland – Joleka		train-train collision	4† 75 wou
Nov 18th	France-UK – Channel Tunnel		tunnelfire	34 wou
Mar 8th	UK – Watford		train-train collision	1† 22 wou
Feb 1st	US – San Bernardino CA		derailment, LOC chemicals	2† 12 wou
Feb 16th	US – Silver Spring MD		train-train collision	11†
Mar 4th	US – Weyauwega WI		derailment, C_3H_8 fire	
Sep 16th	Switzerland – Courfaivre		train-train collision	30 wou
1997				
Oct 23th	Australia – Beresfield NSW		train-train collision	
Dec 9th	Germany – Hanover		train-train collision	90 wou
Sep 19th	Italy – Piacenza		train-train collision	8† 29 wou
Nov 13th	Switzerland– Appenz ell		train-train collision	17 wou
1998				
Jun 3rd	Germany – Eschede		derailment	101†
Mar 6th	Finland – Jyväskylä		derailment	10† 94 wou
Jul 13th	Netherlands – Weert		tractor-train collision	1† 4 wou
Mar 10th	US – Buffalo MT		bus-train collision	2† 4 wou
Mar 15th	US – Bourbonnais IL		truck-train collision	11† 100 wou
Oct 2nd	US – Hinesville GA		truck-train collision	2†
Nov 1st	Switzerland – Bern		train-train collision	2† >> wou

Date	*TRANSPORT (Train) Country - Location (If Known)*	*Direct Cause*	*Event*	*Deaths-†; Wounded (Wou)*
1999				
Dec 3rd	Australia – Glenbrook NSW		derailment	7†
Dec 30th	Canada – Mount St Hilaire Quebec		train-train collision, oil fire	2†
Mar 2nd	Germany – Göttingen		fire	-
Oct 5th	UK – Paddington London		train-train collision	31† 400 wou
Mar 28th	US – Tennga GA		bus-train collision	>>† wou
Aug 20th	US – Brookings SD		derailment	1† 1 wou
2000				
Jul 13th	Canada – Wainwright Alberta		truck-train collision	2 wou
Mar 8th	Japan – Tokyo Nakameguro		train-train collision	5† 63 wou
Jan 4th	Norway – Asta Amot		train-train collision	19†
Nov 11th	Austria – Kaprun		tunnel fire	155†
Oct 17th	UK – Hatfield Hertfordshire	shattered rails	derailment	4† 102 wou
Nov 1st	UK – Bristol		train-train collision	
Jun 6th	Switzerland – Hüswil		train-train collision	1†
2001				
Mar 27th	Belgium – Pécrot	language	train-train collision	8† 12 wou
Feb 7th	Canada – Toronto Ontario		derailment	2 wou
Apr 12th	Canada – Stewiacke Nova Scotia		derailment	22 wou
Dec 18th	Greece – Orestiada		derailment	1†
Jul 2nd	Netherlands – Eindhoven	filling gas	explosion	10 wou
May 15th	UK – Great Heck North Yorkshire		auto-train-train collision	10† 80 wou
May 15th	US – Toledo OH	runaway		
Jul 18th	US – Baltimore MD		derailment tunnel fire	
Sep 11th	US – Marshall TX		train-train collision	
Sep 13th	US – Wendover UT		train-train collision	more wou
2002				
Oct 13th	Australia – Benalla Victoria		truck-train collision	3† 1 wou
Sep 9th	Germany – Bad Münder		derailment, LOC C_3H_5ClO	
Nov 6th	France – Nancy	fire		12† 9 wou
Jul 20th	Italy – Rometta Marea		derailment	8†
Aug 20th	Netherlands – Amersfoort	leakage CH_2CHCN	LOC acrylonitril	

(*Continued*)

(Continued)

Date	*TRANSPORT (Train) Country - Location (If Known)*	*Direct Cause*	*Event*	*Deaths-†; Wounded (Wou)*
May 10th	UK – Potters bar London		derailment	7† 11 wou
Nov 24th	UK – Southall	derailment		no wou
Jan 18th	US – Minot SD		derailment, LOC NH_3	1† > wou
Apr 18th	US – Cresent City Fl		derailment	4† 142 wou
Apr 23th	US – Placentia CA		train-train collision	2† 22 wou
May 30th	US – Hempfield Township PA		auto-train collision	2† 2 wou
Jul 29th	US – Kensington MD		derailment	95 wou
Sep 15th	US – Farragut TN		derailment, LOC H_2SO_4	
Sep 27th	US – Jamaica NY		derailment	1†
Feb 21th	Switzerland – Chiasso		train-train collision	2† 3 wou
2003				
Jan 31th	Australia – Waterfall NSW		derailment	7†
Feb 3rd	Australia – Melbourne Victoria	runaway		
Jun 11th	Germany – Schrozberg		train-train collision	6† 25 wou
Mar 20th	Netherlands – Roermond		train-train collision	1† 7 wou
Jun 3rd	Spain – Albacete Valencia		train-train collision	19†
Jul 7th	UK – Evesham Worcestershire		bus-train collision	3†
Aug 3rd	UK – Romney		bus-train collision	1† 4 wou
Jun 20th	US – Commerce CA	runaway		
Oct 12th	US – Chicago IL		derailment	45 wou
Dec 18th	US – Alexandria VA		derailment	
Oct 14th	Switzerland – Zurich		train-train collision	1† 45 wou
Aug 7th	Switzerland – Gsteigwiller		train-train collision	1† 63 wou
2004				
Nov 15th	Australia – Berajondo Queensland		derailment	
May 21th	Netherlands – Amsterdam		train-train collision	20 wou
Jun 11th	Netherlands – Terneuzen	mobile crane	overturned	no wou
Sep 30th	Netherlands – Roosendaal		train-train collision	20 wou
Nov 22th	Netherlands – Arnhem	leakage $(CH_3)_3OCH_3$	LOC	25 wou
Feb 15th	UK – Tebay Cumbria	runaway		4†
Nov 6th	UK – Ufton Nervet Berkshire		auto-train collision	7† 100 wou

Date	*TRANSPORT (Train) Country - Location (If Known)*	*Direct Cause*	*Event*	*Deaths-†; Wounded (Wou)*
Jun 28th	US – Macdona TX		train-train collision, LOC Cl_2	6† 51 wou
Nov 11th	US – San Antonio TX		derailment	1†
Nov 29th	US – Richland FL		train-train collision	1† 3 wou
Sep 10th	Sweden – Nosaby		truck-train collision	2† 47 wou
2005				
Aug 5th	Australia – Cheakamus British Col		derailment, LOC NaOH	
Aug 1st	Greece – Kilkis		truck-train collision	1†
Jan 7th	Italy – Crevalcore		train-train collision	17†
Oct 23th	Italy – Apulia Bari		derailment	
Apr 25th	Japan – Amagasaki Hyögo		derailment	107† 549 wou
Dec 25th	Japan – Shonai Yamagate		derailment	5† 32 wou
Sep 5th	Netherlands – Amersfoort		train-train collision	10 wou
Nov 3rd	Netherlands – Wijhe		truck-train collision	1† 31 wou
Jul 26th	Austria – Gramatneusiedl		train-train collision	13 wou
Jan 6th	US – Granitville SC		train-train collision	9† 250 wou
Jan 26th	US – Glendale CA		auto-train-train collision	11† 100 wou
May 5th	US – Galt IL		derailment	
Jul 10th	US – Anding MS		train-train collision	2†
Aug 2nd	US – Raleigh NC		truck-train collision	2†
Sep 17th	US – Chicago IL		derailment	83†
Oct 15th	US – Texarkana AR		train-train collision, LOC $C_3H_8O_2$	1†
Feb 28th	Sweden – Ledsgard		derailment, LOC Cl_2	
2006				
Apr 28th	Australia – Victoria		truck-train collision	2† 28 wou
May 25th	Australia – Lismore Victoria		truck-train collision	
Feb 17th	Canada – Montreal	bridge	derailment	
Sep 22th	Germany – Lathem Emsland		collision	22†
Oct 11th	France – Zoufftgen Metz		train-train collision	5† 20 wou
Oct 17th	Italy – Rome		metro-metro collision	1† 50 wou
Dec 13th	Italy – Avio		train-train collision	2†
Jul 14th	Luxemburg		fire	31 wou
Jun 24th	Netherlands – Maastricht		train-train collision	41 wou
Nov 29th	Netherlands – Rotterdam		train-train collision	
Nov 21th	Netherlands – Arnhem		train-train collision	31 wou
Jul 3rd	Spain – Valencia		derailment	41† 47 wou

(*Continued*)

(Continued)

Date	*TRANSPORT (Train) Country - Location (If Known)*	*Direct Cause*	*Event*	*Deaths-†; Wounded (Wou)*
Aug 21th	Spain – Villada		derailment	6† 36 wou
Jan 6th	US – Possum Point VA		derailment	
Jan 16th	US – Brooks KY			
Jun 14th	US – Madera CA		train-train collision	
Jul 1st	US – Abington PA		train-train collision	36 wou
Oct 20th	US – New Brighton PA	bridge	derailment, C_2H_5OH fire	
Nov 9th	US – Baxter CA	runaway		2†
Nov 30th	US – North Baltimore OH		derailment	3 wou
May 17th	Switzerland – Thun		derailment	3†
2007				
Jun 5th	Australia – Kerang		truck-train collision	11† 23 wou
Jun 15th	Italy – Sardinia		train-train collision	3†
Mar 13th	Netherlands – Amsterdam		train-train collision	1 wou
Feb 23th	UK – Grayrigg Cumbria		derailment	1†
Jul 16th	US – Lakeland FL		auto-train collision	4†
Jul 17th	US – Plant City FL		auto-train collision	1†
Oct 3rd	US – Port Wentworth GA		tractor-train collision	
Nov 9th	US – Anacostia rivier	bridge	derailment	
Nov 30th	US – Chicago		train-train collision	> wou
2008				
Nov 27th	Australia – Cardwell Queensland		truck-train collision	2† > wou
Dec 31th	Canada – Villeroy Quebec		derailment, LOC C_3H_8	
Apr 26th	Germany – Landrückentunnel		sheep-train contact	17 wou
Mar 7th	Greece – Larissa		derailment	28 wou
Oct 11th	Netherlands – Gouda		train-train collision	
Feb 1th	UK – Barrow-upon-Soar	bridge	derailment	1 wou
Mar 1th	UK – Tring	loss of container		
Oct 6th	UK – Whatley Quarry Somerset	runaway		
Feb 5th	US – Boswell IN		auto-train collision	2† 1 wou
Mar 17th	US – Marysville WA		derailment	
Mar 25th	US – Canton Junction MA	runaway		150 wou

Date	*TRANSPORT (Train) Country - Location (If Known)*	*Direct Cause*	*Event*	*Deaths-†; Wounded (wou)*
May 28th	US – Woodland MA		train-train collision	1† 12 wou
Sep 12th	US – Chatworth CA		train-train collision	25† 135 wou
Oct 14th	US – Decator AL		derailment	1†
Nov 20th	US – Kent OH	bridge	derailment	
Nov 22th	US – Clarendon TX		derailment	
Jun 5th	Canada – Oshawa Ontario		derailment	
Sep 12th	Germany– Lössnitzgrundbahn		train-train collision	
Sep 16th	Ireland – Dublin		bus-tram collision	21 wou
Dec 15th	US – Marysville WA		auto-train collision	1 wou
2009				
Jan 1st	Australia – Innisfall Queensland		truck-train collision	1† 6 wou
Nov 16th	Ireland – Wicklow		derailment	
Jun 29th	Italy – Viareggio		derailment, gas explosion	32†
May 29th	Netherlands – Zwolle		train-train collision	1 wou
Jun 28th	Netherlands – Halfweg		bus-train collision	7 wou
Sep 24th	Netherlands – Barendrecht		train-train collision	1† 1 wou
Jan 7th	US – Cincinnati OH		derailment, LOC C_3H_8	
Jun 19th	US – Rockford IL		derailment, fire C_2H_5OH	1† > wou
Jun 22th	US – Washington WA		metro-metro collision	9†
Jul 9th	US – Canton Township MI		auto-train collision	5†
Nov 24th	US – Houston TX		derailment	
2010				
Feb 15th	Belgium – Buizingen		train-train collision	18†
Feb 25th	Canada – St Charles de Bellechasse		derailment	4 wou
Mar 30th	Canada – Pickering Ontario		derailment	
Jan 4th	Finland – Helsinki		derailment	
Apr 12th	Italy– Merano Zuid Tirool	landslide		9† 28 wou
Mar 24th	Norway – Alnabu Oslo	runaway		3† 4 wou
Jan 4th	UK – Carrbridge		derailment	
Feb 12th	US – Farragut WA		derailment	1 wou
Mar 15th	US – HoustonTX		bus-metro collision	20 wou

(Continued)

Date	*PROCESS INDUSTRY Country – Location, Company (If Known)*	*Direct Cause*	*Event*	*Deaths-†; Wounded (Wou)*
1990				
	Japan – Sodegraura	H_2	explosion	10† 7 wou
Nov 15th	Portugal – Porto de Leixhos	C_3H_8	gas cloud explosion	14† 76 wou
Mar 20th	UK – Stanlow Shell	$C_6H_3Cl_2NO_2$	internal explosion	1† 5 wou
Jul 25th	UK – Birmingham	Cl_2CO	emission	60 wou
Jul 5th	US – Channelview TX Arco	flammable substances	explosion, fireball	17† 5 wou
Jul 19th	US – Cincinnati OH	C_8H_{10}	gas cloud explosion	
1991				
Dec 13th	Netherlands – Botlek DSM	maintenance storage tank	$C_6H_5CO_2H$	7† 3 wou
Mar 12th	US – Seadrift TX	$(CH_2)_2O$	gas cloud explosion	1†
May 1st	US – Serlington LA		explosion	8† 120 wou
Jun 17th	US – Charleston SC	reactor mixture	internal explosion	
Jul 14th	US – Kensington GA	C_4H_6	gas cloud explosion	
Jan 5th	Switzerland – Nyon	Cl_2	emission	
1992				
Jan 23rd	Germany – Schkopau	Cl_2	emission	186 wou
Jul 8th	Netherlands – Uithoorn Cindu	raising temperature	explosion	3† 11 wou
Jul 21st	UK – Bradford Allied Colloids	storage chemicals	fire	33 wou
Sep	UK – Castleford Hickson & Welch	storage tank	fire	5†
Jul 28th	US – Westlake LA	welding failure	explosion	32 wou
1993				
Feb 2nd	Germany – Giesheim	reactor mixture	toxic LOC	
	Ireland – Ringaskiddy	chemicals	explosion	
1994				
Apr 8th	Netherlands – Zaandam Eurofill		explosion	1† 7 wou
May 27th	US – Bel Pre OH	CHs	explosion	3†
Dec 13th	US – Port Neal IA	NH_4NO_3	explosion	4† 18 wou
1995				
Oct 9th	UK – Wilton Teeside BASF	CH_2CHCH_3	fire	-
Apr 10th	US – Savannah GA Powell Duffryn	H_2S	fire	300 wou

Date	*PROCESS INDUSTRY Country – Location, Company (If Known)*	*Direct Cause*	*Event*	*Deaths-†; Wounded (Wou)*
1996				
Feb 1st	US – Martinez CA	H_2 plant	fire	2 wou
1997				
May 2nd	Japan – Yokkaichi Mie	erosion, corrosion	C_2H_4 explosion	
Jan 28th	Netherlands– Botlek Kemira	cleaning	explosion $TiCl_4$	1† 3 wou
Jun 22nd	US – Deer Park TX		fire explosion	30 wou
	US – Burnside	dryer	LPG	2† 2 wou
1998				
Sep 25th	Australia – Longford Vic	gas	fire, explosion	2† 8 wou
Sep 21st	France – Toulouse	NH_4NO_3	explosion	29† 2500 wou
Jan 7th	US – Mustang NV Sierra Chemical Co		2 explosions	4† 6 wou
Mar 4th	US – Pitkin LA Sonat Exploration Co	overpressure vessel	fire	4†
Mar 27th	US – Hansville LA Union Carbide	O_2 deficiency	suffocation	1† 1 wou
Apr 8th	US – Pattersen NJ Morton International	runaway reaction	explosion	9 wou
1999				
Jun 8th	Germany – Wuppertal	NaOH	explosion	91 wou
Oct	Netherlands – Geleen DSM	HCN	emission	-
Feb 19th	US – Allentown PA Concept Sciences	NH_2OH, high T high c	explosion	5† 4 wou
2000				
May 29th	UK – Grangemouth BP	current breakdown		
Jun 7th	UK – Grangemouth BP	breakdown in steam system		
Jun 10th	UK – Grangemouth BP	leakage catalytic cracker	CH emission	
Mar 27th	US – Pasadena TX Phillips	$CH_2(CH)_2CH_2$	explosion	1† 69 wou
2001				
Sep 21st	France – Toulouse	NH_4NO_3	explosion	30† 5000 wou
Feb 11th	US – St James	$C_6H_5C_2H_3$	fire	-
Mar 13th	US – Augusta GA BP Amaco	overpressure reactor	LOC hot plastic	3†
Jul 7th	US – Tusla OK	As	emission	138 wou
May 1st	US – Brazoria County TX 3rd Coast Ind		fire	

(*Continued*)

(Continued)

Date	*PROCESS INDUSTRY Country – Location, Company (If Known)*	*Direct Cause*	*Event*	*Deaths-†; Wounded (Wou)*
Aug 14th	US – Festus MO DRC Enterprises	hose rupture loading	LOC Cl_2	66 wou
Sep 11th	US – Dayton OH	failed cooling	explosion	-
Oct 13th	US – Pascagoula MS First Chemical Co	distillation column	explosion fire	
Oct 13th	US – Cincinnati OH Environmental	distillation column	explosion fire	
	US – OH	chemicals	explosion	17 wou
	US – PA	dynamite	explosion	1† 3 wou
	US – NV		fire, explosion	5 wou
2002				
	US – MS	dryer, rubber	dust explosion	4† 8 wou
	US – Austin TX	feeding	explosion	5 wou
2003				
Apr 1st	Netherlands – Geleen DSM	start up		3† 2 wou
Jan 2nd	US – Gnadenhutten OH Catalyst Sys	vacuum dryer	explosion $(C_6H_5)_2C_2O_4$	1 wou
Jan 13th	US – Rosharon TX BLSR Operating Ltd	discharge waste condensation	gas cloud explosion	3† 4 wou
Jan 29th	US – Kinston NC West Pharmaceutical	fine dust under roof	dust explosion	6† 12 wou
Apr 11th	US – Louisville KY DD Williamson	overpressure vessel	LOC NH_3	1†
Jul 20th	US – Baton Rouge LA Honeywell	leakage	LOC $SbCl_5$ Cl_2	4 wou
Sep 21st	US – Miami Township OH Isotech/Sigma	distillation tower NO	Explosion LOC CO	1 wou
Nov 17th	US – Glendale AZ DPC Enterprises	loading	LOC Cl_2	
2004				
Jan 1st	Australia – Moomba	brittleness of metal	gas cloud explosion	
Apr 12th	US – Dalton GA MFG Chemial	triallyl cyanurate	emission	154 wou
Apr 23rd	US – Illiopolis IL Formosa Plastic	overheating reactor	LOC C_3H_5OH	154 wou
Apr 23rd	US – Illiopolis IL Formosa Plastic co		explosion	5† 2 wou
Aug 19th	US – Ontario CA Sterigenics	sterilisation chamber	$(CH_2)_2O$ explosion	

Date	*PROCESS INDUSTRY Country - Location, Company (If Known)*	*Direct Cause*	*Event*	*Deaths-†; Wounded (Wou)*
Oct 1st	US – Washington DC	dust	dust explosions	14†
Dec 3rd	US – Houston TX Marcus Oil & Chemical	tank	explosion	1 wou
2005				
Dec 10th	Germany – Münchsmünster	C_6H_{14}	gas cloud explosion	20 wou
Mar 15th	Netherlands – Groningen Perkinelmer	$NaBH_4$	explosion	1† 1 wou
Jan 26th	US – Perth Amboy NJ Acetylene Service		explosion	3†
Jun 20th	US – Point Comfort TX Formosa Plastics		collision, gas cloud explosion	16 wou
Jun 24th	US – St Louis MO Praxair	gas cylinders	fire	-
	US – Forth Worth TX	solvents	explosion	4 wou
2006				
Mar 20th	Japan – Nigata	methylcellulose	explosion	17 wou
Mar 8th	Netherlands – Rotterdam Nerefco		explosion	1 wou
Jan 31st	US – Morgantown NC Synthron Inc	batch process	emission $CH_2COO(CH_2)_3CH_3$	1† 12 wou
Apr 29th	US – TX	ethyleen unit	fire, explosion	
May 5th	US – Raleigh MS Partridge Raleigh	welding production tank	explosion, fire	3† 1 wou
Jun 14th	US – Bellwood IL Universal Form Co	mixing department	explosion, fire	1† 5 wou
Jul 17th	US – Valley Center KS Barton Solvents	solvents	fire, explosion	
Oct 29th	US – Des Moines IA Barton Solvents	filling tank	explosion $CH_3COOC_2H_5$	
Nov 22nd	US – Danvers MA CAI/ Arnel Chemical	ink, paint	explosion	
Dec 19th	US – Jacksonville FL T2 Laboratories Inc.	runaway reaction	explosion $(CH_3C_5H_4)Mn(CO)_3$	4† 13 wou
2008				
Jul 3rd	Australia – Varanus Island	gas	explosion	-
	Canada – Toronto Sunrise Propane		explosion	2† 54 wou
Nov 28th	Netherlands – Rotterdam – Vopak		explosion	2 wou
Feb 7th	US – Port Wentworth GA Imperial Sugar	packing department	dust explosion	14† 38 gew
Jun 11th	US – Houston TX Goodyear Tire-Rubber	maintenance	heat exchanger NH_3	1† 7 wou

(*Continued*)

(Continued)

Date	*PROCESS INDUSTRY Country - Location, Company (If Known)*	*Direct Cause*	*Event*	*Deaths-†; Wounded (Wou)*
Jul 29th	US – Tomahawk WI Packaging Co	welding tank	explosion H_2	3† 1 wou
Oct 11th	US – Petrolia PA INDSPEC Chemical Co	transshipment smoking H_2SO_4	emission H_2SO_4	
Nov 12th	US – Chesapeake VA Allied terminals		tank rupture	4 wou
2009				
Jul 11th	Netherlands – Nijmegen Kelko	carboxylmethyl-cellulose	fire	1†
May 4th	US – West Carrollton OH Veolia Env.	fumes	fire, explosion	22 wou
Jun 9th	US – Garner NC ConAgra Foods		explosion	4† 12 wou
Dec 7th	US – Belvidere IL NDK Crystal Inc		explosion	2 wou
	US – Columbus WI Columbus Chem Ind		explosion	3 wou
2010				
Jan 23rd	US – Belle WV DuPont Co	leaky hose	emission CCl_2O	1†
Jan 30th	US – Belle WV DuPont co		emission CH_3Cl	

Date	*UP-DOWNSTREAM Country - Location, Company (If Known)*	*Installation*	*Direct Cause*	*Event*	*Deaths-†; Wounded (Wou)*
1990					
Apr 1st	Australia – Sydney BORAL	LPG opslag	LPG	BLEVE	
	Australia – St Peters	LPG opslag	LPG	explosion	-
May 14th	Netherlands – Botlek Esso	refinery	coating tank	explosion	1†
Aug	North Sea – Fulmar Alpha Shell	platform		explosion	3 wou
Jan 24th	US – Gulf of Mexico	pipeline		condensate LOC	
Mar 3rd	US – North Blenheim NY	pipeline	C_3H_8	gas cloud fire	2† 7 wou
Apr 1st	US – Warren PA	refinery	LPG	explosion fire	
May 6th	US – Gulf of Mexico	pipeline		major oil LOC	
May 30th	US – Gulf of Mexico Keyes Marine 303	platform	blow-out		
Sep 16th	US – Bay City TX	tanker	gasoline	explosion	1†
Nov 3rd	US – Chalmette LA	refinery	CHs	gas cloud explosion	
1991					
Aug 21st	Australia – Coode Island	storage tanks	C_3H_3N	internal explosion	
Dec 10th	Germany– Noord Rijn	refinery	corrosion – erosion	explosion	
Apr 11th	Italy – Genua Haven	tanker		grote oil LOC	
Aug 23rd	Norway – North Sea Sleipner Field	platform	structural failures	sunken platform	
Aug	UK – North Sea Fumar A	platform		explosion	3 wou
Jan 12th	US – Port Arthur	refinery		fire	2 wou
Mar 3rd	US – Lake Charles LA	refinery	hot oil, water	internal explosion	
Apr 13th	US – Sweeny TX	refinery	CHs	explosion	
Nov 3rd	US – Beaumont TX	refinery		fire	
1992					
Nov 1st	Australia – North West Shelf	platform	stability legs		
Nov 9th	France – La Mede	refinery	CHs	gas cloud explosion	-
Sep 1st	Greece – Elefsina	refinery	LPG	explosion emission	21† 20 wou
Oct 16th	Japan – Sodegaura	refinery	CHs	gas cloud explosion	10† 7 wou
Jun 30th	Netherlands – Lijnden	LPG storage		fire	
Dec 3rd	Spain – La Coruna Aegean Sea	tanker		major oil LOC	
Aug 29th	US – Gulf of Mexico Blake IV	platform	blow-out	fire	1 wou

(*Continued*)

(Continued)

Date	*UP-DOWNSTREAM Country - Location, Company (If Known)*	*Installation*	*Direct Cause*	*Event*	*Deaths-†; Wounded (Wou)*
Aug 31st	US – Gulf of Mexico	pipeline		major oil LOC	
Oct 8th	US – Wilmington CA		erosion-corrosion	H_2 explosion	
1993					
Jan 5th	UK – Shetland Island Braer	tanker		major oil LOC	
Aug 2nd	US – Baton Rouge LA	refinery	creepage	HC fire	
	US – Jacksonville	terminal	tank overflow	fire	1†
1994					
Feb 25th	Japan – Kawasaki	refinery		fire	
	Japan – Ueda	tank		explosion	1† 3 wou
Jul 4th	UK – Pembroke	refinery		fire	
Jul 24th	UK – Milford Haven Texaco	refinery	CHs	explosion	26 wou
Jan 12th	US – Gulf of Mexico Rowan Odessa	platform		fire	1†
1995					
Oct 16th	US – Rouseville	refinery	CHs	fire	
1996					
Nov 16th	Australia – Gulf of St Vincent Maersk	platform	failing platform leg	sunken platform	
	Italy – Paese	terminal	LPG	BLEVE	-
Feb 28th	Netherlands – Rotterdam CMI	storage co.	chemicals	fire	
Feb 15th	UK – Milford Haven Sea Empress	tanker		major oil LOC	
Jan 24th	US – Gulf of Mexico Sundowner 15	platform	blow-out	fire	
Jun 6th	US – Simpsonville Sacramento CA	pipeline	corrosion	emission diesel	
1997					
	US – Gulf of Mexico Pool Ranger 4	platform		sunken platform	
Jan 4th	US – Gulf of Mexico Pride 1001[E]	platform	blow-out	fire	
Jan 27th	US – Martinez CA	refinery	pipe rupture CHs	fire explosion	1† 46 wou
Dec 2nd	US – St Helena CA	pipeline	corrosion	emission gasoline	
1998					
	Atlantic OceanWest Rigmar 151	platform		sunken platform	

Date	*UP-DOWNSTREAM Country - Location, Company (If Known)*	*Installation*	*Direct Cause*	*Event*	*Deaths-†; Wounded (Wou)*
Jun 9th	Canada – Saint John	refinery		explosion, fire	1†
Nov 10th	Canada Nova Scotia Odyssey	tanker		major oil LOC	
Oct 6th	France – Berre l'Etang	refinery	corrosion CHs	fire	
Jul 3rd	Netherlands – Pernis Shell	refinery		$(CH_3)_3OCH_3$	1† 14 wou
Jul	UK – North Sea Glomar Arctic IV	platform		explosion	2†
Jan 26th	US – Gulf of Mexico	pipeline		condensaat LOC	
Apr 9th	US – Albert City IA	storage tank		BLEVE C_3H_8	2† 7 wou
Jul 17th	US – Gulf of Mexico Nabors Rig 269	platform	blow-out		3† 13 wou
Sep 29th	US – Gulf of Mexico	pipeline		major oil LOC	
Dec 3rd	US – Gulf of Mexico Petronius A	hoisting module	falling load	explosion	
1999					
Dec 11th	France – Gulf of Biscay Erika	tanker		major oil LOC	
Feb 19th	Greece – Tessaloniki	refinery	HCs	explosion fire	-
Jan 27th	US – Winchester KY	pipeline	material fatigue	emission oil	
Feb 9th	US – Knoxville TN	pipeline	material fatigue	diesel emission	
Feb 23rd	US – Martinez CA Tosco Avon	refinery	maintenance LOC	fire	4† 1 wou
Mar 3rd	US – Hunt TX	pipeline	corrosion	emission gasoline	
Mar 24th	US – Prince Wbaai Exxon Valdez	tanker		major oil LOC	
Mar 25th	US – Richmond CA	refinery	CHs	gas cloud explosion	
Apr 7th	US – Prince Georges MD	pipeline	corrosion	emission petroleum	
Jul 23rd	US – Gulf of Mexico	pipeline		major oil LOC	
Sep 9th	US – Gulf of Mexico NFX A	platform		blow-out fire	
2000					
Dec 31st	Norway – North Sea	platform			
Jan 19th	US – Gulf of Mexico	platform		major LOC	
Jan 21st	US – Gulf of Mexico	pipeline		major oil LOC	

(*Continued*)

(Continued)

Date	*UP-DOWNSTREAM Country - Location, Company (If Known)*	*Installation*	*Direct Cause*	*Event*	*Deaths-†; Wounded (Wou)*
Aug 19th	US – Carlsbad NM El Paso pipeline	pipeline	corrosion	explosion	10†
	US – Douglas WY		C_3H_8 lek	BLEVE	-
	US – TX	truck	C_3H_8	explosion	2† 1 wou
2001					
Apr 9th	Netherlands Dutch Antilles – Aruba	refinery	oil	fire	
Apr 16th	UK – Humber Estuary Killingholme	refinery	erosion corrosion gas	explosion, fire	185 wou
Mar 1st	US – Gulf of Mexico Ensco 51	platform	blow-out		
Apr 23rd	US – Carson City CA	refinery	leakage	fire	
Apr 28th	US – Lemont	refinery	CHs	fire	
Jul 13th	US – Gulf of Mexico Marine VI	platform	blow-out		
Jul 17th	US – Delaware City DE Motiva	refinery	maintenance tank	fire	1† 8 wou
Aug 14th	US – Lemont IL	refinery	CHs	fire	
Sep 21st	US – Lake Charles	refinery	CHs	fire	3 wou
	US – LA	refinery	leak	explosion	2 wou
2002					
Apr 14th	Canada – Brookdale Manitoba	pipeline	corrosion	fire	
Sep 1st	Greece – Eleusis	pipeline	oil	explosion, fire	14† 30 wou
Nov 7th	Netherlands – Vlissingen Total	refinery		fire	3 wou
Dec 12th	Netherlands – Europoort Kuwait Petrol	refinery	leak desulfuri-zation	gas cloud explosion	1† 1 wou
Nov 13th	Spain – Galicia Prestige	tanker		major oil LOC	
Feb 22nd	UK – Dundee Glomar Artic IV	platform	maintenance	explosion	3†
Aug 9th	US – Gulf of Mexico Ocean King	platform	blow-out	fire	
2003					
Jan 6th	Canada – Fort McMurray Alberta	refinery	CHs	explosion, fire	1 wou
Sep 28th	Netherlands – Geertruidenberg Amer	power plant	scaffolding	maintenance	5†
Mar 1st	US – Gulf of Mexico	platform		major LOC	
May 21st	US – Gulf of Mexico	platform		major LOC	

Date	*UP-DOWNSTREAM Country - Location, Company (If Known)*	*Installation*	*Direct Cause*	*Event*	*Deaths-†; Wounded (Wou)*
Sep 11th	US – Gulf of Mexico Parker 14-J	platform		failing crane	>> wou
2004					
Jul 30th	Belgium – Gellingen	pipeline	gas	explosion	24† 132 wou
May 4th	Netherlands – Vlissingen Total	refinery		explosion	3 wou
Aug 5th	Netherlands – Bergeijk Diffutherm	storage	bitumen tar plasma	explosion	4 wou
Nov 28th	Norway – Snorre A	platform	gas blow-out		
Apr 8th	US – Gallup NM Giant Industries	refinery	maintenance	LOC gas fire	4 wou
Apr 11th	US – Gulf of Mexico	platform		major LOC	
2005					
May 31st	Netherlands – Warfum NAM	CH_4	maintenance	gas cloud explosion	3† 2 wou
Dec 11th	UK – Buncefield Hertfordshire Oil	tank storage	oil overfilling	gas cloud explosion	60 wou
Mar 23rd	US – Texas City TX BP	refinery	start up CHs	explosion, fire	15† 180 wou
Jul 10th	US – Gulf of Mexico BP Thunderhose	platform	ballast system		
Sep 24th	US – Gulf of Mexico	platform		LOC condensation	
Nov 5th	Norway – North Sea	platform		LOC	
2007					
Apr 12th	Norway - North Sea Bourbon Dolphin	supplying	capsized		8†
Nov 5th	US – Delaware City DE Valero	refinery	maintenance reactor	N_2 suffication	2†
2006					
Apr 23th	Norway – North Sea Maersk Giant	platform	blow-out		
Feb 16th	US – Sunray TX Valero Refinery	refinery	leakage	fire C_3H_8	
Aug 16th	US – Pascagoula MS	refinery	oil	fire	
	US – Amarillo CO Valero McKee	refinery			
2008					
Feb 18th	US – Big Spring TX Refinery	start-up	C_3H_8	explosion	2 wou
summer	US – Corpus Christi TX CITGO	refinery		fire, emission HF	1 wou

(*Continued*)

(Continued)

Date	*UP-DOWNSTREAM Country - Location, Company (If Known)*	*Installation*	*Direct Cause*	*Event*	*Deaths-†; Wounded (Wou)*
2009					
Aug 21st	Australia – Montara Timor sea	platform	H_2S condensate	LOC	
	France – Dunkirk	refinery	fire		1† 5 wou
Feb 13th	Netherlands – Botlek Kuwait Petroleum	refinery		fire	
Jun 4th	Norway – North Sea Ekofisk	platform		ship/platform	
Jan 12th	US – Woods Cross UT Silver Eagle	refinery	leaking tank	gas cloud explosion	
Oct 31st	US – Carnes MS	tank storage		explosion	1†
Nov 4th	US – Woods Cross UT Silver Eagle	refinery	failing pipe	explosion	
2010					
Apr 2nd	US – Anacortes WA Tesoro	refinery	failing heat exchanger		
Apr 13th	US – Weleetka OK	tank storage		explosion	2†
Apr 29th	US – Gulf of Mexico Deepwater Horizon	platform	failing centralisation	explosion	11† 17 wou

Date	*NUCLEAR INDUSTRY Country - Location*	*Direct Cause*	*Event*
1990			
1991			
Feb 8th	Japan – Mihama	leakage	LOC radioactivity
Nov 17th	US – Scriba		fire shutdown
1992			
Aug 2nd	Canada – Pickering reactor 1	heavy water leakage	LOC H^3
Jan 20th	India – Tarapur Maharashtra	rod	LOC radioactivity
Mar 31st	India – Bulandshahr	fire	near meltdown
Feb 22nd	Japan – Fukushima		high pressure steam LOC
Mar	Russia – Leningrad	leakage	LOC radioactivity
Apr 6th	Russia – Tomsk	mechanical failure	explosive emissive radioactivity
Feb 3rd	US – Bay City TX	pump	shutdown
Feb 27th	US – Buchanan NY	failing system	shutdown
Mar 2nd	US – Soddy-Daisy TN	failing installation	shutdown
Apr 21st	US – Southport	failing generator	shutdown
Dec 25th	US – Newport MI	maintenance turbine	shutdown
1993			
1994			
Dec 10th	Canada – Pickering reactor 2	leakage cooling water	LOC heavy water
1995			
Feb 2nd	India – Kota Rajasthan	leakage	LOC waste water
Dec	Japan – Tsuruga	damage reactor	LOC Na^{22}
Jan 14th	US – Wiscasset ME	broken steam tube	shutdown
1996			
	UK – Scotland	leakage	LOC waste water
Feb 22nd	US – Millstone 1 2 CT	leakage pumps	shutdown
Sep 2nd	US – Crystal River 3 FL	failing equipment	shutdown
Sep 5th	US – Clinton IL	failing pump	shutdown
Sep 20th	US – Senaca 1 2 IL	failing water system	shut down
1997			
Mar 11th	Japan – Tokaimura		fire explosion
	Russia – St Petersburg	maintenance	LOC radioactive water
Sep 9th	US – Bridgman 1 2 MI	condenser system	
1998			
	Russia – St Petersburg	lek	
Feb	UK – Sellafield	leakage nucleair filter	radiation exposure
1999			
Dec 27th	France – Blayais	flooding	shut down

(*Continued*)

(Continued)

Date	*NUCLEAR INDUSTRY Country - Location*	*Direct Cause*	*Event*
Jun 18th	Japan – Shiga	runaway nuclear reaction	explosion
Sep 30th	Japan – Ibaraki Prefecture	adding too enriched U	nuclear reaction
Sep 30th	Japan – Tokai Mura	too much U in tank	chain reaction, contamination
Oct 4th	South Korea – Wolsung 3	cleaning	exposure radioactive gas
Oct	Ukraine – Pripyat	maintenance	exposure radiation
May 25th	US – Waterford CT	steam leakage	shutdown
Sep 29th	US – Lower Alloways Creek NJ	leakage	LOC toxic gas
2000			
Oct	UK – Sellafield	switch gear	power breakdown
2001			
	UK – Glasgow Hunterston B	LOC	radioactive groundwater
	UK – Dumfries Chapelcross	falling reactor rods	exposure operator
	UK – Sellafield	leakage tank	Tc^{99} LOC groundwater
Mar 6th	UK – Sellafield	leakage	Pu^{239} contamination workplace
May 6th	UK – Sellafield	leakage	Pu^{239} contamination workplace
Jul 6th	UK – Sellafield	overstroming	Pu^{239} contamination workplace
Sep	UK – Lancaster Heysham 1	reactor-coating cracks	
2002			
Jun 11th	Canada – Bruce reactor 6	maintenance	cracked reactor rods
Jan 21st	France – Manche	failing safety	shutdown
Apr 19th	Hungary – Paks	splinted rod	emissive J^{131}
Oct 22nd	India – Kalpakkam	leakage	LOC Na^{22}
	Japan – Onagawa	Leakage	exposure radioactivity
Jan	UK – Dungeness	Leakage	
Mar 11th	UK – Lancaster Heysham 1	failing installation	falling rod
May	UK – Dunbar Torness	metal fatigue pump	failing gas circulation
Nov 6th	UK – Didcot Harwell	leakage	Am^{241} contamination workplace
Nov 12th	UK – Caithness Dounreay	leakage	contamination hands, shoes
Feb 16th	US – Davis Besse NNP, Oak Harbor OH	corrosion pipes	shutdown
Jan 15th	US – Bridgman MI		fire
Mar 5th	US – Davis Besse NNP, Oak Harbor OH	corrosion reactor vessel	shut down

Date	*NUCLEAR INDUSTRY Country - Location*	*Direct Cause*	*Event*
2003			
2004			
May 16th	France – Lorraine Cattenom-2	electrical cables	fire, shut down
Aug 9th	Japan – Mihama 3	failing inspection	non-radioactive steam explosion
Apr 14th	UK – Bradwell	opening wrong valve	LOC radioactivity
Jun 9th	UK – Hartlepool	leakage	LOC H^3
2005			
	Japan – Fukushima	leakage	LOC radioactive steam
	UK – Sellafield	manual manipulation	major accident
	UK – Sellafield	maintenance	major accident
Feb 13th	UK – Sellafield	maintenance	contamination head, hands
Apr 20th	UK – Sellafield	leakage	LOC radioactivity
Jun 16th	US – Braidwood IL	leakage H^3	contamination groundwater
Nov	US – Braidwood IL	leakage H^3	contamination groundwater
Mar 6th	US – Erwin TN	leakage U^{235}	shutdown
Jul 25th	Sweden – Forsmark	shortcut cooling system	near meltdown
2006			
2007			
Jan 7th	UK – Sizewell	leakage	contamination
Jan 10th	UK – Sellafield	leakage	contamination
May 20th	UK – Caithness Dounreay	leakage	Pu^{239} Tc^{99} contamination
2008			
Dec 21st	Canada – Darlington Clarington	valve storage tank	LOC H^3
Jul 13th	France – Tricastin	leakage	LOC radioactive waste water
Dec	Japan – Hamaoka	leakage	LOC radioactive water
Sep 19th	UK – Sellafield	leakage	LOC radioactivity workplace
	UK – Dungeness B	critical nuclear reaction	placing rods
	UK – Sellafield	leakage	contamination
	UK – Sellafield	leakage	contamination
Apr 1st	UK – Sellafield	LOC cooling water	
May 28th	UK – Sellafield	leakage	contamination
Jul 24th	UK – Sellafield	leakage	LOC radioactive material
Nov 23rd	UK – Dungeness B	fire	closing reactor

(*Continued*)

(Continued)

Date	*NUCLEAR INDUSTRY Country - Location*	*Direct Cause*	*Event*
Dec 27th	UK – Capenhurts Urenco	absent safety	outside safety envelope
Nov 21st	US – Three Mile Island Harrisburg PA	maintenance	contamination radioactive dust
2009			
2010			
Aug 9th	France – Gravelines	filed retreav ng rods	chaotic rods
Jan 7th	US – Buchanan NY	automatic shut down	LOC H^3 steam
Feb 1st	US – Montpellier VT	leakage	LOC H^3

Appendix 2

High-tech-high-hazard safety, 1950–2010

First Period	*Major Accident*	*Theories, Models, Metaphors*	*Safety Management*	*General Management*
1930–1950				Quantitative, operations research, decision-making based upon mathematical, statistical models
1950–1970	Windscale (UK, 1957) Feyzin (Fr, 19'66) Aberfan (UK, 1966) Pernis (NL, 1968)		Loss prevention; Association of British Chemical Manufacturers (UK, 1964) Process safety techniques; Hazop, FMEA, FTA, (US, 1960, 1962, UK, 1963)	Modern management, company is open system, managing as a decision-making and information processing activity
1970–1990	Flixborough (UK, 1974) Beek (Nl, 1975) Seveso (Italy, 1976) Alfaques (Spain, 1978) Three Mile Island (US, 1979) Mexico City (Mex, 1984) Bhopal (India, 1984) Clapham Junction (UK, 1988) Piper Alpha (UK, 1988)	System safety; Johnson (US, 1970) Organisational culture; Turner (UK, 1971) Lose-tightly coupled organisation; Reeves (UK, 1972) WASH 1400 (US, 1975) Cause-effect diagram; Nielsen (Den, 1971) Changes, non-routine conditions causing accidents; Petersen (US, 1971), Johnson (US, 1973a, b) Gas cloud explosions; Nettleton (UK, 1976, 1977) Disasters, organisational incubation period; Turner (UK, 1976, 1978) Safety culture; INSAG (Int, 1988) Layers of protection analysis; AIChE (US, 1989) High reliability; Berkeley (US, 1989)	Self-regulating system, from detailed descriptions to goal regulation; Robens report (UK, 1972) MORT; Johnson (US, 1973a, b) Loss control management; Bird (US, 1974), Bird and Loftus (US, 1976)	Quality management; Deming, Juran (US, 1951, 1982) Humanisation of labour; Swain (US, 1973) Management systems are organisation specific; Schein (US, 1972) Typology of organisational structures; Mintzberg (Can, 1979) In search of excellence; Peters (US, 1982) Change masters; Kanter (US, 1984) Images of organisations; Morgan (US, 1986) Anglo-Saxon and Rhineland management models (1989)

Second Period	*Major Accident*	*Theories, Models, Metaphors*	*Safety Management*	*General Management*
1990–2010	Bijlmer (NL'92) Faro (Por, 1992) Eindhoven (NL, 1996) Eschede (Ger, 1998) Sellafield (UK, 1998–2008) Paris (Fr, 2000) Gellingen (Bel, 2004) Buncefield (UK, 2005) BP Texas (US, 2005) BP Macondo (US, 2010)	Tripod; Groeneweg (NL, 1992) Complexity decisions; Sagan (US, 1993) Normalisation of deviations; Vaughan (US, 1996) Operator, technical, management culture; Schein (US, 1996) Man-made disasters – reprint; Turner (UK, 1997) Swiss cheese – final version; Reason, (UK, 1997) Drift to danger; Rasmussen (Den, 1997) Bowtie; Visser (NL, 1998) Normal accidents – reprint; Perrow, (US, 1999) Practical drift; Snook (US, 2000) Resilience engineering; Hollnagel (Sw, 2004)	Sloppy management; Turner (US, 1994) Sense-making; Weick (US, 1995) Third phase, safety management, culture; Hale (NL, 1998) ARAMIS, I-risk; (EC, 2002) Socio-technical approach of safety; Wilpert (Ger, 2002)	Organisational learning; Senge (US, 1990) EFQM/INK management model (EU, NL, 1991) National cultures; Hofstede (NL, 1991) Organisational culture; Schein (US, 1992) Single- and double-loop learning; Argyris (US, 1992) Precautionary principle (Int. Rio Declaration, 1992) Six Sigma; Welch (US, 1995) Corporate Social Responsibility; SER (NL, 2000)

Can: Canada, Bel: Belgium, Den: Denmark, EC: European Council, EU: European Union, Fr: France, Ger: Germany, Gr: Greece, Int: International, Mex: Mexico, NL: The Netherlands, Por: Portugal, Sw: Sweden, UK: United Kingdom, US: United States

Appendix 3

Occupational safety 1800s–2010

Third Period	*Theories*	*Models, Metaphors*	*Safety Management*	*General Management*
1800s–1910	External factors; Heijermans (NL, 1905) External factors; Eastman (US, 1910),	Safety Technique; Thackrah (UK, 1930), Factory Act (UK, 1844), Westerouwen (NL, 1893), Calder (UK, 1899) Road to happiness, Safety First movement; US Steel (US, 1906)	Accidents are part of the job	Classical management
1910–1930	Externa; factors; Home Office (UK, 1911) Accident proneness; Greenwood and Wood (UK, 1919) External factors; DeBlois (US, 1926)	The 3 E's: engineering, education, enforcement; National Safety Council (US, 1914) Hazard is energy, the probabilistic approach; DeBlois (US, 1926) Accident costs 1:4; Heinrich (US, 1927) Accident causes 88:10:2; Heinrich (US, 1928) Accident mechanism, iceberg, 1:29:300; Heinrich (US, 1929)	Selection, training workers in 'safety idea', safety committees; Cowee (US, 1916) Safety engineer, standardisation procedures; Williams (US, 1927) Safety = efficient production; American Engineering Council (US, 1928) Good management is better than good tools; Heinrich (US, 1929)	Scientific management, observations, measurements, registration, selection, training workers, standards, procedures, cooperation between management and workers; Taylor (US, 1911)
1930–1950	External factors; Vernon (UK, 1936)	Accident process, dominoes; Heinrich (US, 1941) Epidemiological triangle; Haddon, Gordon; (US, 1949)	Management supports safety initiatives, analyses causes, develops and implements solutions; DeBlois, Heinrich (US, 1926, 1931) Accident prevention is similar to quality control; Heinrich (US, 1941) Management shows safety leadership; Armstrong et al. (US, 1945) Managing safety as a ladder; Heinrich; (US, 1950a)	Behavioural management, human relations, behaviour motivation, leadership; (US, 1930s)

(*Continued*)

(Continued)

Third Period	*Theories*	*Models, Metaphors*	*Safety Management*	*General Management*
1950–1970	Task dynamics, man–machine interaction; Winsemius (NL, 1951) Man–machine systems; Singleton (UK, 1967a, 1969)	Human factors, ergonomics; Swain (US, 1964), Singleton (UK, 1960) Hazard-barrier-target; Gibson, Haddon (US, 1961) Damage iceberg 1:100:500; Bird and Germain (US, 1966) Multi-causality of accidents, disrupted information flow; Hale and Hale; (UK, 1970), Dunn (UK, 1972) Organisational domino; Bird (US, 1974)		Modern management, company is open system, managing as a decision-making and information processing activity
sixth Period	*Theories*	*Models, Metaphors*	*Safety Management*	*General Management*
1970–1990	Prospective study, ergonomics system design, sloppy management; Powell et al. (UK, 1971) Risk homeostasis Wilde (Can, 1982)	10 preventive strategies; Haddon (US, 1973) Positive feedback; Sulzer and Santamaria (US, 1980) System model; Shannon (UK, 1980) Safety climate; Zohar (Israel, 1980a) Deviation model; Kjellén and Larsson (Sw, 1981) IJEM risk factors; Faverge (ILO, 1983) Information model; Saari (Fin 1984) Accident epidemiology; Stout (US, 1987) Human behaviour; Hale (NL, 1987) Behaviour classification; Rasmussen et al. (Den, 1987) Safety culture, INSAG (Int, 1988)	Safety management, multi-causality, audits, participative safety; Petersen (US, 1971, 1975, 1978) Self-regulating system, from detailed to goal regulation; Robens report (UK, 1972) ISRS; Bird and Germain (US, 1985) 100 low-cost solutions, Kogi et al. (ILO, 1989)	Quality Management; Juran (US, 1951; Deming 1982) Humanisation of labour: Swain (US, 1973) Management systems are organisation specific; Schein (US, 1972) Typology of organisational structures; Mintzberg (Can, 1979) Total Quality Management; Deming (US, 1982) In search of excellence; Peters (US, 1982) The change masters; Kanter (US, 1984)

				Images of organisations; Morgan (US, 1986) Anglo-Saxon and Rhineland management models (1989)
1990–2010		Tripod; Groeneweg (NL, 1992) Hearts; Lawrie et al. (UK, NL, 2006) Bowtie; Visser (NL, 1998) Safety maturity; Keil Center (UK, 1999) Occupational Risk Model; (NL, Gr, Den, UK, 2003)	3rd phase, safety management and culture; Hale and Hovden (NL, 1998) ISRS audit; Bird and Germain (US, 1990) Health & Safety audit system; HSE UK, 1991) Solutions exchange, Swuste (NL, 1996)	Organisational learning; Senge (US, 1990) EFQM/INK management model (EU, NL, 1991) National cultures; Hofstede (NL, 1991) Organisational culture; Schein (US, 1992) Single, double loop learning; Argyris (US, 1992) Six Sigma; Welch (US, 1995) Corporate social responsibility; SER (NL, 2000)

Can: Canada, Den: Denmark, Gr: Greece, Int: International, NL: The Netherlands, Sw: Sweden

References

Abdolhamidzadeh B Abbasi T Rashtchian D Abbasi S (2011). Domino effect in process-industry accidents. An inventory of past events and identification of some patterns. *Journal of Loss Prevention in the Process Industries* 24:575–593

Abeytunga P (1990). The SOLUTIONS Database, a means to share solutions. In: *Proceedings of the 23rd International Congress on Occupational Health*. International Commission on Occupational Health, Montreal

Agricola G (1556). *De Re Metallica*. Translated by Hoover H and Hoover L (1950). Dover Publications, New York

AGS (2009). *Adviesraad Gevaarlijke Stoffen. Strategie in de kennisinfrastructuur voor veilige chemie en energie (Advisory Board Hazardous Substances. Strategy of the knowledge structure for safety in chemical industries and energy)*. AGS, Den Haag

AGS (2010). *Adviesraad Gevaarlijke Stoffen. Risicoberekening volgens voorschrift. Een ritueel voor vergunningverlening (Advisory Board Hazardous Substances. Risk calculations by prescription: Rituals for granting permits and land-use planning)*. AGS, Den Haag

Aldrich M (1997). *Safety First: Technology, Labour and Business in the Building of American Safety 1870–1939*. John Hopkins University Press, Baltimore

Ale B (2002). Risk assessment practices in The Netherlands. *Safety Science* 40:105–126

Ale B (2003). *Risks and safety*. A historical outline. Technical University Delft. Faculty of Technology, Policy and Management (in Dutch). Dutch board for Occupational Health and Safety (1992). Advice on limit values for genotoxic carcinogens

Ale B Baksteen H Bellamy L Bloemhoff A Goossens L Hale A Mud M Oh J Papazoglou I Post J Whinston J (2008). Quantifying occupational risk: The development of an occupational risk model. *Safety Science* 46(2):176–185

Allen F Garlick A Hayns M Taig A (1992). *The Management of the UKAEA Working Group on the Risks to Society from Potential Major Accidents with an Executive Summary*. Elsevier Applied Science, London

Amalberti R (1993). Safety in flight operations. In: Wilpert B & Qvale T (Eds.). *Reliability and Safety in Hazardous Work Systems. Interdisciplinary Study Group New Technology and Work (NeTWork). Approaches to Analysis and Design*. Lawrence Erlbaum Associates Publisher, Hove UK

Amalberti R (2001). The paradox of almost totally safe transportation system. NeTWork proceedings. *Special Issue Safety Science* 37(2–3):109–126

Amalberti R (2002). Revisiting safety and human factors paradigms to meet the safety challenges of ultra-complex as safe systems. In: Wilpert B Fahlbruch B (Eds.). *System*

Safety. Challenges and Pitfalls of Interventions. Interdisciplinary Study Group New Technology and Work (NeTWork), p. 265–276. Pregamon, Oxford

Amalberti R Auroy Y Aslanides M (2004). *Understanding Violations and Boundaries*. Canadian Healthcare Safety Symposium, Edmonton, AL

American Engineering Council (1928). *Safety and Production. An Engineering and Statistical Study of the Relationship between Industrial Safety and Production*. Harper & Brothers Publishers, New York

Andreas H (1979). Veiligheid en systeembenadering (Safety and a system approach). *De Veiligheid* 55(11):605–609

Andriessen J (1974). Waarom wil men veilig werken? I-III (Why one wants to work safely? I-III). *De Veiligheid* 50(6):251–258, 50(7/8):315–320, 50(9):381–384

Andriesen S Swaan J (1991). *Beleidsevaluerend Onderzoek Naar de Melding en Registratie van Arbeidsongevallen* (Policy evaluating survey on notification and registration of occupational accidents). Rapport S119. Ministerie van Sociale Zaken en Werkgelegenheid, Den Haag

Aneziris O Papazoglou I Baksteen H Mud M Ale B Bellamy L Hale A Bloemhoff A Post J Oh J (2008). Quantifying RA for falling from height. *Safety Science* 46(2):198–220

Aneziris O Papazoglou L Kallianiotis D (2010). Occupational risk of tunnelling construction. *Safety Science* 48(8):964–972

Anonymous (1979). Democratisering en humanisering van de arbeid. Arbeidsomstandighedenbeleid van Minister Albeda zoals neergelegd in de Memorie van Toelichting op de Begroting van het ministerie van Sociale Zaken voor 1980 (Democratisation and humanisation of labour. Working conditions policy of Minister Alberda, as laid down in the Explanatory Memorandum to the Budget of the Ministry of Social Affairs for 1980). *De Veiligheid* 55(10):505–508

Anonymous (1980a). Zorg voor menswaardige arbeid. Arbeidsomstandighedenbeleid van Minister Albeda zoals neergelegd in de Memorie van Toelichting op de Begroting van het ministerie van Sociale Zaken voor 1981 (Care for decent work. Working conditions policy of Minister Alberda, as laid down in the Explanatory Memorandum to the Budget of the Ministry of Social Affairs for 1981). *De Veiligheid* 56(10):495–499

Anonymous (1980b). *U Speelt te Wild Met uw Kinderen en Verrekt een Spier – Ga Drie Plaatsten Terug* (You are playing too wildly and strain a muscle – go back three places). Mare, weekblad Rijksuniversiteit Leiden 18 september, pp. 9–11

Anonymous (1982). Het Swain veiligheidsproject (The Swain safety project). *De Veiligheid* 58(11):21–24

Anonymous (1983). Veiligheid tot welke prijs (Safety at what price). *De Veiligheid* 59(4):189–191, 199

Anonymous (1889a). *Inquiry into the Life and Labour of the People in London. Charles Booth (1886–1903)*. http://booth.lsc.ac.uk/static/a/3.html

Anonymous (1889b). Voorkoming van ongevallen in fabrieken en werkplaatsen in Nederland (Prevention of accidents in factories and workshops in The Netherlands). *De Ingenieur* 4(43):370–372

Anonymous (1890). Middelen ter voorkoming van ongevallen (Means to prevent accidents). *De Ingenieur* 5(2):17–18

Anonymous (1891). Het Veiligheidsmuseum te Amsterdam (The safety museum of Amsterdam). *De Ingenieur* 9(41):378

Anonymous (1892). Verslagen van de Inspecteurs van den Arbeid in het Koninkrijk der Nederlanden Over 1892. (Reports of the inspectors of labour in the Kingdom the

Netherlands 1892). Departement van Waterstaat, Handel en Nijverheid. De Gebroeders van Cleef,'s-Gravenhage
Anonymous (1893). Economische kroniek, voorkomen van ongevallen (Economical chronicle, prevention of accidents). *De Economist* 42(1):340–348
Anonymous (1896). Economische kroniek. voorkomen van ongevallen (Economical chronicle, prevention of accidents). *De Economist* 45(1):323–349
Anonymous (1897a). De Veiligheidswet (The safety law). *De Ingenieur* 13:153
Anonymous (1897b). Nederlandse Vereniging ter voorkoming van ongelukken (Dutch Association to prevent accidents). *De Ingenieur* 13:153
Anonymous (1909). Adres in zake het ongeval-Oldigs bij de maatschappij 'Electra' te Amsterdam (Address of the accident-Oldigs at the company 'Elektra' Amsterdam). *De Ingenieur* 24(10):227–229
Anonymous (1910a). *Industrial Accidents, a Selected List of Books*. Carnegie Library of Pittsburg
Anonymous (1911a). Reviews on Eastman, accidents and the law. *Michigan Law Review* 1910:9(1):81–82; *Harvard Law Review* 1911:24(3):250–252; *The Yale Law Journal* 1910:20(1):83–84; Deibler F (1911). *The Journal of Political Economy* 19(2):140–141; Nearing S (1911). *Annals of the American Academy of Political and Social Science* 37(1):233–234; Abbot G (1911). *The American Journal of Sociology* 17(3):403–405; *Columbia Law Review* 1911:11(4):388–389; Martin J (1911). *The Academy of Political Science* 26(2):315–317
Anonymous (1911b). Het ongeval op Staatsmijn 'Emma' op 26 April 1911 (The accident at the state mine Emma at April 26th 1911). *De Ingenieur* 26(21):553
Anonymous (1913). *Accident Prevention, Safety First*. The United Gas Improvement Company, Philadelphia
Anonymous (1914a). Opening van het Veiligheidsmuseum (Opening of the safety museum). *De Ingenieur* 29(26):506–507
Anonymous (1914b). *Safeguards for Machine Tools and Power Presses. Machinery's Reference Series nr 140*. The Industrial Press, New York
Anonymous (1915). Industrial accident statistics. *Science* 42(1077):238–239
Anonymous (1920). *Coal Mining Hazards*. The Travelers Insurance Company, Hartford, CN
Anonymous (1921). *Safety in Building Constructions*. The Travelers Insurance Company, Hartford, CN
Anonymous (1926a). Industrial Psychology. *Nature* 118(2969):462
Anonymous (1926b). The greatest industrial story ever told. *Safety* 12(5):133–140
Anonymous (1928a). The award of the Harriman medals. *Safety* 14(1):5–22
Anonymous (1928b). Our museum. *Safety* 14(5):131–135
Anonymous (1930). News and views. *Nature* 126(3174):320–324
Anonymous (1931). Science and the human factor. *Nature* 127(3213):805–807
Anonymous (1931–1939). Accident proneness in the scientific literature: Safety work in chemical industries. *Nature 1931* 128(3218):1–3
Anonymous (December 1935). Health and safety in industry. *Nature* 14:928–929
Anonymous (1936). Onderzoek naar de psychologie van arbeiders die dikwijls door een ongeval zijn getroffen (Research to the psychology of workers, frequently been hit by accidents). *De Veiligheid* 13(6):81–84; 13(7):98–100
Anonymous (1931–1939). Accident proneness in the scientific literature: Safety work in chemical industries. *Nature* 1931:128(3218):1–3; Accidents and their prevention. *Nature*

March 1943:17:409; Health and safety in industry. *Nature* December 1935:14:928–929; Accidents. *Nature* March 1937:13:446; Health and safety of industrial workers. *Nature* 1937:139(3525):857–859; Scientific aspects of industrial accidents. *Nature* 1937:140(3544):559–561; The toll of accidents. *Nature* 1939:September 16:504

Anonymous (March 1937). Accidents. *Nature* 13:446

Anonymous (1937a). Hoe en waarom gebeuren ongevallen (How and why accidents do occur) *De Veiligheid* 14(3):34–40

Anonymous (1937b). Georganiseerde bestrijding van ongevallen (Organised fight for accidents). *De Veiligheid* 14(4):58–63

Anonymous (1937c). Een goed boek over veiligheid en het voorkomen van ongevallen (Book review of Vernon 1936, Accidents and their prevention). *De Veiligheid* 14(6):81–86

Anonymous (1937d). Health and safety of industrial workers. *Nature* 139(3525):857–859

Anonymous (1937e). Scientific aspects of industrial accidents. *Nature* 140(3544):559–561

Anonymous (September 1939). The toll of accidents. *Nature* 16:504

Anonymous (1939a). Veiligheid begint van bovenaf (Safety starts at the top). *De Veiligheid* 16 (5):72–75

Anonymous (1939b). De cost gaet voor de baet uyt (You must lose a fly to catch a trout). *De Veiligheid* 16(11):146–148

Anonymous (1940a). Het Veiligheidsmuseum, jaarverslag 1939 (The safety museum, annual report 1939). *De Ingenieur* 55(29):A240

Anonymous (1940b). Psychotechniek en veiligheid (Psychotechnique and safety). *De Veiligheid* 17(7):106–107

Anonymous (March 1943). Accidents and their prevention. *Nature* 17:409

Anonymous (1949). Procedures for performing a failure mode effect and critical analysis. November 9, United States Military Procedure MIL-P-1629

Anonymous (1952). Accident proneness. *British Medical Journal* 2(4785):656–657

Anonymous (1953). Mijlpalen, 60 jaar veiligheidswerk (Milestone, 60 years of safety work). *De Veiligheid* 29(1):1–2

Anonymous (1958). Home safety project. Final report California State Department of Public Health.

Anonymous (1962). Basis voor effectieve veiligheidsoverdracht door de baas. (Basis for effective safety transfer by the boss). *De Veiligheid* 38(1):3–4

Anonymous (1964). Accident proneness. *The Canadian Medical Association Journal* 90:646–647

Anonymous (1965). ‘n Amerikaans oordeel over Europese fabrieken (An American judgement of European factories). *De Veiligheid* 41(2):19–20

Anonymous (1968). Grondbeginselen van de preventie in de bedrijven (Fundamentals of prevention in companies). *De Veiligheid* 44(3):63

Anonymous (1973). Psychotechniek bij de PTT. Catalogus bij gelijknamige tentoonstelling. Mededeling uit het Rijksmuseum voor de Geschiedenis der Natuurwetenschappen te Leiden (Psychotechnique at the National Post, Cataloque of the National Museum on the History of Science), nr 145. 9/3

Anonymous (1974). De ontwikkeling van veiligheids- en gezondheidsaspecten in het ontwerp en de uitvoering van machines en installaties (Development of safety and health aspects in the design and manipulation of machinery and installations). *De Veiligheid* 50(7/8):313

Anonymous (1977a). *America & Lewis Hine, Photographs 1904–1940*. Aperture, New York

Anonymous (1977b). Protecting production or workers British Society for the Social Responsibility in Science – BSSRS. *Nature* 270:93
Anonymous (1979). Democratisering en humanisering van de arbeid. Arbeidsomstandighedenbeleid van Minister Albeda zoals neergelegd in de Memorie van Toelichting op de Begroting van het ministerie van Sociale Zaken voor 1980 (Democratisation and humanisation of labour. Working conditions policy of Minister Alberda, as laid down in the Explanatory Memorandum to the Budget of the Ministry of Social Affairs for 1980). *De Veiligheid* 55(10):505–508
Anonymous (1980a). Zorg voor menswaardige arbeid. Arbeidsomstandighedenbeleid van Minister Albeda zoals neergelegd in de Memorie van Toelichting op de Begroting van het ministerie van Sociale Zaken voor 1981 (Care for decent work. Working conditions policy of Minister Alberda, as laid down in the Explanatory Memorandum to the Budget of the Ministry of Social Affairs for 1981). *De Veiligheid* 56(10):495–499
Anonymous (1980b). U speelt te wild met uw kinderen en verrekt een spier – ga drie plaatsten terug (You are playing too wildly and strain a muscle – go back three places). *Mare*, weekblad Rijksuniversiteit Leiden 18 September, 9–11
Anonymous (1982). Het Swain veiligheidsproject (The Swain safety project). *De Veiligheid* 58(11):21–24
Anonymous (1983). Veiligheid tot welke prijs (Safety at what price). *De Veiligheid* 59(4):189–191, 199
Anonymous (2010). *Biography Edwards Deming*, Wikipedia, consulted January 22, 2010
Antonsen S (2009). Safety culture and the issue of power. *Safety Science* 47(2):183–191
Arbous A Kerrich J (1951). Accident statistics and accident proneness. *Biometrics* 7(4):340
Archer B (1965). Safety preventing industrial accidents. *Design* 202:58–63
Argyris C (1976). Single-loop and double-loop models in research on decision making. *Administrative Science Quarterly* 21(3):363–375
Argyris C (1992). *On Organisational Learning*. Blackwell, Cambridge
Arlidge J (1892). *The Hygiene, Diseases and Mortality of Occupations*. Percival, London
Armstrong D (1949). Accident prevention. *Public Health Reports* 64(12):355–362
Armstrong T Blake R Bloomfield J Boulet C Gimbel M Homan S Keefer W Page R (1945). *Safety Organisation*. Prentice-Hall Inc., New York
Armstrong T Blake R Bloomfield J Boulet C Gimbel M Homan S Keefer W Page R (1953). *Industrial Safety*, 2nd edition. Prentice Hall Inc., New York
Arturson G (1981). The Los Alfaques disaster a boiling-liquid expanding-vapour explosion. *Burns* 7(4):233–250
Arxiu Nacional de Catalunya (1987). *Safety Posters 1925–1936*. Generalitat de Catalunya Departement de Culture, Barcelona
Ashe S (1917). *Organisation in Accident Prevention*. McGraw-Hill Book Co, New York
Ashford N (1976). *Crisis in the Workplace: Occupational Disease and Injury. A Report to the Ford Foundation*. MIT Press, Cambridge, MA
Association of British Chemical Manufacturers (1964). *Safety and Management, a Guide for the Chemical Industry*. Heffer & Sons Ltd, Cambridge
Aven T (2012). Foundational issues in risk assessment and risk management. *Risk Analysis* 32(10):1647–1656
Baker A (1982). The role of physical simulation experiments in the investigation of accidents. *Journal of Occupational Accidents* 4:33–45
Baker S Haddon W (1974). Reducing injuries and their results: A scientific approach. *The Millbank Memorial Fund Quarterly. Health and Safety Society* 52(4):377–389

Bakker M Berkers E (1995). Techniek ter discussie. In: Lintsen H Bakker M Homburg E Lente D Van Schot J Verbong G (Eds.). *Techniek in Nederland. De wording van een moderne samenleving 1800–1890 (Technique at discussion, In: Technique in The Netherlands. The Making of a Modern Society 1800–890). Deel VI Techniek en samenleving.* Walburg Pers, Zutphen

Bank R van de (2008). *Persoonlijke mededeling (Personal Correspondence)*

Baram M (1998). Process safety management and the implications of organisational change. In: Hale A Baram M (Eds.). *Safety Management, the Challenge of Change. Interdisciplinary Study Group New Technology and Work (NeTWork)*, pp. 191–205. Pergamon, Oxford

Baram M (2009). Globalization and workplace hazards in developing nations. *NeTWork Proceedings Special Issue Safety Science* 47(6):756–766

Barlow R Proschan F (1975). *Statistical Theory of Reliability and Life Testing.* Holt Rinehart Winston Inc., New York

Barnett R Brickman D (1986). Safety hierarchy. *Journal of Safety Research* 17:49–55

Beaumont P Else D (1990). Sharing solutions – the Share Program. *Journal of Health and Safety* 4:15–20

Becht S Franke R Geißelmann A Hahn H (2009). An industrial view of process intensification. *Chemical Engineering and Processing* 48:329–332

Beck U (1986). *Risikogesellschaft. Auf dem Weg in eine andere Moderne.* Suhrkamp, Frankfutt/Main

Bedford T (1951). Obituary H.M. Vernon. *British Medical Journal* 8(2):92–93

Beek F ter Fransen R Kerklaan P Swuste P (1982). Cyanamid. *Risicobulletin* 4(1):5–8

Belke J (2001). Chemical accident risk US industry. A preliminary analysis of accident risk data from US hazardous chemical facilities. Proceedings of the Loss Prevention & Safety Promotion in the Process Industries, Stockholm June 19–21, pp. 31275–1314

Bell J Healey N (2006). *The Causes of Major Hazards Incidents and How to Improve Risk Control and Health and Safety Management: A Review of the Existing Literature.* Reports 2006/117. Health and Safety laboratory, Buxton, Derbyshire

Bellamy L Ale B Geyer T Goossens L Hale A Oh J Mud M Bloemhoff A Papazoglou I Whinston J (2007). Storybuilder – A tool for the analysis of accident reports. *Reliability Engineering & System Safety* 92(6):735–744

Bellamy L Ale B Whinston J Mud M Baksteen H Hale A Papazoglou I Bloemhoff A Damen M Oh J (2008a). The software tool storybuilder and the analysis of the horrible stories of occupational accidents. *Safety Science* 46(2):186–197

Bellamy L Brouwer W (1999). AVRIM2, a Dutch major hazard assessment and inspection tool. *Journal of Hazardous Materials* 65:191–210

Bellamy L Geyer T Wilkinson J (2008b). Development of a functional model which integrates human factors, safety management systems and wider organisational issues. *Safety Science* 46(3):461–492

Bellamy L Mud M Damen M Baksteen H Aneziris O Papazoglou I Hale A Oh J (2010). Which management system failures are responsible for occupational accidents? *Safety Science Monitor* 14(1)

Bellhouse G (1920). *Accident Prevention and 'Safety First'.* Manchester University Press, Manchester

Benavides F Benach J Martinez J Gonzales S (2005). Description of fatal accident rates in five EU countries. *Safety Science* 43(8):497–502

Benner L (1975). Accident investigation. Multilinear events sequencing methods. *Journal of Safety Research* 7(2):67–73

Benner L (1985). Rating accident models and investigation methodologies. *Journal of Safety Research* 16:105–126

Bergen A van (2000). De jacht op de dode ogenblikken (The hunt for dead moments). *Arbeidsomstandigheden* 77(5):10–13

Bergsma J (1974). Het voorkomen voor zijn (To prevent). *De Veiligheid* 50(6):263–266

Berman D (1978). *Death on the Job*. Monthly Review Press, New York

Bernstein P (1996). *Against the Gods: The Remarkable Story of Risk*. John Wiley & Sons, New York.

Berg A van den (1985). The Multi-Energy Method, a framework for vapour cloud explosion blast prediction. *Journal of Hazardous Materials* 12:1–10

Beyer D (1916). *Industrial Accident Prevention*. Houghton Milner Co, Boston

Beyer D (1917). Accident prevention. *Annals of the American Academy of Political and Social Science* 70:238–243

Binneveld H (1991). *Een Zaak van Vertrouwen (A Matter of Trust). Arbeidsinspectie 1890–1990*. SDU, Den Haag

Bird F (1974). *Management Guide to Loss Control*. Institute Press, Loganville, GA

Bird F (1978). Materiële schade (Material damage). *De Veiligheid* 54(5):199–201

Bird F Germain G (1966). *Damage Control, A New Horizon in Accident Prevention and Cost Improvement*. American Management Association, The Comet Press

Bird F Germain G (1985). *Practical Loss Control Leadership*. International Loss Control Institute, Loganville, GA

Bird F Germain G (1990). *Practical Loss Control Leadership*. International Loss Control Institute, Loganville, GA

Bird F Loftus R (1976). *Loss Control Management*. International Loss Control Institute, Institute Press, Loganville, GA

Bjordal E (1980). Zijn risicoanalyses verouderd? (Are risk analyses out of date?). *De Veiligheid* 56(12):627–629

Blaauw G (1950). Het wekken van de veiligheidsgedachte in de bedrijven bij toezichthoudend personeel en arbeiders (Creating a safety notion in companies among foremen and workers). Voordracht voor de Commissie van overleg. *De Veiligheid* 26:159–164

Blair B (2004). Keeping Presidents in the nuclear dark. Episode 2. The SIOP option that wasn't. *The Defence Monitor* 33(3):1–2

Blake R (1963). *Industrial Safety*, 3rd edition. Prentice Hall Inc., New York

Blake R Armstrong T Bloomfield J Boulet C Gimbel M Homan S Keefer S Page R (1945). *Industrial Safety*. Prentice Hall Inc., New York

Blanchard R (1917). *Industrial Accidents and Workmen's Compensation*. University of Pennsylvania, Appleton & Co, New York

Blijswijk M van Mutgeert B (1987). Al denkend en pratend worden mensen meer veiligheidsbewust. Swain project DSM (thinking and talking makes people more safety conscious, Swain project at DSM). *Maandblad voor Arbeidsomstandigheden* 63(4):232–237

Bloemen J (1967). Veiligheidsorganisatie in een bedrijf – N.V. Nederlandse Aardolie Maatschappij (Safety organisation in a company – NAM). *De Veiligheid* 43(5):135–142

Boer A den (1967). Veiligheid en wetenschap (Safety and science). *De Veiligheid* 43(9):253–258

Boersma (1974). Opening address by Drs. Boersma (Minister of Social Affairs). In: Buschmann C (Ed.). *Loss Prevention and Safety Promotion in the Process Industry*. Proceedings of the 1st International Loss Prevention Symposium, The Hague 28–30 May.

Boesten A (1978). Risico-aanvaardbaarheid-1 (Risk acceptability-1). *De Veiligheid* 54(11):539–541

Boesten A (1979). Risico-aanvaardbaarheid-2 (Risk acceptability-2). De Veiligheid 55(2):87–88

Bogardus E (1911a). The relation of fatigue to industrial accidents. *The American Journal of Sociology* 17(3):351–374

Bogardus E (1911b). The relation of fatigue to industrial accidents. *The American Journal of Sociology* 17(4):512–539

Boodhoo K Harvey A (2012). *Process Intensification Technologies for Green Chemistry: Engineering Solutions for Sustainable Chemical Processing.* John Wiley and Sons Inc., New York

Boonstra J (1983). Een paradox in veiligheids- en gezondheidszorg (A paradox in safety and health care). *Tijdschrift voor Sociale Gezondheidszorg* 61(13):450–457

Boonstra J (1983/4). Veiligheid tot welke prijs (safety at what price). *Risicobulletin* 5(5–6):15–18, 6(1):12–16, 6(3):8–13, 6(5):12–15

Booth C (1889). *Life and Labour of the People.* McMillan, London

Borg ter B (1939). *De maatregelen ter beveiliging van het hoogovenbedrijf te IJmuiden sociaal hygiënisch beschouwd. (prevention of the Hoogoven steel Works IJmuiden from a social and hygienic perspective) Proefschrift faculteit Geneeskunde Rijksuniversiteit Utrecht.* Noord-Hollandsche Uitgevers Maatschappij, Amsterdam

Boskma P (1977). Definitie van het risicoprobleem (Definition of the risk problem). *De Veiligheid* 53(5):237–238

Boudri H (1979). Ergonomie en veiligheid, een veelzijdige benadering (Ergonomics and safety, a multifaceted approach. *De Veiligheid* 55(4):183–184

Bradley P Baxter (2002). Fires, explosions and related incidents at work in Great Britain in 1998/1999 and 1999/2000. *Journal of Loss Prevention in the Process Industries* 15(5):365–372

Brehmer B (1991). Modern information technology: Timescales and distributed decision making. In: Rasmussen J Brehmer B Leplat J (Eds.). *Distributed Decision Making. Interdisciplinary Study Group New Technology and Work (NeTWork)*, pp. 193–210. John Wiley & Sons, Chichester

Brehmer B (1993). Cognitive aspects of safety. In: Wilpert B Qvale T (Eds.). *Reliability and Safety in Hazardous Work Systems. Interdisciplinary Study Group New Technology and Work (NeTWork)*, pp. 23–42. Approaches to analysis and design. Lawrence Erlbaum Associates-Publishers, Hove UK

Breivik G (2011). Risk a plague or a joy? Some reflections on the dual nature of risk in present society. *Safety Science Monitor* 15(1):1–8

Brinke W ten Bannink B (2004). *Dutch Dikes and Risk hikes. A Thematic Policy Evaluation of Risks of Flooding in the Netherlands.* RIVM, Bilthoven. Report 500799002

Broadhurst A (1971). Factory legislation and safety training. *Industrial and Commercial Training* 4(1):38–40

Brouwer J Moerman P (2005). *Angelsaksen versus Rijnlanders. Zoektocht naar overeenkomsten en verschillen in Europees en Amerikaans denken (Anglo-Saxons Versus and Rhinelanders. A Search for Similarities and Diffences in European and American Ways of Thinking).* Garant, Apeldoorn

Brugmans I (1958). De arbeidende klasse in Nederland in de 19e eeuw, 1813–1870 (the working class in The Netherlands in the 19th century, 1813–1870). In: *Aula boeken.* Het spectrum, Utrecht

Bruin M de Swuste P (2008). Analysis of hazard scenarios for a research environment in an oil and gas exploration and production laboratory. *Safety Science* 46(2):261–271

Brundtland G (1987). Report of the world commission on environment and development: Our common future. Transmitted to the General Assembly as an Annex to document A/42/427 – Development and International Co-operation: Environment

Bryden R Gradinger S Dick N Paul T (2016). *Re-Energising the Life-Saving Rules*. Society of Petroleum Engineers, Dallas, April 11. doi: 10.2118/179289-MS

BSI (1999). *British Standard Institute, Bristol BSI-OHSAS 18001*. Occupational Health and Safety Assessment Series, Bristol

Buitelaar W Vreeman R (1985). *Vakbondswerk en Kwaliteit van de Arbeid. Voorbeelden van Werknemersonderzoek in Nederland. (Union and Quality of Work. Examples of Workers' Research in the Netherlands)*. Proefschrift TUDelft, SUN Nijmegen

Burdorf L Swuste P Kromhout H (1997). De komst en betekenis van de arbeidshygiëne in Nederland. *Tijdschrift voor Sociale Gezondheidszorg* 75:81–88

Burggraaf J Groeneweg J (2016). Managing the human factor in the incident investigation process. *Proceedings of the SPE International Conference and Exhibition on Health, Safety, Security, Environment and Social Responsibility in Stavanger, Norway, Ref. SPE 179207*. Society of Petroleum Engineers, London.

Burggraaf J Guldenmund F Groeneweg J (2020). Everything under control, check your variation! A Shewhartian view on process safety and other applications. *Safety Science* (submitted)

Burnham J (2008). Accident Proneness (Unfallneigung): A classic case of simultaneous discovery/construction in psychology. *Science in Context* 21(1):99–118

Bus J Swuste P (1999). Voortdurend op jacht naar Murphy (Constantly on the hunt for Murphy). *Maandblad voor Arbeidsomstandigheden* 76(3):30–33

Busch C (2018). Heinrich's local rationality: Shouldn't 'new view' thinkers ask why things made sense to him? Thesis MSc in Human Factors and System Safety, Lund University, Lund

Buschmann C (1972). Gevaarlijke stoffen en wetgeving (Hazardous substances and the law). *De Veiligheid* 48(3):81, (4):131, (5):177–179

Buschmann C (Ed.) (1974). Loss Prevention and Safety Promotion in the Process Industries. In: *Proceedings of the first International Loss Prevention Symposium, May 28th -30st Royal institute of Engineers (KIvI) and Royal Netherlands Chemical Society (KNCV)*. Elsevier, Amsterdam

Calder J (1899). *Prevention of Factory Accidents. Being an Account of Manufacturing Industry and Accident and a Practical Guide to the Law on Safe-Guarding, Safe-Working, and Safe-Construction of Factory Machinery, Plant and Premises*. Longmans, Green and Co, London

Cambon J Guarnieri F Groeneweg J (2006). Towards a new tool for measuring Safety Management Systems performance. In: Hollnagel E Rigaud E (Eds.). *Proceedings of the 2nd Symposium on Resilience Engineering*. Juan-Les-Pins, France, November 8–10, 2006

Carozzi L Stocker A (1932). The medical aspects of industrial accidents. *International Archives of Occupational and Environmental Health* 4(1):14–41

Carroll A (2008). *A History of Corporate Social Responsibility. The Oxford Handbook of Corporate Social Responsibility*. Oxford University Press, Oxford

Carson P Mumford C (1979). Major hazards in the chemical industry Part II their identification and control. *Journal of Occupational Accidents* 2:85–98

Carson R (1962). *Silent Spring*. Houghton Mifflin, Boston

Cates A (1992). Shell Stanlow fluoraromatic explosion -20 March 1990: Assessment of the explosion and of blast damage. *Journal of Hazardous Materials* 32(1):1–38

CCPS (1992). *Center for Chemical Process Safety (CCPS), Plant Guidelines for Technical Management of Chemical Process Safety*. American Institute of Chemical Engineers, New York

CCPS (2000). *Centre for Chemical Process Safety. Guidelines for Chemical Process Quantitative Risk Analysis*, 2nd edition. AIChE, New York

CCPS (2001). *Center for Chemical Process Safety, Simplified Process Risk Assessment*. American Institute of Chemical Engineers, New York

Chaplin C (1936). *Modern Times*. United Artists, Beverley Hills, CA

Chaplin R Hale A (1998). An evaluation of the use of the International Safety Rating System (ISRS) as intervention to improve the organisation of safety. In: Hale A Baram M (Eds.). *Safety Management, the Challenge of Change*. NeTWork Workshop, Bad Homburg; Elsevier Science, Oxford

Chemical Industries Association (2008). *Process Safety Leadership in Chemical Industries*. CIA, London

Chemical Safety Board, finished research projects, consulted 1990–2010, Januari 2016

Chhokar J Wallin J (1984). Improving safety through applied behaviour analysis. *Journal of Safety Research* 15:141–151

Choudhry R Fang D Mohamed S (2007). The nature of safety culture: A survey of the state of the art. *Safety Science* 45(10):993–1012

Clarke A (1973). *Profiles of the Future. An Inquiry into the Limits of the Possible*. Popular Library, New York

Clarke D (2005). Human redundancy in complex hazardous systems. A theoretical framework. *Safety Science* 43(9):655–677

Clarke D (2008). A Tribute to Trevor Kletz. *Journal of System Safety* 44(6). www.system-safety.org/ejss/past/novdec2008ejss/spotlight1_p1.php

Clarke S (2006). Contrasting perceptual, attitudinal and dispositional approaches to accident involvement in the workplace. *Safety Science* 44(6):537–550

Cleveland R Cohen H Smith M Cohen A (1979). *Safety Program Practice in Record Holding Plants*. DHEW-NIOSH Publications no. 79–136, NIOSH Morgantown

Clough I (2009). *100 Largest Losses 1972–2009: Large Property Damage Loss in the Hydrocarbon Industry*. Marsh Global Energy Risk Engineering, London

Cobben L Esch S van Roos J de Smelik J Span H Vossen J (1976). *Gevaren-analyse van de chemische bedrijven van D.S.M. in de westelijke mijnstreek in relatie tot de omgeving. Commissie Cobben in opdracht van de provincie Limburg (Hazard Analysis of Chemical Companies of DSM in the Western Mine Region in Relation to the Environment. Commission Cobben Commissioned by the Province of Limburg)*. Commissie der Koningin in de provincie Limburg, Sittard

Coglianese C Finkel A Zaring D (2009). *Import Safety*. University of Pennsylvania Press, Philadelphia.

Cohen A (1977). Factors in successful occupational safety training. *Journal of Safety Research* 9(4):168–178

Cohen A Smith M Anger W Self (1979). Protective measures against workplace hazards. *Journal of Safety Research* 11(3):121–131

Cohen A Smith M Cohen H (1975). *Safety Practices in High Versus Low Accident Rate Companies*. US Department of Health, Education, and Welfare, Centre of Disease Control, NIOSH, Cincinnati

Cohen H Jensen R (1984). Measuring the effectiveness of an industrial lift truck safety training program. *Journal of Safety Research* 15:125–135

Collins J Pizatella T Etherton J Trump T (1986). The use of simulation for developing safe workstation design for mechanical power presses. *Journal of Safety Research* 17:73–79

Collis E Greenwood M (1921). *The Health of the Industrial Worker*. Churchill, London

Comeche S (1979). Veiligheidsproblemen bij het ontwerpen van machines (Safety problems during the design of machines). *De Veiligheid* 55(3):109–113

Compes P (1982). Perspective of accident research by safety science. *Journal of Occupational Accidents* 4:105–119

Conrad J (Ed.) (1980). *Society Technology and Risk Assessment*. Academic Press, London

Cooke R (2009). A brief history of quantitative risk analysis. *Resources* 8–9

Cooper M (2017). *Personal Communication*

Copius Peereboom J (1941). Een nieuw schema voor het rubriceren van ongevalsoorzaken (A new scheme for the categorisation of accidents). *De Veiligheid* 18(12):162–176

Coppola A (1984). Reliability engineering of electronic equipment: A historical perspective. *IEEE Transactions on Reliability* R-33(1):29–35

Coronel S (1876). Beroepshygiëne (industrial hygiene). *De Economist, tijdschrift voor alle standen* pp. 189–220

Corvalan C Driscoll T Harrison J (1994). Role of migrant factors in work related fatalities in Australia. *Scandinavian Journal of Work Environment and Health* 20:364–370

Covello V Mumpower J (1985). Risk analysis and risk management: A historical perspective. *Risk Analysis* 5(2):104–120

Covello V Lave L Moghissi A Uppuluri V (1987). *Uncertainty in Risk Assessment, Risk Management and Decision Making. Advances in Risk Analysis*. Plenum Press, New York

Cowee G (1911). *Geology of Andover Granite*. Massachusetts Institute of Technology, Department of Mining and Metallurgy, Boston

Cowee G (1916). *Practical Safety Methods and Devices, Manufacturing and Engineering*. D. van Nostrand Company, New York

Cowee G (1931). *Common Stocks and the Next Bull Market*. Fort hill Press, Boston

Cowee G (1938). *Bankers and Brokers Blanket Bonds*. The Spectator, Philadelphia

Cowee G (1942). *Cumulative Liability and its Elimination in Connection with Fidelity Coverage*. Fort Hill Press, Boston

Cowee G (1960). *The Ups and Downs of Common Stocks*. Vantage Press, New York

Cozzani V Gubinelli G Salzano E (2006). Escalation thresholds in the assessment of domino accidental events. *Journal of Hazardous Materials* A129:1–21

Cozzani V Tugnoli A Salzano E (2007). Prevention of domino effects: From active and passive strategies to inherent safer design. *Journal of Hazardous Materials* A139:209–219

CPR 14E (1979). *Methods for the Calculation of Physical Effects of the Release of Dangerous Materials (Liquids and Gases). Yellow Book*. Directorate General of Labour, Den Haag (later PGS 2 en inmiddels vervangen door het handboek risicoberekeningen BEVI

CPR (1982). *Commissie voor Preventie van Rampen door Gevaarlijke Stoffen*. Experimenten met Chloor (Experiments with chlorine), Den Haag

CPR 12E (1988). *Commissie voor Preventie van Rampen door gevaarlijke stoffen*. Methods for Determining and Processing Probabilities. Red Book. The Hague.

CPR 16 (1989). *Commissie voor Preventie van Rampen door gevaarlijke stoffen*. Methods for the Determination of Possible Damage to People and Objects Resulting from Releases of Hazardous Materials. Green Book, The Hague.

CPR 18E (1999). *Commissie voor Preventie van Rampen door gevaarlijke stoffen*. Guideline for Quantitative Risk Assessment. Purple Book, Den Haag

Cremer and Warner (1980). *An Analysis of the Canvey Report*. Cremer and Warner, London

Cremer and Warner (1982). *Risk Analysis of Six Potentially Hazardous Industrial Objects in the Rijnmond Area*. Londen (de zogenoemde COVO-studie). Cremer and Warner, London

Creyghton J (1949). Dispositie tot ongevallen (Disposition for accidents). *De Veiligheid* 25:115–117

Creyghton J (1952). Jan ongeluk (Accident John). *De Veiligheid* 28:9–11

Crowl D Louvar J (2002). *Chemical Process Safety – Fundamentals with Applications*, 2nd edition. Prentice Hall, Upper Saddle River, NJ

CSB (2007). *Chemical Safety Board, Chemical Safety and Hazard Investigation Board Investigation Report no. 2005–04-i-tx Refinery Explosion and Fire, 15 Killed, 180 Injured*

Cullen W (1990). *The Public Inquiry into the Piper Alpha Disaster*. Department of Energy. Her Majesty's Stationary Office, London

Daft R Weick K (1984). Model of organisations as interpretation systems. *Academy of Man Review* 9(2):284–295

Damen M (2011). *Handleiding WebORCA, Werken met de Occupational Risk Calculator (Manual WebORCA, Working with the Occupational Risk Calculator)*. RIGO, Amsterdam

Dantzig D van (1956). Economic decision problems for flood prevention. *Econometrica* 24:276–287

Dantzig D van Kriens J (1960). Report of the Delta Committee, Part 3, Section II.2 *The Economic Decision Problem Concerning the Security of the Netherlands Against Storm Surges*. Delta Committee, The Hague (in Dutch)

Darbra R Palacios A Casal J (2010). Domino effects in chemical accidents: Main features and accident sequences. *Journal of Hazardous Materials* 183:565–573

DeBlois L (1919). Supervision as a factor in accident prevention. *NSC Proceedings* 8:132–136

DeBlois L (1926). *Industrial Safety Organization for Executives and Engineer*. McGraw-Hill Book Company, New York

DeBlois L (1927). Has the industrial accident rate declined since 1913? *Proceedings of the Casualty Actuarial Society* XIV:84–96

Dechy N Bourdeaux T Ayrault N Kordek M Le Coze J (2004). First lessons of the Toulouse ammonium nitrate disaster, 21 September 2001, AZF plant France. *Journal of Hazardous Materials* 111(1–3):131–138

DeLong T (1970). *Fault Tree Manual*. Texas A&M University, College Station, TX

Delvosalle C Fievez C Pipart A Debray B (2006). ARAMIS project. A comprehensive methodology for the identification of reference accident scenarios in process industries. *Journal of Hazardous Materials* 130(3):200–219

Deming W (1982). *Out of Crisis, Quality, Productivity and Competitive Position*. Cambridge University Press, Cambridge

Denson W (1989). The history of reliability prediction. *IEEE Transactions on Reliability* 47(3-SP):SP-321–328

Department of Employment (1975). *The Flixborough Disaster. Report of the Court of Inquiry*. Her Majesty's Stationery Office, London

Desta J Heuvel P van den (1987). *Veertig jaar Veiligheidskunde in verenigingsverband (Forty years of Dutch safety Association)*. Kluwer, Deventer

DGA (1979). Directoraat-Generaal van de Arbeid. *Hazard ad Operability Study, Why? When? How? Publicatie R no. 3 E*. Ministerie Sociale Zaken, Voorburg

DGA (1981). *Storingsanalyse. Waarom? Wanneer? Hoe? (Failure Analysis, Why, When, How?) V2*. Directoraat-Generaal van de Arbeid van het Ministerie van Sociale Zaken, Voorburg

DGA (1984a). Directoraat-Generaal van de Arbeid. *Instrumentele beveiligings- en gevaardetectiesystemen in de procesindustrie. Enkele principes en grondslagen (Instrumental Safety and Hazard Detection Systems in the Process Industry. Some Principles and Foundations). Publicatie V 6.* Ministerie Sociale Zaken en Werkgelegenheid, Voorburg

DGA (1984b). Directoraat-Generaal van de Arbeid. *Procesveiligheidsanalyse, aanzet tot het opsporen van inherente procesgevaren (Process Safety Analysis, Incentive to Detect Inherent Process Hazards). Publicatie V 7.* Ministerie Sociale Zaken en Werkgelegenheid, Voorburg

DGA (1989). Directoraat-Generaal van de Arbeid. *Procedures in de procesindustrie. Voorbeelden en voorstellen met betrekking tot het ontwikkelen, invoeren en beheren van procedures in de procesindustrie (Procedures in the Process Industry. Examples and Proposals of the Development, Implementation and Control of Procedures in the Process Industry). Publicatie V 18.* Ministerie Sociale Zaken en Werkgelegenheid, Voorburg

Dianous V de Fiévez C (2006). ARAMIS project: A more explicit demonstration of risk control through the use of bow-tie diagrams and the evaluation of safety barrier performance. *Journal of Hazardous Materials* 130(3):220–233

Dickens C (1854). *Hard Times*. Republished as Penguin Classic 1993

Dien Y (1998). Safety and application of procedures, or 'how do they' have to use operating procedures in nuclear power plants. *NeTWork Proceedings. Special issue Safety Science* 29(3):179–187

Dik J Hale A Poort R Roden N van Roggeveen V Schaardenburgh K van Swuste P Veld R in t Verwer C (2008). De onbalans in verantwoordelijkheid voor de veiligheid in Nederland (The imbalance in responsibility for safety in the Netherlands). *Tijdschrift voor Toegepaste Arbowetenschap* 21(2):55–60

Doherty J (1981). *Women at Work 153 Photographs by Lewis Hine*. Dover Publications, New York

Donaghy R (2009). One-death-is-too-many. Secretary of State report UK

Döös M Backstrom T Sundstrom Frisk C (2004). Human actions and errors in risk handling – an empirical grounded discussion of cognitive action-regulation levels. *Safety Science* 42(2):185–204

Döös M Laflamme L Backström T (1994). Immigrants and occupational accidents: A comparative study of the frequency and types of accidents encountered by foreign and Swedish citizens at an engineering plant in Sweden. *Safety Science* 18(1):15–32

Dop G (1967). Bedrijfsongevallen, oorzaken, classificatie, preventie (Occupational accidents, causes, classification, prevention). *De Veiligheid* 43(1):13–19; (2):37–40

Dop G (1977). Betrouwbaarheid en onbetrouwbaarheid van technische systemen I, II, III (Reliability and unreliability of technical systems I, II, III). *De Veiligheid* 53(1):3–9; 53(2):53–58; 53(3):111–117

Dop G (1979). Onbetrouwbaarheid van een technisch system (unreliability of technical systems I, II, III). *De Veiligheid* 55(1):19–25, 55(5):249–253, 55(11):593–597, 55(12):657–662

Dop G (1981). De carastrofe theorie 1–3 (The catastrophe theory 1–3). *De Veiligheid* 57(4):165–168, 57(5):217–220, 57(6)273–276

Dop G (1984). Bedrijfszekerheid en technische veiligheid (Company reliability and technical safety). *De Veiligheid* 60(9):433–437

Dop G (1985). Toeval en kansberekening (Coincidence and probability calculation). *De Veiligheid* 61(6):199–201

Dort R van (2014). 50 jaar gasexplosies in de Nederlandse industrie (50 years of gas explosions in the Dutch industry). *Tijdschrift voor toegepaste Arbowetenschap* 27(2):36–49

Dowell III A (2011). Is it really an independent protection layer? *Process Safety Progress* 30(2):126–131

Drimmelen D Swuste P Woude W van der Hoefnagels W Musson Y Hale A (1989). An example of solution-directed workplace analysis: "Pneumatic chipping of pile heads". Proceeding of the International Symposium of the Research Section of the ISSA "Vibration at Work" Vienna, Austria, April 19–21, pp. 168–174

Drupsteen L (2014). Improving organisational safety through better learning from incidents and accidents. PhD thesis Aalborg University, Denmark

Drupsteen L Groeneweg J Zwetsloot G (2013). Critical steps in learning from incidents: Using learning potential in the process from reporting an incident to accident prevention. *International Journal of Occupational Safety and Ergonomics* 19(1):63–77

Duell M (2016). Young girl seen in an iconic Aberfan rescue photo returns to the site of her flattened primary school with her grandson to lay a wreath marking 50 years since the disaster that killed 144. Mail online October 21. www.dailymail.co.uk/news/article-3858518/The-lives-entire-generation-extinguished-reached-prime-Aberfan-falls-silent-remember-144-victims-disaster-village-school-buried-landslide-50-years-ago.html

Duffield S (2003). *Major Accident Prevention Policy in the European Union: The Major Accident Hazard Bureau (MAHB) and the Seveso Directive*. IChemE, London, symposium series no 149

Dunn J (1972). A safety analysis technique derived from skill analysis. *Applied Ergonomics* 3(1):30–36

Dunn R (1929). *Labour and Automobiles*. New York International Publishers, New York

DuPont (1927). *Life of Eleuthere Irenee du Pont from Contemporary Correspondence 1778–1791*, 12 volumes. University of Delaware Press, Newark

Duyvis J (1979). Raakvlakken bedrijfsveiligheid – privéveiligheid (Common ground of industrial and private safety). *De Veiligheid* 55(6):309–310

Dyreborg J Hannerx H Tüchsen F Spangenberg S (2010). Disability retirement among workers involved in large construction projects. *American Journal of Medicine* 53(6):596–600

Eastman C (1908). The American way of distributing industrial accident losses, a criticism. *American Economic Association Quarterly* 10(1):119–134

Eastman C (1910). *Work-Accidents and the Law. The Pittsburgh Survey*. Charities Publications Committee, New York

Eastman C (1911). Three essentials for accident prevention. *Annals of the American Academy of Political and Social Science* 38(1):98–107

Eberts R Salvendy G (1986). The contribution of cognitive engineering to safe design of CAM. *Journal of Occupational Accidents* 8:49–67

EC (1989). *European Council Directive 89/391/EEC of 12 June 1989 on the Introduction of Measures to Encourage Encouragement Improvements in the Safety and Health of Workers at Work*. EC, Luxemburg

EFQM (2019). *European Foundation for Quality Management*. efqm.org consulted December 23

Eindhoven J van (1984). Wetenschappelijke benadering van risico, een metacontoverse (Scientific approaches of risk, a meta controverse). *Wetenschap en Samenleving Maart* 17–22

Eisner H Leger J (1988). The international safety rating system in South African mining. *Journal of Occupational Accidents* 10:141–160

Elling R (1991). Veiligheidsvoorschriften in de industrie (Safety procedures in the industry). Doctoral Thesis University of Twente, Faculty of Philosophy and Social Science. Publication WNW nr 8, Netherlands

Ellis L (1975). A review of research on efforts to promote occupational safety. *Journal of Safety Research* 7(4):180–189

Engels F (1844). *The Conditions of the Working Class in England.* Leipzig, gepubliceerd in het Engels 1887 Lovell, New York, London

Ericson II C (1999). Fault Tree Analysis, a history. In: *Proceedings of the 17th International System Safety Conference.* Springer-VerlagBerlin, Heidelberg

Ericson II C (2006). A short history of system safety. *Journal of System Safety* 42(3)

Etchells J (2005). Process intensification: Safety pros and cons. *Process Safety and Environmental Protection* 83(2):85–89

European Commission (1989). Directive concerning the execution of measures to promote the improvement of the safety and health of workers at their work and other subjects (Framework Directive). *Official Journal of EC*, June 12

Fadier E Garza C de la (2006). Safety design: Towards a new philosophy. *Safety Science* 44(1):55–73

Farewell V Johnson T (2010). Woods and Russel, Hill and the emerge of medical statistics. *Statistics in Medicine* 29 (14):1459–1476 Published online 2010 May 14. doi: 10.1002/sim.3893

Farmer E (1925). The method of grouping by differential tests in relation to accident proneness. *Industrial Fatigue Research Board* (Annual Report 1925):43–45

Farmer E (1940). Accident proneness and accident liability. *Occupational Psychology* 14(3):121–131

Farmer E (1942). The personal factor in accidents. *Medical Research Council. Industrial Health Research Board.* Emergency report no. 3. His Majesty's Stationary Office, London

Farmer E Chambers E (1926). A psychological study of individual differences in accident rates. *Industrial Health Research Board.* Report no. 38. His Majesty's Stationary Office, London

Farmer E Chambers E (1929). A study of personal qualities in accident proneness and proficiency. *Medical Research Council. Industrial Health Research Board.* Report no. 55. His Majesty's Stationary Office, London

Farmer E Chambers E (1939). A study of accident proneness among motor drivers. *Medical Research Council. Industrial Health Research Board.* Report no 84. His Majesty's Stationary Office, London

Farmer E Chambers E Kirk F (1933). Test for accident proneness. *Medical Research Council. Industrial Health Research Board.* Report nr 68. His Majesty's Stationary Office, London

Faverge F (1983). Accidents, human factor. In: Parmeggiani L (ed.). *Encyclopaedia of Occupational Health and Safety*, 3rd revised edition. International labour Office, Geneva

Faverge J (1967a). *Research in Belgian Coal Mines.* Communauté Européenne de Charbon et d'Acier, Luxembourg

Faverge J (1967b). *Research in French Coal Mines.* Communauté Européenne de Charbon et d'Acier, Luxembourg

Faverge J (1970). L'homme agent d'infiable et de fiabilite du processus industriel. *Ergonomics* 13(3):301–327

Fawcett H (1965a). *Chemical Booby Traps. Safety Industrial & Engineering Chemistry,* pp. 89–90. ACS Publications, Washington, DC

Fawcett H (1965b). The literature on chemical safety. *Journal of Chemical Education* 42(10):A815–A818; 42(11):A897–A899

Feggetter A (1982). A method for investigating human factor aspects of aircraft accidents and incidents. *Ergonomics* 25:1065–1075

Fellner D Sulzer B (1984). Increasing industrial safety practices and conditions through posted feedback. *Journal of Safety Research* 15:7–21

Fetter Z (1947). Verslag van de 25e vergadering van de Commissie van overleg inzake veiligheid en hygiëne (Report on the 25th meeting of the Consultative Commision on safety and hygiene). *De Veiligheid* 23:84–92

Ficq C (1976). *Rapport Over de Explosie bij DSM Beek (L)*, 7 November 1975. Gaswolk-explosie in Naftakraker II. Korps Rijkspolitie Maastricht, Dienst Bewaking en Beveiliging DSM, Dienst Stoomwezen, Arbeidsinspectie

Fine W (1971). *Mathematical Evaluation for Controlling Hazards*. Naval Ordnance Laboratory, white oak, Silver Spring, Maryland

Fischhoff B (1995). From risk perception to risk communication: Risk perception and communication unplugged: 20 years of process. *Risk Analysis* 15(2):137–145

Fischhoff B (2019). Tough calls. How we make decisions in the face of uncertainty and incomplete information. *Scientific American* 321(3):74–79

Fischhoff B Furby L Gregory R (1987). Evaluating voluntary risk of injury. *Accident Analysis & Prevention* 19(1):51–62

Fischhoff B Lichtenstein S Slovic P Derby S Keeney R (1981). *Acceptable Risk*. Cambridge University Press, Cambridge

Fischhoff B Watson S Hope C (1984). Defining risk. *Policy Sciences* 17:123–139

Fleming M (2001). *Safety Culture Maturity Model*. Keil Centre, HSE, HMSO, Norwich

Fletcher J (1978). Total loss control. *De Veiligheid* 54(5):203–207

Flin R Mearns K O'Connor P Bryden R (2000). Measuring safety climate identifying common features. *Safety Science* 34:177–192

Fox D Hopkins B Kent W (1987). The long term effects of a token economy on safety. Performance in open-pit mining. *Journal of Applied Behaviour Analysis* 20:215–224

Fraser J (1951). *Psychology*. Sir Isaac Pitman & Sons LTD, London

Frederik W (1951). Statistische analyse van industriële ongevallen als basis voor ongevallenpreventie (Statistical analysis of industrial accidents as a basis for accident prevention). Voordracht voor Commissie van overleg. *De Veiligheid* 27:52–58

Freedman R (1994). *Kids at Work. Lewis Hine and the Crusade Against Child Labour*. Clarion Books, New York

Freeman J (2018). *Behemoth*. W.W. Norton & Company, New York

Freud S (1901). *Psychopathology of Everyday Life*, translated by A Brill (1914). MacMillan Co, New York

Frijters A Swuste P (2008). Safety assessment in design and preparation phase. *Safety Science* 46(2):272–281

Froggatt P Smiley J (1964). The concept of accident proneness: A review. *British Journal of Industrial Medicine* 21(1):1–12

Fuller C (2005). As assessment of the relation between behaviour and injury in the workplace: A case study in professional football. *Safety Science* 43(4):213–224

Garbowski M Roberts K (2016). Reliability seeking virtual organisations: Challenges for high reliability organisations and resilience engineering. *Safety Science* (in press). http://dx.doi.org/10.1016/j.ssci.2016.02.016

Gardner D Cross J Fonteyn P Carlopio J Shikdar A (1999). Mechanical equipment accidents in small manufacturing businesses. *Safety Science* 33(1):1–12

Gauchard G Mur J Touron C Benamghar L Dehaene D Perrin P Chau N (2006). Determinants of accident proneness a case – control study in railway workers. *Occupational Medicine* 56(3):187–190

Gerritsen W (1957). De optimale samenwerking tussen bedrijfsarts en veiligheidsinspecteur (Optimum cooperation between occupational physician and safety inspector). *De Veiligheid* 33:118–120

Gerven T van Stankiewicz A (2009). Structure, energy, synergy, time – the fundamentals of process intensification. *Industrial and Engineering Chemistry Research* 48(5):2465–2474

Geus A de (1997). *De Levende Organisatie. Over Leven en Leren in een Turbulente Omgeving (The Living Organisation. About Life and Learning in a Turbulent Environment).* Scriptum Management, Schiedam

Gezondheidsraad (1965). *Rapport Betreffende de Algemeen Statistische Berichtgeving van Ongevallen aan Werknemers Overkomen. (Report on the General Statistical Reporting of Accidents Happening to Workers – in Dutch).* Health Council, The Hague

Gezondheidsraad (1975). Dutch Health Council. *Kerncentrales en volksgezondheid (Nuclear power plants and public health).* Den Haag

Gezondheidsraad (1978). Dutch Health Council. *Canvey – An Investigation of Potential Hazards from Operations in the Canvey Island/Thurrock Area.* Den Haag

Gezondheidsraad (1979). Dutch Health Council. *Advies Over de Carcinogeniteit van Chemische Stoffen (Advice on Carcinogenicity of Chemical Substances).* Den Haag

Gezondheidsraad (1980). Dutch Health Council. *Advies Inzake Carcinogene Stoffen (Advice on Carcinogenic Substances), Werkgroep van Deskundigen.* Den Haag

Gezondheidsraad (1995). Dutch Health Council. *Niet Alle Risico's Zijn Gelijk. Kanttekeningen bij de Grondslag van de Risicobenadering in het Milieubeleid (Not all Risks are Equal, Comments on the Basis of the Risk Approach in Environmental Policy).* Den Haag, Gezondheidsraad, publicatie nr. 1995/06

Gezondheidsraad (1996). Dutch Health Council. *Risico, Meer Dan een Getal (Risk is More than a Number). Commissie Risicomaten en Risicobeoordeling.* Rapport 03, Den Haag

Gezondheidsraad (2006). Dutch Health Council. *Gevolgen van rampen voor de gezondheid op middellange en lange termijn* (The Medium and Long-term Health Impact of Disasters). Health Council of the Netherlands, The Hague, publication no. 2006/18

Gibson J (1961). The Contribution of Experimental Psychology to the Formulation of the Problem of Safety – A Brief for Basic Research. *Behavioural Approaches to Accident Research*, pp. 77–89. Association for the aid of crippled children, New York, included in: Haddon W Suchman E Klein D (Eds.) (1964). *Accident Research, Methods and Approaches.* Harper & Row, New York

Giele J (1981). *Een Kwaad Leven. De Arbeidsenquête van 1887 (A Bad Life, The Industrial Enquête of 1887): Amsterdam, Maastricht, De Vlasindustrie, Tilburg en Eindverslag.* Link, Nijmegen

Gilbreth F Gilbreth L (1917). *Applied Motion Studies.* Sturgis and Walton, New York, NY

Glaser B Strauss A (1967). *The Discovery of Grounded Theory: Strategies for Qualitative Research.* Aldine, Chicago

Glendon I (2008). Safety Culture Snapshot of a developing concept. *Journal of Occupational Health and Safety Australia and New Zealand* 24(3):179–189

Glendon A Stanton N (2000). Perspective on safety culture. *Safety Science* 34(1–3):193–214

Gómez-Castro F Segovia-Hernández J (2019). *Process Intensification.* De Gruyter, Berlin

Gómez-Mares M Záratev L Casal J (2008). Jet fires and the domino effect. *Fire Safety Journal* 43(8):583–588

Gool R van (2016). *Veertig Jaar Geleden Zat Heel Schoonebeek Onder de Olie (Forty Years Ago the Whole of Schoonebeek was Covererd with Oil).* RTV Drente November 8th

Goossens L (1981). Veiligheidskunde aan de Technische Hogeschool Delft. *Tijdschrift voor Sociale Geneeskunde* 59(9):312–316

Goossens L Cooke R (1997). Application of some risk assessment techniques: Formal expert judgement and accident sequence precursor. *Safety Science* 26(1–2):35–47

Goossens L Cooke R (2001). Expert judgement elicitation in risk assessment. In Linkov I Palma-Oliveira J (Eds.). *Assessment and Management of Environmental Risks, NATO Science Series (Series IV: Earth and environmental series)*, vol 4. Springer, Dordrecht. doi: 10.1007/978-94-010-0987-4_45

Goossens L Cooke R Hale A Rodic-Wiersma L (2008). Fifteen years of expert judgment at TUDelft. *Safety Science* 46:234–244. doi: 10.1016/j.ssci.2007.03.002

Gordon J (1949). Epidemiology of accidents. *American Journal of Public Health* 39:504–515, opgenomen in: Haddon W Suchman E Klein D (Eds.) (1964). *Accident Research, Methods and Approaches*. Harper & Row, New York

Gorissen A Schroder C (1996). Beroepsziekten en ongevallen in Nederland een schatting van de omvang en incidentie op basis van Duitse cijfers (Occupational diseases and accidents in the Netherlands, an estimation of the size and incidence based upon German figures). *Tijdschrift voor Sociale Gezondheidszorg* 74:251–258

Gorter R (1935). Warum Unfälle, Dombrowsky (book review why accidents). *De Veiligheid* 12(11):116–117

Gorter R (1946). Moderne inzichten over de bestrijding van ongevallen (Modern insights in the prevention of accidents. *De Veiligheid* 22(9):58–59

Gorter R (1947). Veiligheid na de oorlog (Safety after the war). *De Veiligheid* 23:50–52

Greenwood E (1934). *Who Pays*? Doran & Co, New York

Greenwood M Wood H (1919). The incidence of industrial accidents upon individuals with special reference to multiple accidents. *Industrial Fatigue Board*, report nr 4. Her Majesty's Stationary Office, London

Greisler D (1999). William Edward Deming: The man. *Journal of Management History* 5(8):434–453

Griffiths R (1981). *Dealing with Risk. The Planning, Management and Acceptability of Technological Risk*. Manchester University Press, Manchester

Groeneweg J (1988). Schuldcultuur afschaffen (Abolish blame culture). *Nammogram* 29(2):8–10

Groeneweg J (1992). Controlling the controllable, the management of safety. Proefschrift Rijksuniversiteit Leiden, DSWO Press, Leiden

Groeneweg J (2002). *Controlling the Controllable, Preventing Business Upsets*, 5th revised edition. Global Safety Group, Leiden

Groeneweg J Hudson P Vandevis T Lancioni G (2010). Why improving the safety climate doesn't always improve the safety performance. Proceedings of the SPE International Conference on Health, Safety and Environment in Oil and Gas exploration. Rio de Janeiro. SPE 127152, Society of Petroleum Engineers London

Groeneweg J Lancioni G Metaal N (2003). Tripod: Managing organisational components of business upsets. In: Bedford T van Gelder P (Eds.). *Safety and Reliability*, pp. 707–712. Proceedings of the 2003 ESREL Conference, Maastricht

Groeneweg J Lancioni G Metaal N Verhoeve K (2004). *Tripod: Professionalism versus amateurism in the management of safety*. Proceedings of the 7th SPE International Conference on Health, Safety and Environment in Oil and Gas exploration. Calgary, Canada Ref. SPE 86828, Society of Petroleum Engineers London, pp. 937–943

Groeneweg J Roggeveen V (1998). Tripod: Controlling the human error component in accidents. In: Lydersen S Hansen G Sandtorv H (Eds.). *Safety and Reliability*, pp. 809–816. Proceedings of the European Conference of Safety and Reliability, ESREL '98, Rotterdam

Groeneweg J Ter Mors E (2016). The influence of communicating on safety measures on risk-taking behavior. Proceedings of the SPE International Conference and Exhibition on Health, Safety, Security, Environment and Social Responsibility in Stavanger, Norway, Ref. SPE 129255, Society of Petroleum Engineers, London

Groeneweg J Wagenaar W (1989). *Ongevalspreventie bij de NAM, een Geïntegreerde Aanpak (Accident Prevention at NAM, an Integrated Approach)*. Werkgroep Veiligheid, Leiden, Rijksuniversiteit te Leiden, R-89/30

Grondstrom R Jarl T Thorson J (1980). Serious occupational accidents – an investigation of causes. *Journal of Occupational Accidents* 2:283–289

Groothuizen T (1976). De Explosie in Flixborough (the Explosion at Flixborough). *De Veiligheid* 52(2):41–44

Grossmann I Westerberg A (2000). Research challenges in process systems engineering, *AIChE Journal* 46(9):1700–1703.

Grote G (2012). Safety management in different high-risk domains – all the same? *Safety Science* 50:1983–1992

Guarnieri M (1992). Landmarks in the history of safety. *Journal of Safety Research* 23:151–158

Guastello S (1991). Some further evaluations if the International Safety Rating System. *Safety Science* 14:253–259

Guastello S (1993). Do we really know well our occupational prevention program work. *Safety Science* 16(3–4):445–463

Guillaume E (2011). Identifying and responding to weak signals to improve learning from experiences in high-risk industry. Doctoral thesis, Delft University of Technology

Guldenmund F (2000). The nature of safety culture: A review of theory and research. *Safety Science* 34(1–3):215–257

Guldenmund F (2007). The use of questionnaires in safety culture research: An evaluation. *Safety Science* 45:723–743

Guldenmund F (2009). De organisatorische driehoek als basis voor gedragsverandering (The organisational triangle as basis for behavioural changes). *Tijdschrift voor Toegepaste Arbowetenschap* 22(4):142–145

Guldenmund F (2010). Misunderstanding safety culture and its relation to safety management. *Risk Analysis* 30(10):1466–1480

Guldenmund F Arbntzen K Vriends S (1999). *Meting Veiligheidsbeleving Personeel Oxystaalfabriek 1 (Measurement f Safety Pereption Oxysteel Plant-1)*. Vakgroep Veiligheidskunde, TUDelft, Delft

Guldenmund F Booster P (2005). De effectiviteit van structurele maatregelen op veiligheid, een casusbeschrijving (The effectiveness of structural measures on safety, a case study). *Tijdschrift voor Toegepaste Arbowetenschap* 18(2):38–44

Guldenmund F Hale A Goossens L Betten J Duijn N (2006). The development of an audit technique to assess the quality of safety barrier management. *Journal of Hazardous Materials* 130(3):234–241

Guldenmund F Swuste P (2001). Veiligheidscultuur toverwoord of onderzoeksobject (Safety culture, a magic word, or an object of research). *Tijdschrift voor Toegepaste Arbowetenschap* 14(4):2–8

Gun R (1993). Role of regulation in accident prevention. *Safety Science* 16(1):47–66

Gundlach H (2002). Certification, a tool for safety regulation? In Hale A R Hopkins, A Kirwan B (Eds.). *Changing Regulation: Controlling Hazards in Society,* pp. 233–252. Pergamon, Oxford

Haavik T Antonsen S Rosness R Hale A (2016). HRO and RE: A pragmatic perspective. *Safety Science* (in press). http://dx.doi.org/10.1016/j.ssci.2016.08.010

Haddon W (1963). A note concerning accident theory and research with special reference to motor vehicle accidents. *Annals of the New York Academy of Science* 107:635–646

Haddon W (1968). The changing approach to the epidemiology, prevention, and amelioration of trauma: The transition to approaches etiologically rather than descriptive based. *American Journal of Public Health* 58(8):1431–1438

Haddon W (1970). On the escape of tiger: An ecologic note. *American Journal of Public Health* 60(12):2229–2234

Haddon W (1973). Energy damage and the ten countermeasure strategies. *Human Factors* 15(4):355–366

Haddon W (1974). Advances in the epidemiology of injuries as a basis for public policy. *Public Health Reports* 95(5):411–421

Haddon W (1980a). The basic strategies for reducing damage from hazards of all kind. *Hazard Prevention* 16(11):8–12

Haddon W (1980b). Advances in the epidemiology of injuries as a basis for public policy. *Public Health Reports* 95(5):411–421

Haddon W Suchman E Klein D (Eds.) (1964). *Accident Research, Methods and Approaches*. Harper & Row New York

Haimes Y (1991). Total risk management. *Risk Analysis* 11(2):169–171

Häkkinen K (1982). The progress of technology and safety in materials handling. *Journal of Occupational Accidents* 4:157–163

Hale A (1978). The role of government inspectors of factories with particular reference to their training needs. PhD thesis, University of Aston, Birmingham

Hale A (1985). *De Menselijke Paradox in Technologie en Veiligheid (The Human Paradox in Technology and Safety. Inaugural Lecture)*. Oratie THDelft Delftse Universitaire Pers

Hale A (1987). Opleidingen in veiligheid (Safety education). *Risicobulletin* 9(4):11–12

Hale A (2000). Culture confusion. *Safety Science* 34(1):1–14

Hale A (2001a). Conditions of occurrence of major and minor accidents. *Journal of the Institution of Occupational Safety and Health* 5(1):7–21

Hale A (2001b). Regulating airport safety, the case of Schiphol. NeTWork proceedings. Special issue. *Safety Science* 37(2–3):127–149

Hale A (2002). Risk contours and risk management criteria for safety at major airports, with particular reference to the case of Schiphol. *Safety Science* 40(1–4):299–323

Hale A (2003). Safety Management in Production. *Human Factors & Ergonomics* 13(3):185–201

Hale A (2005). Safety management, what do we know, what do we believe we know, and what do we overlook? *Tijdschrift voor Toegepaste Arbowetenschap* 18(3):58–66

Hale A (2006). *Method in Your Madness: System in Your Safety*. Afscheidsrede TUDelft, Delft

Hale A (2014). Editorial. Foundation of safety science: A postscript. *Safety Science* 67:64–69

Hale A Ale B Goossens L Heijer T Bellamy L Mud M Roelen A Baksteen H Post J Papazoglou I Bloemhoff A Oh J (2007a). Modelling accidents for prioritizing prevention. *Reliability Engineering and Safety Systems* 92:1701–1715

Hale A Glendon A (1987). *Individual Behaviour in the Control of Danger Industrial Safety Series Volume 2*. Elsevier Science Publisher, Amsterdam

Hale A Goossens L Poel I van der (2002). Oil and gas industry regulation: From detailed technical inspection to assessment of safety management. In: Kirwan B Hale A Hopkins A (Eds.). *Changing Regulations, Controlling Risk in Society. Interdisciplinary Study Group New Technology and Work (NeTWork)*, pp. 79–106. Pregamon, Oxford

Hale A Guldenmund F (2010). Veiligheidsverbetering wat werkt wanneer? (Safety interventions, what is working when?) *Tijdschrift voor Toegepaste Arbowetenschap* 23(2):50–56

Hale A Guldenmund F Goossens L Karczewski J Duijm N Hourtoulou D Le Coze J Plot E Kontic D Gerbec M (2005). Management influences on major hazard prevention: The ARAMIS audit. *Proceedings of the European safety and reliability Conference*, ESREL 2005, volume 1, p 767–773

Hale A Guldenmund F Loenhout P Oh J (2010). Evaluating safety management and culture interventions to improve safety: Effective interventions. *Safety Science* 48:1026–1035

Hale A Hale M (1970). Accidents in perspective. *Occupational Psychology* 44:115–122

Hale A Hale M (1972). *A Review of the Industrial Accident Research Literature of the National Institute of Industrial Psychology*. Her Majesty's Stationery Office, London

Hale A Heijer T (2006). Defining resilience (pp. 35–40), Is resilience really necessary? The case of railways (pp. 125–148). In: Hollnagel E Woods D Leveson N (Eds.). *Resilience Engineering, Concepts and Precepts*. Ashgate, Aldershot

Hale A Heming B Carthey J Kirwan B (1997). Modelling of safety management systems. *Safety Science* 26(1/2):121–140

Hale A Hovden J (1998). Management and culture third age of safety. A review of approaches to organizational aspects of safety, health and environment. In: Feyer A Williamson A (Eds.). *Occupational Injury, Risk, Prevention and Intervention*. Taylor & Francis, London

Hale A Kirwan B Kjellén U (2007). Editorial. In: Hale A Kirwan B Kjellén U (Eds.). *Safety by design. NeTWork proceedings special issue Safety Science* 45(1–2):3–327; Safety by design where are we now. NeTWork proceedings Special issue. *Safety Science* 45(1–2):305–327

Hale A Storm W (1996). Is certificering van arbo-deskundigen een voldoende flexibel middel voor kwaliteitsborging (Is certification of working condition experts flexible enough guarantee for a quality assurance). *Tijdschrift voor Toegepaste Arbowetenschap* 9(4):55–61

Hale A Swuste P (1997). Avoiding square wheels International experience in sharing solutions. *Safety Science* 25(1–3):3–14

Hale A Swuste P (1998). Safety rules: Procedural freedom or action constraint? *Safety Science* 29(3):163–177

Hale A Walker D Walters N Bolt H (2012). Developing the understanding of underlying causes of construction fatal accidents. *Safety Science* 50:2020–2027

Hall T (1995). *The Quality Systems Manual: The Definitive Guide to the ISO 9000 Family and Ticket*. John Wiley and Sons, Chichester

Halpern J (1989). Cognitive factors influencing decision making in a highly reliable organization. *Industrial Crisis Quarterly* 3:143–158

Hämäläinen P (2010). Global estimates of occupational accidents and fatal work related diseases. Thesis, Tampere University of Technology

Hämäläinen P Takala J Saarela K (2006). Global estimates on occupational accidents. Safety *Science* 44(2):137–156

Hanekamp J Vera-Navas G Verstegen S (2005). The historical roots of precautionary thinking: The cultural ecological critique and 'The Limits to Growth'. *Journal of Risk Research* 8(4):295–310

Hanken A Andreas H (1980). Systeemleer en veiligheid (System theory and safety). *De Veiligheid* 56(12):603–609

Hard W (1907). Making steel and killing men. *Everybody's Magazine*, November

Hard W (1910). Injured in the course of duty. Industrial accidents, how they happen, how they are paid for and how they ought to be paid for. *Reprints from Everybody's Magazine*. The Ridgway Co, New York

Harms C (1966). Ontwikkeling en toepassing van beveiligingen bij bewerkingsmachines (Development and application of safety of machinery). *De Ingenieur* 46:G51–54

Harms-Ringdahl L (1987a). Safety analysis in design. Evaluation of a case study. *Accident Analysis & Prevention* 19(4):305–317

Harms-Ringdahl L (1987b). *Safety Analysis*. Elsevier Applied Science, London

Harms-Ringdahl L (1993). *Safety Analysis, Principles and Practice in Occupational Safety*. Elsevier Applied Science, London

Harms-Ringdahl L (2004). Relationship between accident investigation risk analysis and safety management. *Journal of Hazardous Materials* 111(1–3):13–19

Harper A Cordery J Klerk de N Sevastos P Geelhoed E Gunson C Robinson L Sutherland M Osborn D Colquhoun J (1996b). Curtin Industrial safety trial: Management behaviour and program effectiveness. *Safety Science* 24(3):173–179

Harper A Gunson C Robinson L Klerk de N Osborn D Geelhoed E Sutherland M Colquhoun J (1996a). Curtin Industrial safety trial: Methods and safe practice and housekeeping outcomes. *Safety Science* 24(3):159–172

Harrisson D Legendre C (2003). Technological innovations, organizational change and workplace accident prevention. *Safety Science* 41(4):319–338

Hart H 't (1950). Propaganda en organisatie van het veiligheidswerk in de US. *De Veiligheid* 28:51–56

Hart H 't (1952). Propaganda en organisatie van het veiligheidswerk in de US (Propaganda and organisation of safety in the US). *De Veiligheid* 28:51–56

Hart H 't (1966). De ontwikkeling, keuring en gebruik van beveiliging bij bewerkingsmachines (The development, control and use of machinery). *De Ingenieur* 78(46):G55–61

Harvey B (1979). Flixborough 5 years later. *The Chemical Engineer* 349:697–698

Hasle P Kines P Andersen L (2009). Small enterprise owners' accident causation attribution and prevention. *Safety Science* 47(1):9–19

Hasselt J (1907). Het ongevallen-risico van machines (Accident risks of machines). *De Ingenieur* 22(32):604–606

Hattem B van (1980). Het veiligheids- en gezondheidswerk in de industriebond NVV (Safety and health activities of the NVV labour union). *De Veiligheid* 56(11):583–587

Hauer E (1980). Bias by selection Overestimation of the effectiveness of countermeasures. *Accident Analysis & Prevention* 12:113–117

Hauer E (1983). Reflection on methods of statistical inference in research on the effect of countermeasures. *Accident Analysis & Prevention* 15(4):275–285

Hawkins F (1987). *Human Factors in Flight*. Gover Technical Press, Aldershot, Hant, UK

Hawkins S (1992). An introduction to task dynamics. In Docherty G Ladd D (Eds.). *Gesture, Segment, Prosody. Laboratory Phonology*, pp. 9–25. Cambridge University Press. doi: 10.1017CBO9780511519918.002

Health Council of the Netherlands (1979). *Advisory Report on the Carcinogenicity of Chemical Substances*. The Hague. (in Dutch)

Heertje A (2009). De economische waarde van veiligheid (The econom value of safety). *Tijdschrift voor Toegepaste Arbowetenschap* 22(4):114–116

Heijermans L (1905) *Gezondheidsleer voor Arbeiders (Hygiene for Workers)*. Brusse Rotterdam

Heijermans L (1907) *Het Onderwijs in de Bedrijfshygiëne. Voordracht Gehouden te Delft bij de Opening van den Cursus in de Sociale en Technische Hygiëne, Georganiseerd Door de Sociaal Technische Vereniging van Democratische Ingenieurs en Architecten voor Studeerenden aan de Technische Hoogeschool en Belangstellenden (Education in Occupational Hygiene. Presentation held During the Start of a Course in Social and Technical Hygiene, Organised by the Social Technical Society of Democratic Engineers and Architects, studying at the Technical Highschool and Other Persons Interested)*. Brusse, Rotterdam

Heijermans L (1908). *Handleiding tot de Kennis der Beroepsziekten (Textbook on the Knowledge of Occupational Diseases)*. Brusse, Rotterdam

Heinrich H (1927). The "incidental" cost of accidents. *National Safety News* (February 1927):18–20

Heinrich H (1928). The origin of accidents. *National Safety News* (July 1928):9–12 & 55

Heinrich H (1929). The foundation of a major injury. *National Safety News* January 19(1):9–11, 59

Heinrich H (1931). *Industrial Accident Prevention, a Scientific Approach*, 1st edition. McGraw-Hill Book Company, London

Heinrich H (1932). The safety engineer aids the life underwriter. *National Safety News* (Augustus 1932):21–22

Heinrich H (1938a). Accident cost in the construction industry. *NSC Transactions*:374–377.

Heinrich H (1938b). It's up to the foreman! *Ind. Supervisor* 4:4–5 & 14

Heinrich H (1941). *Industrial Accident Prevention, a Scientific Approach*, 2nd edition. McGraw-Hill Book Company, London

Heinrich H (1942a). Men in motion. *Ind. Supervisor* 10:4–5 & 11

Heinrich H (1942b). The foreman's place in the safety program. *NSC Transactions* 1:191–194

Heinrich H (1945a). Key men in industry: Part 1. *Ind. Supervisor* 13(1):4–6

Heinrich H (1945b). Key men in industry: Part 2. *Ind. Supervisor* 13(2):12–14

Heinrich H (1945c). Key men in industry: Part 3. *Ind. Supervisor* 13(3):12–14

Heinrich H (1950a). *Industrial Accident Prevention*, 3rd edition. McGraw Hill Book Company New York, Toronto & London

Heinrich H (1950b). The human element in the cause and control of industrial accidents. *NSC Transactions* 30:7–10

Heinrich H (1951). The safety engineer and home safety. *NSC Transactions* 13:6–8

Heinrich H (1956). Recognition of safety as a profession. *NSC Transactions* vol 24:37–40

Heinrich H Blake, R (1956). The accident cause ratio – 88:10:2. *National Safety News* (May, 1956):18–22

Heinrich H Granniss E (1959). *Industrial Accident Prevention*, 4th edition. McGraw Hill Book Company, New York, Toronto & London

Heinrich H Petersen D Koos N (1980). *Industrial Accident Prevention*, 5th edition. McGraw Hill Book Company New York, Toronto & London

Helderslot D (2009). Remembering Flixborough. *Journal of Chemical Safety*:46–47

Hendrickx L (1991). How versus how often: The role of scenario information and frequency information in risk judgment and risky decision making. Doctoral thesis Rijksuniversiteit Groningen

Henstra D (1992). Risicoklassificatie door middel van een nomogram (Risk classification using a monogram). *NVVK-Info* 1(2):39–4–2

Hermans H (2007). *Een Monster Loert . . . De Collectie Historische Gezondheidsaffiches van de Universiteit van Amsterdam*. Vossiuspers, Amsterdam

Hermansson H (2005). Consistent risk management: Three models outlined. *Journal of Risk Research* 8(7–8):557–568

Herold J (1945). Dispositie tot Ongevallen. De Betekenis van een Bedrijfsopleiding (Disposition to Accidents, The Impact of Company Training). Proefschrift KUNijmegen Ernest van Aelst, Maasticht.

Hidden A (1989). *Investigation into the Clapham Junction Railway Accident*. Department of Transport. Her Majesty's Stationary Office, London

Hine L (1908). Child Labour, reporter of the Child Labour Committee. Onderdeel van de catalogus van de Rotterdamse tentoonstelling in het Nederlandse Fotomuseum. Lewis Hine. Fundación MAPFRE, Instituto de Cultura, Madris 2010

HMSO (1990). *Railway Accident. Report of the Inquiry into the Collision at Bellgrove Junction on 6 March 1989*. Department of Transport. Her Majesty Stationary Office, London

Hoeff J van der (1970). Moet veiligheid beloond worden? Ongevallenbestrijding in de Praktijk (Must safety be rewarded? Accident prevention in practice). *De Veiligheid* 46(6):213–214

Hoffman F (1909). Industrial accidents and industrial diseases. *Publications of the American Statistical Association* 11(88):567–603

Hofstede G (1978). The poverty of management control philosophy. *Academy of Management Review* 3(3):450–461

Hofstede G (1991). *Cultures and Organizations: Software of the Mind*. McGraw Hill, London

Hohnen P Hasle P (2011). Making work environment audible. A critical study on certified safety management systems in Denmark. *Safety Science* 49(7):1022–1029

Høiset S Hjertager B Solberg T Malo K (2000). Flixborough revisited – an explosion simulation approach. *Journal of Hazardous Materials* A77(1–3):1–9

Hollander A Vliet L van (1992). Melden en registreren van bedrijfsongevallen en beroepsziekten (Notification and registration of occupational accidents and occupational diseases). *Risicobulletin* 14(4):10–11

Hollnagel E (1983). Human error. *Position Paper NATO Conference on Human Error*, Bellagio Italy

Hollnagel E Cacciabue P Bagnara S (1994). The limits of automation in air traffic control and aviation Bulletin International. *Journal of Human-Computer Studies* 40:561–566

Hollnagel E Pedersen O Rasmussen J (1981). *Notes on Human Performance Analysis*. Risø-M-2285, Risø National Laboratory, Roskilde, Denmark

Hollnagel E Woods D Leveson N (2006). *Resilience Engineering, Concepts and Precepts*. Ashgate, Aldershot

Holmes N Triggs T Gifford S Dawkins A (1997). Occupational injury risk in a blue collar small business industry: Implications for prevention. *Safety Science* 25(1–3):67–78

Home Office (1911). *Report of the Departmental Committee on Accidents in Places Under the Factory and Workshop Act*. HMSO, London

Hoorn W van Meulman J Vincent B (1980). *Geschiedenis van de Psychologie (The History of Psychology)*. Collegedictaat Rijksuniversiteit Leiden, Psychologisch Instituut, Leiden

Hope E Hanna W Stallybrass C (1923). *Industrial Hygiene and Medicine*. Baillière, Tindall and Cox, London

Holy Scripture (1996). *New Living Translation*, Copyright 1996, 2004, 2015. Tyndale House Foundation, Carol Stream, Illinois

Hopkins A (1999a). For whom does safety pay? The case of major accidents. *Safety Science* 32(2–3):143–153

Hopkins A (1999b). *Managing Major Disasters*. Allen & Unwin, St Leonards NSW

Hopkins A (1999c). The limits of normal accidents theory. *Safety Science* 32(2–3):93–102

Hopkins A (2000a). *Lessons from Longford: The Esso Gas Plant Explosion*. CCH Australia Limited, Sydney

Hopkins A (2000b). A culture of denial sociological similarities between the Moura and Gretley mine disasters. *Journal of Occupational Health Studies* 16(1):29–36

Hopkins A (2001). Was 3 Mile Island a normal accident? *Journal of Contingencies and Crisis Management* 9(2):65–72

Hopkins A (2006). What are we to make of safe behaviour programs? *Safety Science* 44(7):583–597

Hopkins A (2008). *Failure to Learn: The BP Texas City Refinery Disaster*. CCH Australia Limited, Sydney

Hopkins A (2012). *Disastrous Decisions. The Human and Organisational Causes of the Gulf of Mexico Blowout*. CCH Australia Limited, Sidney

Hopkins A (2014). Issues in safety science. *Safety Science* 67:6–14

Horn-van Nispen, M. ten (2001), Johan van Veen. *Tijdschrift voor Waterstaatsgeschiedenis* 10

Hovden J (2002). The development of new safety regulations in the Norwegian oil and gas industry. In: Kirwan B Hale A Hopkins A (Eds.). *Changing Regulations, Controlling Risk in Society. Interdisciplinary Study Group New Technology and Work (NeTWork)*, pp. 57–77. Pregamon, Oxford

Hovden J Albrechtsen E Herrera A (2010). Is there a need for new theories models and approaches to occupational accident prevention? *Safety Science* 48(8):950–956

Hovden J Sten T (1984). The workers as a safety resource in modern production systems. *Journal of Occupational Accidents* 6(1–3):213–214

Hovden J Tinmannsvik R (1990). Internal control. A strategy for occupational safety and health experiences from Norway. *Journal of Occupational Accidents* 12(1):21–30

HSC (1976). *Health and Safety Commission, Advisory Committee on Major Hazards, First Report*. Her Majesty's Stationery Office, London

HSC (1979). *Health and Safety Commission, Advisory Committee on Major Hazards, Second Report*. Her Majesty's Stationery Office, London

HSE (1976). *Health and Safety Executive, Success and Failure in Accident Prevention*. Her Majesty's Stationary Office, London

HSE, Health and Safety Executive (1978). *Canvey – An Investigation of Potential Hazards from Operations in the Canvey Island/Thurrock Area: Complete Report*. Stationery Office Books, London

HSE, Health and Safety Executive (1981). *Second Canvey Report*. Her Majesty's Stationery Office, London

HSE, Health and Safety Executive (1985). *Monitoring Safety. Occasional Paper Series OP9*. HMSO, London

HSE, Health and Safety Executive (1991). *Successful Health and Safety Management*. HMSO, Richmond, Surrey

HSE, Health and Safety Executive (2001). *Reducing Risks, Protecting People. HSE's Decision Making Process*. HMSO, Norwich

HSE, Health and Safety Executive (2003a). *Out of Control, why Control Systems go Wrong and how to Prevent Failure*. HMSO, London

HSE, Health and Safety Executive (2003b). *Major Investigation report BP Grangemouth 29th May-10th June 2000*. A public report prepared by the HSE on behalf of the competent authority, HSE, SEPA, HMSO, Colegate, Norwich

HSE, Health and Safety Executive (2005). *Trends and Context to Rates of Workplace Injuries Report 386*. Her Majesty's Stationary office, Norwich

HSE, Health and safety Executive (2009). *Underlying Causes of Construction Fatal Accidents – A Comprehensive Review of Recent Work to Consolidate and Summarize Existing Knowledge, Phase 1 Report*. Construction Division. Her Majesty's Stationary office, Norwich

HSE, Health and Safety Executive (2010). *Refinery Fire at Feyzin*. 4th of January 1966. www.hse.gov.uk/comah/sragtech/casefeyzin66.htm consulted March 19th 2018

HSE, Health and Safety Executive (2019). *The History of HSE* website hse,gov.uk consulted October 19th

Hubbard R Neil J (1985). Major and minor accidents at the Thames barrier construction site. *Journal of Occupational Accidents* 7:147–164

Hudson P (2007). Implementing a safe culture in a major multinational. *Safety Science* 45(6):697–722

Hudson P (2010a) *Private Communication*

Hudson P (2010b). *Safety Science. It's not Rocket Science, it's Much Harder*. Inaugural address Delft University of Technology, September 24th

IAEA (1991). International Atomic Energy Agency. *Safety Cultures*. Safety series nr 75-INSAG-4. IAEA, Vienna

IChemE (1987). The Feyzin Disaster. *Loss Prevention Bulletin 077*, October p 1–4

IJsendoorn P van (2016). Een verhaal van liefde, overleven en gemeenschap. *De Volkskrant* 21 october

ILO (1988). International Labour Office. *Major Hazard Control, a Practical Manual*. ILO, Geneva

ILO (1992). International Labour Office. *Introduction to Work Study*, 4th edition. International Labour Office, Geneva

ILO (2001). International Labour Office. *Guidelines on Occupational Safety and Health Management Systems*. International Labour Office, Geneva

INSAG (1988). International Nuclear Safety Advisory Group. *Basic Safety Principles for Nuclear Power Plants*. Safety series nr 75-INSAG-3. IAEA, Vienna

INSAG (1999). International Nuclear Safety Advisory Group. *Management of Operational Safety in Nuclear Power Plants* (INSAG-13). International Atomic Energy Agency, Vienna

Irwin A (1984). Risicoanalyse Wie kiest er in wetenschap en technologie? (Risk analysis, who is chosing in science and technology?). *Wetenschap en Samenleving Maart* 7–9

ISO 16732–1 (2012). International Organisation for Standardisation. *Fire Safety Engineering – Guidance on Fire Risk Assessment*. International Organization for Standardization, Geneva, Switzerland

Janssens M Brett J Smith F (1995). Cross cultural research Testing the variability of coorporation wide safety policy. *Academy of Management Journal* 38(2):364–382

Johnson S (2006). *The Gost Map*. Penguin Books, London

Johnson S (2007). The predictive validity of safety climate. *Journal of Safety Research* 38(5):511–521

Johnson W (1970). *New Approaches to safety in industry*. Industrial and Commercial Techniques LTD, London

Johnson W (1973a). Sequences in accident causation. *Journal of Safety Research* 5(2):54–57

Johnson W (1973b). *The Management Oversight and Risk Tree – MORT, Including Systems Developed by the Idaho Operations Office and Aerojet Nuclear Company*. US Atomic Energy Commission, Division of Operational Safety – SAN 821–2/UC-41

Jones D (1963). *Danger of Coal Slurry being Tipped at the Rear of the Pantglas School*. Letter of the Borough & waterworks Engineer D. Jones to Mr. T Ritchie District Public Works Superintendent, Reservoir House Treharris, July 24th Cited at Parliament notes Hansard October 26th 1967 vol 751 cc 1909–2014

Jong M de Poll K (1984). Ergonomische normen en richtlijnen voor aangedreven handgereedschap (Ergonomic norms ad guidelines for empowered tools). *De Veiligheid* 60(12):609–612

Jongen M Swuste P (2014). Regelt veiligheidscultuur procesveiligheid? (Does safety culture regulates process safety?) *Tijdschrift voor Toegepaste Arbowetenschap* 27(1):22–25

Jongen M Swuste P (2017). Onderwijs in procesveiligheid. Kunnen we het niveau handhaven? (Education in process safety. Can we maintain a standard?) *Tijdschrift voor Toegepaste Arbowetenschap* 30(2):62–65

Joseph G (2003). Recent reactive incidents and fundamental concepts. *Journal of Hazardous Materials* 104(1–3):65–73

Judt T (2012). *Thinking the Twentieth Century*. Penguin books, London

Juran J (1951). *Quality Control Handbook*. McGraw-Hill New York

Juran J Barish N (1955). *Case Studies in Industrial Management*. McGraw Hill Book Company New York

Kampen J van Beek A van der Groeneweg J (2014). The value of safety indicators. *SPE Economics and Management* 5(5):131–140.

Kampen J van Lammers M Steijn W Guldenmund F Groeneweg J (2010). The effectiveness of 48 safety interventions according to safety professionals. *Chemical Engineering Transactions* 79. AIDIC S.R.L., Italy

Kanter R (1984). *The Change Masters, Corporate Entrepreneurs at Work*. George Allen & Unwin, London

Kaplan S Garrick B (1981). On the quantitative definition of risk. *Risk Analyses* 1(1):11–27

Kasperson R Renn O Slovic P Brown H Emel J Goble R Karperson J Ratick S (1988). The importance of the media and the social amplification of risk: The social amplification of risk; A conceptual framework. *Risk Analysis* 8(2):177–178

Keefer D (1945). Accident costs. In: Armstrong T Blake R Bloomfield J Boulet C Gimbel M Homan S Keefer W Page R (Eds.). *Industrial Safety*. Prentice Hall Inc., New York

Keenan V Kerr W Sherman W (1951). Psychological climate and accidents in the automobile industry. *Journal of Applied Psychology* 35(2):108–111

Keller W Modarres M (2005). A historical overview of probabilistic risk assessment developments and its use in the nuclear power industry: A tribute to the late Professor Norman Carl Rasmussen. *Reliability Engineering and System Safety* 89(3):271–285

Kellogg P (1909). *The Pittsburgh Survey*. Charities Publications Committee, New York

Kelly K Cardon N (1991). *The Myth of 10–6 as a Definition of Acceptable Risk*. 84th annual meeting air and waste management association, Vancouver, B.C., Canada, 16–21 June. Downloaded on 14 July 2013 from http://heartland.org/sites/all/modules/custom/heartland_migration/files/pdfs/17603.pdf

Kemeny J (1979). *Report of the Presidential Commission on the Accident at Three Mile Island*. The need for change: The legacy of TMI, Washington DC

Kerckhove J van de (1991). Tegen een eenzijdige benadering van het menselijk falen. Machines moeten werken, mensen moeten denken (Against a one-sided approach towards human failure. Machines must work, people must think). *Maandblad voor Arbeidsomstandigheden* 67(5):313–318

Kerklaan P (2006). De lange houdbaarheid van de ongevallenwet in Nederland 1901–1967 The long tenability of the law on industrial injuries in The Netherland 1901–1967). *Tijdschrift voor Sociale en Economische Geschiedenis* 3(4):64–90

Kerklaan P Smid T Mechelen W Houwaart E van (2002). De stempel van de arbeid (The mark of labour). *Tijdschrift voor Gezondheidswetenschappen* 5:321–329

Keyser V de (1979). De menselijke factor en het ongeval (The human factor and the accident). *De Veiligheid* 55(12):633–637

Keyser V de Qvale T Wilpert B Ruiz Quintanilla S (Eds.) (1988). *The Meaning of Work and Technological Options*. John Wiley & Sons Ltd, Chichester

Khan F Abbasi S (1998). Models for domino effect analysis in chemical process industries. *Process Safety Progress* 17(2):107–123

Khan F Abbasi S (1999). Major accidents in process industries. *Journal of Loss Prevention in the Process Industries* 12:361–378

Khan F Amyotte P (2002). Inherent safety in offshore oil and gas activities a review of the present status and future directions. *Journal of Loss Prevention in the Process Industries* 15:279–289

Kidam K Hurme M Hassin M (2010). Technical analysis of accidents in chemical process industry. *Chemical Engineering Transactions* 19:451–456

Kinnersley S Roelen A (2007). The contribution of design to accidents. NeTWork proceedings Special issue. *Safety Science* 45(1–2):31–60

Kinney G Wiruth A (1976). *Practical Risk Analysis for Safety Management*. Naval Weapons Center, China Lake, California

Kirchsteiger C (1999). Trends in accidents disasters and risk sources in Europe. *Journal of Loss Prevention in the Process Industries* 12:7–17

Kirchsteiger C (2002). Towards harmonising risk-informed decision making: The ARAMIS and compass projects. *Journal of Loss Prevention in the Process Industries* 15(3):199–203

Kirchsteiger C Christou M Papadakis A (1998). *Risk Assessment ad Management in the Context of Seveso II Directive. Industrial Safety Series, volume 6*. Elsevier, Amsterdam

Kirwan B (1994). *A Guide to Practical Human Reliability Assessment*. Taylor & Francis, Bristol

Kirwan B (1998). Safety management assessment and task analysis – a missing link? In: Hale A Baram M (Eds.). *Safety Management, the Challenge of Change. Interdisciplinary Study Group New Technology and Work (NeTWork)*, pp. 67–91. Pregamon, Oxford

Kirwan B (2001). The role of the controller in the accelerating industry of air traffic management. NeTWork proceedings. Special issue. *Safety Science* 37(2–3):151–185

Kirwan B Hale A Hopkins A (2002). Insight into safety regulation. In: Kirwan B Hale A Hopkins A (Eds.). *Changing Regulations, Controlling Risk in Society. Interdisciplinary Study Group New Technology and Work (NeTWork)*, pp. 1–12. Pregamon, Oxford

Kjellén U (1982). An evaluation of safety information systems at 6 firms. *Journal of Occupational Accidents* 3:273–288

Kjellén U (1984a). The deviation concept in occupational accident control I definition and classification. *Accident Analysis & Prevention* 16(4):289–307

Kjellén U (1984b). The role of deviations in accident causation. *Journal of Occupational Accidents* 6:117–126

Kjellén U (1984c). The deviation concept in occupational accident control II data collection and assessment of significance. *Accident Analysis & Prevention* 16(4):307–323

Kjellén U (1987). Simulating the use of computerised injury near miss information in decision making. *Journal of Occupational Accidents* 9:87–105

Kjellén U (1996). Improving the efficiency of safety management in industry. In: Menckel E Kullinger B (Eds.). *15 Years of Occupational Accident Research in Sweden*. Swedish Council for Work Life Research, Stockholm

Kjellén U (2000). *Prevention of Accidents Through Experience Feedback*. Taylor & Francis, London

Kjellén U (2002). Transfer of experience from users to design to improve safety in offshore oil and gas production. In: Wilpert B Fahlbruch B (Eds.). *System Safety. Challenges and Pitfalls of Interventions. Interdisciplinary Study Group New Technology and Work (NeTWork)*, pp. 207–224. Pregamon, Oxford

Kjellén U (2007). Safety in the design of offshore platforms. Integrated safety versus safety as add-on characteristic. *Safety Science* 45(1–2):107–127

Kjellén U Albrechtsen E (2017). *Prevention of Accidents and Unwanted Occurrences. Theory, Methods and Tools in Safety Management*. Taylor and Francis, CRC Press, Boca Raton FL

Kjellén U Hovden J (1993). Reduction risks by deviation control – a retrospection into research strategy. *Safety Science* 16(3–4):417–438

Kjellén U Larsson T (1981). Investigation accidents and reducing risks – a dynamic approach. *Journal of Occupational Accidents* 3:129–140.

Klauw M van der Bakhuys M Stam C Fekke J Nijman S Venema A (2012). *Monitor Arbeidsongevallen in Nederland 2010 (Monitor Occupational Accidents in the Netherlands 2010)*. TNO Kwaliteit van leven, Hoofddorp

Klein J (2009). Two centuries of process safety at DuPont. *Process Safety Progress* 28(2):114–122

Kletz T (1976). Accident data – The need for a new look at the sort of data that are collected and analysed. *Journal of Occupational Research* 1:95–105

Kletz T (1978). 'What You Don't Have Can't Leak'. *Chemistry and Industry* 6:287–292

Kletz T (1981). Benefits and risks-their assessment in relation to human need. In: Griffiths R F (Ed.). Dealing with Risk. University of Manchester Press, Manchester

Kletz T (1982). Human problems with computer control. *Plant/Operation Progress* 1(4):209–211

Kletz T (1984a). The Flixborough explosion 10 years later. *Plant/Operation Progress* 3(3):133–135

Kletz T (1984b). *Cheaper, Safer Plants*. Institute of Chemical Engineers, Rugby

Kletz T (1985a). *An Engineer's View of Human Error*, 1st edition 1985, 2nd edition 1991. Institute Chemical Engineers, Rugby, Warwickshire, UK

Kletz T (1985b). *What Went Wrong*? Gulf Publishing Company, Houston

Kletz T (1985c). Inherently safer plants. *Plant/Operation Progress* 4(3):164–167

Kletz T (1986). Transportation of hazardous substances. *Plant Operation/Progress* 5(3):160–164

Kletz T (1988a). On the need to publish more case histories. *Plant/Operation Progress* 7(3):145–147

Kletz T (1988b). Fires and explosions of hydrocarbon oxidation plants. *Plant/Operation Progress* 7(4):226–230

Kletz T (1988c). *Learning from Accidents in Industry*. Butterworths, London

Kletz T (1989). Good safety procedures can prevent accidents. *Plant Operation Progress* 8(1):1–2

Kletz T (1999). The origins and history of Loss Prevention. *Trans IChemE* 77(B):109–116

Kloman H (1992). Rethinking risk management. *The Geneva Papers on Risk and Insurance* 17(64):299–313

Klunhaar G (1964). De baas en bedrijfsveiligheid (The boss and occupational safety). *De Veiligheid* 40(7/8):25–28; (11):33–36

Knegtering B Pasman H (2009). Safety of the process industries in the 21st century: A changing need of process safety management for a changing industry. *Journal of Loss Prevention in the Process Industries* 22:162–168

Knudsen F (2009). Paperwork at the service of safety? Workers reluctance against written procedures by the concept of 'seamanship'. *Safety Science* 47(2):295–303

Kogi K Phoon W Thurman J (1989). *Low-Cost Ways of Improving Working Conditions 100 Examples from Asia*. International Labour Office, Geneve

Kolkman H (1980). De menselijke fout in beveiligingssystemen (The human error in safety systems). *De Veiligheid* 56(12):621–622

Kolkman H (1981). Normaals de menselijke fout in beveiligingssystemen (Again the human error in safety systems). *De Veiligheid* 57(4):1770178

Kolmogorov A (1956), *Foundations of the Theory of Probability*, 2nd edition. Chelsey, New York

Komaki J Barwick K Scott L (1978) A behavioural approach to occupational safety. Pinpointing and reinforcing safe performance in a food manufacturing plant. *Journal of Applied Psychology* 63(4):434–445

Korstjens, G (1988). Tijdschrift voor toegepaste Arbowetenschap een voldongen feit (Journal of applied Occupational Sciences an accomplished fact). *Maandblad voor Arbeidsomstandigheden* 64(3):73

Kourniotis S Kiranoudis C Markatos N (2000). Statistical analysis of domino chemical accidents. *Journal of Hazardous Materials* 71:239–252

Kraaijvanger T (2014). *De Grootste Kernrampen Ooit (The biggest nuclear disasters ever)*. Scientias.nl consulted January 2016

Kraan C (1981). Factor Mensch. Verslag 7e symposium van de Internationale Sectie Chemische Industrie van de ISSA (The human factor. Report of the 7th symposium of the International section Chemical Industry of ISSA). *De Veiligheid* 57(7/8):329–330

Kraan C Schenke M (1976). Spanningsvelden voor de veiligheidsfunctionaris, door werkgroep 13 (Tention for a safety professional, by working group 13). *De Veiligheid* 52(4):143–145

Kraft J (1950). Zestig jaar veiligheidszorg bij de arbeid (Sixty years of occupational safety). *De Ingenieur* 60(9):A93

Krap J (1890). Voorkoming van ongevallen in fabrieken en werkplaatsen (Prevention of accidents in factories and workshops). *De Ingenieur* 5(6):50

Krause T Hidley J Hodson S (1990). *The Behavior-Based Safety Process: Managing Involvement for an Injury-Free Culture*. Van Nostrand Reinhold, New York.

Krause T Seymour K Sloat K (1999). Long-term evaluation of a behaviour-based method for improving safety performance: A meta-analysis of 73 interrupted time-series replications. *Safety Science* 32(1):1–18

Kromhout H Vermeulen R (2000). Long-term trends in occupational exposure: Are they real? What causes them? What shall we do with them? *Annals of Occupational Hygiene* 44:325–327

Kruithof J (1966). Ongevallenpreventie door mentaliteitsverandering (Accident prevention by mentality changes). *De Veiligheid* 42(1):5–7

Kuiper J (1969). Bedrijfsgeneeskundigen en veiligheidskundigen (Occupational physicians and safety experts). *De Veiligheid* 45(10):293–298

La Porte T Consolini P (1991). Working in Practice but Not in Theory. *Journal of Public Administration Research and Theory* 1(1):19–47

La Porte T Thomas C (1995). Regulatory compliance and the ethics of quality enhancement: Surprises in Nuclear Power Plant operation. *Journal of Public Administration Research and Theory* 5(1):109–137

Laflamme L (1990). A better understanding of occupational accident genesis to improve safety in the workplace. *Journal of Occupational. Accidents* 12:155–165

Laflamme L (1993). Technological improvement of production process and accidents: An equivocal relationship. *Safety Science* 16(3–4):246–266

Laitinen H Marjamäki M Oäivärinta K (1999). The validity of the TR safety observation method on building construction. *Accident Analysis and Prevention* 31:463–472

Laitinen H Ruohomaki I (1996). The effect of feedback and goal setting on safety performance at two construction sites. *Safety Science* 24(1):61–73

Larsson T (2002). Pulverization of risk, privatization of trauma. In: Kirwan B Hale A Hopkins A (Eds.). *Changing Regulations, Controlling Risk in Society. Interdisciplinary Study Group New Technology and Work (NeTWork)*, pp. 15–28. Pregamon, Oxford

Larsson T (2003). Is sme's a safety problem? *Safety Science Monitor* 7(1) online journal

Larsson T Rechnitzer G (1994). Forklift trucks analysis of severe and fatal occupational injuries. *Safety Science* 17(4):275–289

Lateiner A (1958). If we're to stop accidents preventing injuries is not enough. *Industrial Superviser* 26(11):3–5,14

Lateiner A Heinrich H (1969). *Management and Controlling Employee Performance.* Lateiner Publishing, New York

Laukkanen T (1999). Construction work and education: occupational health and safety reviewed. *Construction Management and Economics* 17:53–62

Law F Newell W (1909). *The Prevention of Industrial Accidents, General Pamphlet No 1.* The fidelity and causality company, New York.

Lawley H (1974). Operability studies and hazard analysis. *Chemical Engineering Progress* 70:45–56

Lawley H (1976). Size up plants this way. *Hydrocarbon Processing* 55:247–261

Lawrie M Parker D Hudson P (2006). Investigating employee perception of a framework of safety culture maturity. *Safety Science* 44(3):259–276

Lawton R (1997). Not working to rule understanding procedural violations at work. *Safety Science* 28(2):77–95

Le Coze J (2010). Accident in a French dynamite factory: An example of organisational investigation. *Safety Science* 48:80–90

Le Coze J (2011). A study about changes and their impact on industrial safety. *Safety Science Monitor* 15(2):1–17

Le Coze J (2015). Reflecting on Jens Rasmussen's legacy. A strong problem for a hard problem. *Safety Science* 71:123–141

Le Coze J (2016). Vive la diversié! High Reliability Organisations (HRO) and Resilience Engineering (RE). *Safety Science* (in press). http://dx.doi.org/10.1016/j.ssci.2016.04.006

Le Coze J Pettersen K Reiman T (2014). The foundation of safety science editorial. *Safety Science* 67:1–5

Lee F (1919). The new science of industrial psychology. *Public Health Reports* 34(15): 723–728

Lees F (1980). *Loss Prevention in the Process Industry*. Butterworth Heinemann, Oxford

Lees F (1983). The relative risk from materials in storage and in production. *Journal of Hazardous Materials* 8(2):185–190

Lees F (1996). *Loss Prevention in the Process Industry. Hazard Identification, Assessment and Control*. Butterworth Heinemann, Oxford

Lees R (2014). Aberfan Queens biggest regret. *The Free Library*, retrieved March 8 2020. www.thefreelibrary.com/Aberfan%3a+Queen%27s+%27biggest+regret%27.-a082244332

Leeuwen C van (1982). De veiligheidskundige op weg van fenomeen bestrijding naar systeembenadering (The safety expert, from combating phenomenon to system approach). *De Veiligheid* 58(6):33–34

Leij G van der (1977). Risco analyse (Risk analysis). *De Veiligheid* 53:23–26

Leij G van der (1978). Veiligheid geïntegreerd in de bedrijfsvoering. *De Veiligheid* 54(4):137–142

Leij G van der (1979a). Techniek van het veiligheidsmanagement, boekbespreking Dan Petersen (Technique and safety management, review of Dan Petersens' book). *De Veiligheid* 55(3):129–130

Leij G van der (1979b). Veiligheidsdoorlichting nuttig en haalbaar? (Safety audit, useful and achievable?). *De Veiligheid* 55(5):269–272

Leij G van der Mutgeert B (1977). Risk analysis: industry, government and society, verslag TNO conferentie. *De Veiligheid* 53(4):165–168

Leitner P (1999). Japans post-war economic success: Deming, quality and contextual realities. *Journal of Management History* 5(8):489–505

Lemkowitz A Zwaard A (1988). Veiligheids- en milieu-onderwijs moet in het onderwijspakket (Safety and environmental education must be included in the curriculum). *Chemisch Magazine* (11):708–712

Lemkowitz S (1992). A Unique program for integrating health, safety, environment and social aspects into Undergraduate Chemical Engineering Education. *Plant/Operations Progress* 11(3):140–150.

Lemkowitz S Wilde J de Bibo B (1995). Use and misuse of science in predictive safety and environmental studies. In: Bishop P (Ed.). *Proceedings Environmental Training in Engineering Education*, October 16–19, Ispra, Italy

Le Poole S (1865). Ongelukken in de fabriek (Accidents in the factory). *De Economist* 14:449–457

Leplat J (1984). Occupational accidents research and system approach. *Journal of Occupational Accidents* 6:77–89

Leuftink A (1964). Ergonomie en veiligheid I-III (Ergonomics and safety I-III). *De Veiligheid* 40(1):5–10; (2):29–16; (3):19–26

Li Y (2019). A systematic and quantitative approach to safety management. Doctoral thesis, Delft University of Technology

Li Y Guldenmund FW (2018). Safety management systems: A broad overview of the literature. *Safety Science* 103:94–123.

Lidskog R Sundqvist G (2012). Sociology of risk. In: Roeser S Hillerbrand R Sandin P Petersen M (Eds.). *Handbook of Risk Theory*, volume 2, pp. 1001–1027. Springer, Dordrecht

Liker J Nagamachi M Lifshitz Y (1989) A comparative analysis of participatory ergonomics programs in U.S. and Japan manufacturing plants. *International Journal of Industrial Ergonomics* 3:185–199

Lindhout P Ale B (2009). Language issues, an underestimated danger in major hazard control? *Journal of Hazardous Materials*. doi: 10.1016/j.jhazmat.2009.07.002

Lindhout P Kingston Howlett J Ale B (2010). Controlled readability of Seveso II company safety documents, the design of a new KPI. *Safety Science* 48(6):734–746

Lindsay F (1992). Successful health and safety management. The contribution of management audit. *Safety Science* 15(4–6):387–402

Lingard H Rowlinson S (1998). Behavioural based safety management in Hong Kong's construction industry: The results of a field study. *Construction Management and Economics* 16:481–488

Linhard J (2005). Understanding the return on health, safety and environment investments. *Journal of Safety Research* 36(3):257–260

Lintsen H (1995a). Een land zonder stoom (A country without steam) In: Lintsen H Bakker M Homburg E Lente D Van Schot J Verbong G (Eds.). Technique in The Netherlands. The Becoming of a Modern Society 1800–1890. Part VI Technique and Society. Walburg Pers, Zutphen

Lintsen H (1995b). Een land met stoom (A country with steam). In: Lintsen H Bakker M Homburg E Lente D Van Schot J Verbong G (Eds.). Technique in The Netherlands. The Becoming of a Modern Society 1800–1890. Part VI Technique and Society. Walburg Pers, Zutphen

Lochem J (1943). *Algemene gezondheidsleer (General hygiene)*, 2nd edition. Kosmos, Amsterdam

Löfstedt R Renn O (1997). Examples of what social scientists can do: The Brent Spar controversy, an example of risk communication gone wrong. *Risk Analysis* 17(2):131–136

Luijt P van (1948). Veiligheidspropaganda (Safety propaganda). *De Veiligheid* 24(4):49–51

Maidment D (1993). A changing safety culture on British Rail. *Paper on the 11th NeTWork Workshop on 'The Use of Rules to Achieve Safety'*. Bad Homburg 6–8 May

Maidment D (1997). Responding to public criticism of safety management systems. Is the response always effective and appropriate? In: Hale A Wilpert B Freitag M (Eds.). *After the Event, from Accident to Organisational Learning. Interdisciplinary Study Group New Technology and Work (NeTWork)*, pp. 66–74. Pregamon, Oxford

Malten K (1959). De hygiënische problemen bij het gebruik van kleurstoffen voor het bedrukken van katoen (Hygienic problems during the use of dyes for cotton prints). *De Veiligheid* 35:65–70

Mannan S (Ed.) (2005). *Lees' Loss Prevention in the Process Industry. Hazard Identification, Assessment and Control*, 3rd edition. Butterworth Heinemann, Oxford

Mannan S (Ed.) (2012). *Lees' Loss Prevention in the Process Industry. Hazard Identification, Assessment and Control*, 4th edition. Butterworth Heinemann, Oxford

Manuele F (2002). *Heinrich Revisited, Truisms or Myths*. NSC Press, New York.

Marbe K (1925). Zur praktischen Psychologie der Unfälle und Betriebsunglücke, Verhandlungen der phys.-med. *Gesellschaft zu Würzburg*:172–175

Marsh (2012). *The 100 Largest Losses 1972–2011*, 22nd edition. Marsh & McLennan Co, London

Marshall V (1987). *Major Chemical Hazards*. Ellis Horwood Limited, Chichester

Maslow A (1973). *The Farther Reaches of Human Nature*. Penguin Books, London

Mattila M Rantanen E Hyttinen M (1994). The quality of work environment supervision and safety in building construction. *Safety Science* 17(4):257–268

Mayhew H (1861). *London Labour and the London Poor: Cyclopedia of the Conditions and Earnings of those that Will Work, those that Cannot Work, and those that Will Not Work*. Griffin, Bohn and Company, London Stationary's' Hall Court [reprint by Dover Publications Inc, New York

McAfee P Winn A (1989). The use of incentives/feedback to enhance workplace safety: A critique of the literature. *Journal of Safety Research* 20:7–19

McElroy F (1980). *Accident Prevention Manual of Industrial Operations. Engineering and Technology*. National safety Council, Chicago, Ill

McFarland R (1963). A critique of accident research. *Annals of the New York Academy of Science* 107:686–695

McIntyre G (2000). *Patterns in Safety Thinking, a Literature Guide to Air Transportation Safety*. Ashgate, Aldershot

McIvor A (1987). Employer, the government, and industrial fatigue in Britain, 1890–1918. *British Medical Journal* 44:724–732

McKenna F (1985a). Do safety measures really work? *Ergonomics* 28(2):489–498

McKenna F (1985b). Evidence and assumptions relevant to risk homeostasis. *Ergonomics* 28(11):1539–1542

Meertens D Zwam H van (1976). Een discussiestuk over de toekomst der veiligheidsfunctie, werkgroep 13 (A discussion paper on the future of safety functions). *De Veiligheid* 52(4):113–123

Meiklejohn A (1957). *The Life, Work and Times of Charles Turner Thackrah, Surgeon and Apothecary of Leeds (1795–1833)*. E & S Livingstone Ltd, Edinburgh

Menckel E Carter N (1985). The development and evaluation of accident prevention routines, a case study. *Journal of Safety Research* 16:73–82

Menckel E Kullinger B (1996). *15 Years of Occupational Accident Research in Sweden*. Swedish Council for Work Life Research, Stockholm

Mertens F (2011). *Inspecteren, Toezicht Door Inspecties (Inspection, Supervision Through Inspections)*. Sdu Uitgevers bv Den Haag

Mesritz A Ree v R (1937). *Bedrijfshygiëne en Veiligheidstechniek. Beknopt Leerboek voor het Middelbaar Technisch Onderwijs, voor Studerenden voor het Nijverheids-Akte-Examens en voor in de Prakrijk Werkzame Technici (Occupational Hygiene and Safety Technique, Textbook in Brief for Students Preparing for the Examination of Trade and Industry, and for Engineers Working in Professional Settings)*. De technische boekhandel H. Stam, Amsterdam

Mihailidou E Antoniadis K Assael M (2012). The 319 major industrial accidents since 1917. *International Review of Chemical Engineering* 4(6):529–540

Miller T Hoskin A Yalung D (1987). A procedure for annually estimating wage losses due to accidents in the US. *Journal of Safety Research* 18:101–119

Ministry of Housing (1990). Spatial planning and environment and the Ministry of social affairs and employment. *Note Setting Ionizing Radiation Protection Standards for Labour and Environment: 'Dealing with Risks of Radiation*. "No. 21483' SDU, The Hague (in Dutch)

Ministry of the Interior and Kingdom Relations (2012). *New Perspectives on Dealing with Risks and Responsibilities*. The Hague. (in Dutch)

Mintzberg H (1979). *The Structuring of Organisations, a Synthesis of the Research*. Prentice-Hall. Englewood Cliffs, NJ

Mintzberg H (1983). *Structure in Five: Designing Effective Organisations.* Prentice Hall Inc, Englewood NJ

Mitchell J (1911). Burden of industrial accidents. *Annals of the American Academy of Political and Social Science* 38(1):76–82

Molen H Lehtola M Lappalainen J Hoonakker P Hsiao H Haslam R Hale A Verbeek J (2007). *Interventions for Prevention in the Construction Industry (Review).* The Cochrane Collaboration, Wiley and Sons, London

Monteau M (1983). Control technonoly in occupational safety and health. In: Parmeggiani L (Ed.). *Encyclopaedia of Occupational Health and Safety*, 3rd revised edition. International labour Office, Geneva

Moore P (1968). *Basic Operational Research.* Pitman Publishing, New York

Morgan G (1986). *Images of Organizations.* Sage Publications, London

Mostia W (2010). Why bad things happen to good people. *Journal of Loss Prevention in the Process Industries* 23(6):799–805

Mulder P (2012). *Maslow's Hierarchy of Needs.* Retrieved February 1st 2020 from ToolsHero: www.toolshero.com/psychology/hierarchy-of-needs-maslow/

Muntz E (1932). Accidents and safety work. *Journal of Educational Sociology* 5(7):397–412

MVK Middelbare Veiligheidskunde-(secondary Safety Education) (2008). *Risicobeheersing en Ongevalsmanagement (Risk Control and Accident Management).* Elsevier opleidingen, Amsterdam

Myers G (1915). A study of the causes of industrial accidents. *Publications of the American Statistical Association* 14(117):672–694

Nachreiner F Nickel P Meyer I (2006). Human factors in process control systems: The design of human-machine interface. *Safety Science* 44(1):5–26

Nammogran (1976). Maximale inzet van een ieder bij afsluiten put S457 (Maximum effort of everyone on closing put S457). *Nammogram* 17(11):3–9

Nammogram (1980). Voorbereid op. . . (Prepared for. . .). *Nammogram* 21(3):2–9

National Safety Council (1914). *Referred to in Greenwood (1934). Who Pays?* Doubleday Doran Co Inc. New York

National Safety Council-NSC (2008). www.nsc.org/about/timeline/aspx, consulted November 25, 2008.

Neal A Griffin M Hart P (2000). The impact of organisational climate on safety climate and individual behaviour. *Safety Science* 34:99–109

NEA-OECD Nuclear Energy Agency (2010). *The International Nuclear and Radiological Event Scale (INES): 20 Years of Nuclear Communication.* NEA/COM/7 Paris, Vienna October 14th

Necci A Cozanni V Spadoni G Khan F (2015). Assessment of domino effect State of the art and research needs. *Reliability Engineering and System Safety* 143:3–18

Nettleton M (1976). Alleviation of blast waves from large vapour clouds. *Journal of Occupational Accidents* 1:3–8

Nettleton M (1976/1977). Some aspects of vapour cloud explosions. *Journal of Occupational Accidents* 1:149–158

Nielsen D (1971). *The Cause/Consequence Diagram Method as a Basis for Quantitative Accident Analysis.* Danish Atomic Energy Commission, research Establishment, Risø. Rapport Risø-M-1374

Nielsen D Platz O Runge B (1975). Cause consequence diagram. *IEEE Transactions on Reliabiility* 2 R24(1):8–13

Ninkovich F (1994). *Modernity and Power: A History of the Domino Theory in the Twentieth Century*. Chicago university press, Chicago

Niskanen T Saarsalmi O (1983). Accident analysis in construction of buildings. *Journal of Occupational Accidents* 5:89–98

Nissen M Dijkstra L Oostendorp Y (1996). RI&E bewijst zich nog onvoldoende als beleidsinstrument. ABRIE methode geëvalueerd (RI&E does prove itsself insufficiently as policy instrument. ARIE method evaluated). *Arbeidsomstandigheden* 73(9):422–424

Noesen R (1966). De menselijke factoren en de veiligheid (Human factors and safety). *De Veiligheid* 42(10):275–276

Noort H van (1952). Benzol en verwante stoffen (Benzol and related compounds). *De Veiligheid* 28:88–90

Nossent S Ziekemeyer M Kromhout H Swuste P (1990). Arbeidsomstandigheden en arbeidsomstandighedenbeleid in de rubberverwerkende industrie wel of geen samenhang (Working conditions and working condition policy in the rubber manufacturing industry, coherent or not?). *Tijdschrift voor Toegepaste Arbowetenschap* 3(2):36–39

NPR 5001 (1997). *Nederlandse PraktijkRichtlijn (Dutch Practice Guidance) Model voor een arbo-managementsysteem, NEN-norm (Model for occupational management system, NEN norm)*. NEN, Delft

Nunen K van Swuste P Reniers G et al (2018). Improving pallet mover safety in the manufacturing industry. A Bow-Tie analysis of accident scenarios. *Materials* 11(1955):1–19

Nye D (2013). *America's Assembly Line*. The MIT Press, Cambridge MA

Office of Nuclear Regulations (UK) 2000–2010 www.onr.org.uk/, consulted January 2016

OGP (2010). *International Association of Oil and Gas Producers. Risk Assessment Data Directory Report 434–17 Major Industrial Accidents*. OGP, London

Oirbons J (1981). Arbocommissie voor veiligheid gezondheid en welzijn (Commission for occupational safety health and well-being). *De Veiligheid* 57(11):483–486

Oliver T (1902). *Dangerous Trades from the Legislative, Social and Medical Point of View*. Methmen & Co, London

Olsen S Rasmussen J (1989). The reflective expert and the prenovice: Notes on skill-, rule- and knowledge-based performance in the setting of instruction and training. In: Bainbridge L Ruiz S (Eds.). *Developing Skills with Information Technology. Interdisciplinary Study Group New Technology and Work (NeTWork)*. Wiley, Chichester

Onderzoeksgroep Veiligheid en DuPont (1982). *Research group safety and DuPont. Veiligheid tot welke prijs? Onderzoek naar de invloed van het veiligheidsbeleid van het chemische bedrijf DuPont de Nemours op het handelen en denken van de werknemers en naar de veiligheids- en gezondheidsaspecten van het werk (Safety at what price? Research into the influence of the safety policy of the chemical company DuPont de Nemours on the actions and thinking of the employees and into the safety and health aspects of the work)*. Vakgroep Sociale- en Organisatiepsychologie en de Chemiewinkel van de Rijksuniversiteit, Leiden

Oortmans Gerlings P Hale A (1991). Certification of safety services in large Dutch industrial companies. *Safety Science* 14(1):43–59

Oostendorp Y Zwaard W Gulijk C van Lemkowitz S Swuste P (2016). Introduction of the concept of risk within safety science in The Netherlands focussing on the years 1970–1990. *Safety Science* 85:2015–219

Oosterom N (1979). Humanisering van de arbeid (Humanisation of labour). *De Veiligheid* 55(7/8):382–383

Opland (1979). Illustratie bij het artikel van Reijnders L (1979). Drie visie op veiligheid (Illustration of an article of Reijnders. Three visions on safety). *Risicobulletin* 1(1):5–7

Orton J Weick K (1990). Loosely coupled systems: a reconceptualization. *The Academy of Management Review* 15(2):203–223

Osborne E Vernon H Muscio B (1922). Two contributions to the study of accident causation: the influence of temperature and other conditions in the frequency of industrial accidents, on the relation of fatigue and accuracy to speed and duration of work. *Industrial Fatigue Board*, report nr 19. Her Majesty's Stationary Office Press, Harrow

OVV (2013). *Onderzoeksraad Voor Veiligheid. Veiligheid in Perspectief. Acht jaar Ongevalsonderzoek door de Onderzoeksraad voor Veiligheid 2005–2012 (Eight Years of Accident Investigation by the Dutch Safety Board 2005–2017)*. Grapefish, Voorschoten

Page P Hubbard F Anderson J Henretty P Wallace J Jones E Falconer J McGoldrick J Parker C Eschen G (1910). *Report of Commission Appointed by Governor M. Hay to Investigate the Problems of Industrial Accidents and to Draft a Bill of Employees' Compensation.* E.L. Boardman, public printer, Olympia, Washington

Palmer L (1926). The history of the Safety Movement. *Annals of the American Academy of Political and Social Sciences* 123(1):9–19

Papadakis G Amendola A (Eds.) (1997). *Guidance on the Preparation of a Safety Report to Meet the Requirements of Council Directive 96/82/EC (Seveso II)*. Joint Research Centre, European Commission, Luxembourg

Papazoglou I Ale B (2007). A logical model for quantification of occupational risk. *Reliability Engineering and System Safety* 92(6):785–803

Papazoglou I Bellamy L Hale A Aneziris O Ale B Post J Oh J (2003). I-risk: development of an integrated technical and management risk methodology for chemical installations. *Journal of Loss Prevention in the Process Industries* 16(6):575–591

Parker D Lawrie M Hudson P (2006). A framework for understanding the development of organisational safety culture. *Safety Science* 44(7):551–562

Parker R (1975). *The Flixborough Disaster, Report of the Court of Inquiry*. Department of Employment. Her Majesty's Stationary Office, London

Pasman H (1974). Schadepreventie en veiligheidsbevordering in de procesindustrie (Damage prevention and safety promotion in the process industry). *De Veiligheid* 86(16):311–312

Pasman H (1999). *Risk Control. Chemical Risk Management. Towards Safe Processes and Products for People and Environment.* Inaugural speech, TU Delft (in Dutch)

Pasman H (2000). Risk informed resource allocation policy: Safety can safe costs. *Journal of Hazardous Materials* 71(1–3):375–394

Pasman H (2009). Learning from the past and knowledge management: Are we making progress? *Journal of Loss Prevention* 22(6):672–679

Pasman H (2011). History of English process equipment failure frequencies and the purple book. *Journal of Loss Prevention in the Process Industries* 24(3):208–213

Pasman H (2015). Risk analysis and control for industrial processes – gas, oil and chemicals. *A System Perspective for Assessing Low-Probability, High Consequence Events.* IChemE, Butterworth Heinemann, Oxford

Pasman H Baron R (2002). How is it possible? Why didn't we do anything? A case history! *Journal of Hazardous Materials* 93(1):147–154

Pasman H Duxbury H Bjordal E (1992). Major hazards in the process industry. *Journal of Hazardous Materials* 30(1):1–38

Pasman H Reniers G (2013). Past, present and future of quantitative risk assessment (QRA) and the incentive it obtained from Land-use Planning. *Journal of Loss Prevention in the Process Industries* 28(1):2–9

Pasman H Snijder G (1974). Schadepreventie en veiligheidsbevordering in de procesindustrie (Damage prevention and safety promotion in the process industries). *De Veiligheid* 50(5):211–212

Pasmooij C (1979). Ongunstige arbeidsomstandigheden en mens-factoren in hooggeautomatiseerde systemen. *De Veiligheid* 55(4):161–165

Paté-Cornell E Boykin R (1987). Probabilistic risk analysis and safety regulation in the chemical industry. *Journal of Hazardous Materials* 15(1–2):97–122

Paté-Cornell M (1993). Learning from the Piper Alpha accident: A post-mortem analysis of technical and organizational factors. *Risk Analysis* 13(2):215–232

Patijn J (1903). Toezicht op de veiligheid in mijnen door mijnwerkers I en II (Control of safety in mines by miners I and II). *De Economist* 52(1):559–573 and 52(2):683–704

Patijn R (1945, 1946). De mensch als oorzaak der ongevallen (Men as cause of accidents) I-VI. I Factoren in een fabriek die een ongunstige psychische invloed uitoefenen op den arbeider, waardoor ongelukken kunnen ontstaan (Factors in a factory with an unfavorable influence on workers and might cause accidents). *De Veiligheid* 1945(2):22:14–16; II De geestelijke houding tov de buitenwereld als ongevallenoorzaak (The mental attitude towards the external world as cause of accidents). *De Veiligheid* 1945:22(3):18–20; III Menselijke karaktereigenschappen als ongevallenoorzaak (Character traits as cause of accidents). *De Veiligheid* 1945:22(4):26–28; IV Het temperament als ongelukkenoorzaak (Temperament as cause of accidents). *De Veiligheid* 1946:22(6):43–46; V Het ongelukkenprobleem (The problem of accidents). *De Veiligheid* 1946:22(7):52–53; VI Hoe kan men een brokkenmaker onderscheiden? (How can one discriminate an accident-prone worker?). *De Veiligheid* 1946:22(8):60–61

Pavlov I (1927). *Conditioned Reflexes*. Oxford University Press, heruitgegeven in 1960 en 2003 bij Dover Publications, New York

Pekkarinnen A Anttonen H (1989). The comparison of accidents in a foreign construction project with construction in Finland. *Journal of Safety Research* 20:187–195

Pellanders M (1980). Foto Wessanen meelfabriek, Wormerveer (Photo Wessanen floor mill, Wormerveer)

Peperstraten J van (1992). WEBA-instrument blijkt nog niet populair. *Arbeidsomstandigheden* 68(5):297–299

Perrow C (1984). *Normal Accidents. Living with High-Risk Technologies*. BasicBooks, US

Perrow C (1994). The limits of safety: The enhancement of a theory of accidents. *Journal of Contingencies and Crisis Management* 2(4):212–220

Perrow C (1999). *Normal Accidents. Living with High-Risk Technologies*. Princeton University Press, Princeton NJ

Pesatori A Consonni D Bachetti S Zocchetti C Bonzini M Baccarelli A Bertazzi P (2003). Short- and long-term morbidity and mortality in the population exposed to dioxin after the "Seveso Accident. *Industrial Health* 41:127–138

Pesatori A Consonni D Rubagotti M Grillo P Bertazzi P (2009). Cancer incidence in the population exposed to dioxin after the Seveso accident twenty years of follow-up. *Environmental Health* 8(1):39–50. doi: 10.1186 1476–069X-8–39

Peters H (1969). Veiligheidsorganisatie en – programma in een bedrijf (Safety organisation and program in a company). *De Veiligheid* 45(5):145–150

Peters T Waterman R (1982). *In Search for Excellence, Lessons from America's Best-Run Companies*. Harper & Row Publisher, New York

Petersen D (1971). *Techniques of Safety Management.* McGraw-Hill Book Company, New York

Petersen D (1975). *Safety Management a Human Approach, a Human Approach.* McGraw-Hill Book Company, New York

Petersen D (1978). *Techniques of Safety Management.* McGraw-Hill Book Company, New York

Pettersen K Schulman P (2016). Drift, adaptation and reliability: Towards an empirical clarification. *Safety Science* (in press). http://dx.doi.org/10.1016/j.ssci.2016.03.004

Peuscher W Groeneweg J (2012). A big oil company's approach to significantly reduce fatal incidents. *Proceedings of the SPE/APPEA Conference on Health, Safety and Environment in Oil and Gas Exploration and Production.* Perth, Australia. Ref. SPE 157465, Society of Petroleum Engineers, London.

Pfeifer C Schaeffer M Grether C Stefanski J Tuttle T (1974). An evaluation of policy related research on effectiveness of alternative methods to reduce occupational illness and accidents. *Behavioural, Safety Center.* Westinghouse Electrical Co, Columbia, Maryland

Pidgeon N O'Leary M (2000). Man-made disasters why technology and organizations (sometimes) fail. *Safety Science* 34(1):15–30

Pieters H Hovers J (1960). Zorg voor de veiligheid, een levensvoorwaarde (Care for safety a life condition) (1–8). *De Veiligheid* 36(1):3–6, (2):13–16, (3):4–6, (4):13–16, (5):19–25, (6):9–14, (7):12–20, (8/9):4–9

Pietersen C (2009). *Twenty-Five Years Later: The Two Largest Industrial Disasters Involving Hazardous Substances.* Gelling Publishing, Nieuwerkerk aan den IJssel

Pietersen C Veld B van het (1992). Risk assessment and risk contour mapping. *Journal of Loss Prevention in the Process Industries* 5(1):60–63

Pietersen M (1981). *Het Technisch Labyrinth een Maatschappijgeschiedenis van Drie Industriële Revoluties (The Technical Labyrinth, the History of Society of Three Industrial Revolutions).* Boom, Meppel

Pindur W Rogers S Kim P (1995). The history of management: A global perspective. *Journal of Management History* 1(1):59–77

Pinker S (2018). *Enlightment Now, the Case for Reason, Science, Humanism, and Progress.* Penguin Books, New York

Pinwell G (1866). *Death's Dispensery.* Philidelphia Museum of Art, Philadelphia, PA

Pliny (77). *Naturalis Historia.* Translated in Dutch by Gelder J van Niewenhuis M Peters (2004) T. Plinius, De Wereld, Athenaeum – Polak & Van Gennep, Amsterdam.

Poll K (1983). Ergonomische arbeidsplaatsverbetering (Ergonomic improvement of work places). *De Veiligheid* 59(12):615–618

Poll K (1984). Normen voor tillen (Norms for lifting). *De Veiligheid* 60(5):281–285

Pope W (1976). Systems safety management: Een nieuwe opvatting over interne managementcommunicatie en veiligheid (Systems safety management: A new approach of internal managerial communication and safety). *De Veiligheid* 52(12):487–490

Powell P Hale M Martin J Simon M (1971). *2,000 Accidents, a Shopfloor Study of their Causes on 42 Months' Continuous Observation.* National Institute of Industrial Psychology, London

Poyet C Leplat C (1993). Mixed technologies and management of reliability. In: Wilpert B Qvale T (Eds.). *Reliability and Safety in Hazardous Work Systems. Interdisciplinary Study Group New Technology and Work (NeTWork). Approaches to Analysis and Design,* pp. 131–156. Lawrence Erlbaum Associates Publisher, Hove UK

Purswell J Rumar K (1984). Occupational accident research: Where have we been and where are we going? *Safety Science* 6:219–228

Putman J (1986). Onvoldoende kennis en vaardigheden oorzaak van groot aantal ongevallen (Insufficient knowledge and skills as causes of a great number of accidents). *Maandblad voor Arbeidsomstandigheden* 62(12):746–751

Qureshi Z (2007). *A Review of Accident Modelling Approaches for Complex Socio-Technical Systems*. 12th Australian Workshop on safety (SCS'07), Adelaide

Raafat H (1989). Risk assessment and machine safety. *Journal of Occupational Accidents* 11(1):37–50

Raafat H Abdouni A (1987). Development of an expert system for human reliability analysis. *Journal of Occupational Accidents* 9:137–152

Radandt S (1979). Perspectieven voor de ontwikkeling van de veiligheidstechniek (Perspectives fort he development of safety technique). *De Veiligheid* 55(11):577–578

Rademaker E Suter G Pasman H Fabiano B (2014). Review of past present future. *Loss Prevention Process Safety and Environmental Protection* 92:280–291

Ramamoorthy C Ho S Han Y (1977). *Fault Tree Analysis of Computer Systems*. National Computer Conference 13–16 June, Dallas Texas, p 13–17

Ramazzini B (1700). *De Morbis Artificum Diatriba*. University of Modena, Modena

Ramirez J Pastor E Casal J Amaya R (2015). Analysis of domino effect in pipelines. *Journal of Hazardous Materials* 298:210–220

Rasmussen B Gronberg C (1997). Accidents and risk control. *Journal of Loss Prevention* 10(5–6):325–332

Rasmussen J (1980). What can be learned from human error reports? In: Duncan K (Ed.). *Changes in Working Life*. John Wiley & Sons Ltd, London

Rasmussen J (1982). Human errors. A taxonomy for describing human malfunctioning. *Journal of Occupational Accidents* 4(2–4):311–333

Rasmussen J (1983). Skills, rules, and knowledge; signals, signs, and symbols, and other distinctions in human performance models. *IEEE Transactions on systems man and cybernetics SMC* 13(3):257–266

Rasmussen J (1985). The role of hierarchical representation in decision making and system management. *IEE Transactions on Systems* SMC-15(2):234–243

Rasmussen J (1988a). Human error mechanisms in complex working environments. *Reliability Engineering and System Safety* 22:155–167

Rasmussen J (1988b). Human factor in high risk systems. *IEEE Transactions on systems man and cybernetics SMC* 18:43–48

Rasmussen J (1990). The role of error in organising behaviour. *Ergonomics* 33(10–11): 1185–1199

Rasmussen J (1991). Modelling distributed decision making. In: Rasmussen J Brehmer B Leplat J (Eds.). *Distributed decision making. Interdisciplinary study group New Technology and Work (NeTWork)*, pp. 111–142. John Wiley & Sons, Chichester

Rasmussen J (1993a). Learning from the experience? How? Some research issues in industrial risk management. In: Wilpert B Qvale T (Eds.). *Reliability and Safety in Hazardous Work Systems. Approaches to Analysis and Design. Interdisciplinary Study Group New Technology and Work (NeTWork)*, pp. 43–66. Lawrence Erlbaum Associates Publishers, Hove UK

Rasmussen J (1993b). Diagnostic reasoning in action. *IEE Transactions on Systems* 23(4):981–992

Rasmussen J (1994). Risk management, adaptation, and design for safety. In: Sahlin N Brehmer B (Eds.). *Future Risks and Risk Management*. Kluwer, Dordrecht

Rasmussen J (1997). Risk management in a dynamic society: A modelling problem. *Safety Science* 27(2–3):183–213

Rasmussen J Duncan K Leplat J (Eds.) (1987). *New Technology and Human Error*. John Wiley & Sons, Chichester

Rasmussen J Lind M (1982). A model for human decision making. MP3 2:30 *IEE Transactions on Systems* 270–276

Rasmussen J Reason J (1987). Causes, and human error. In: *New Technology and Human Error*. Wiley, London, 1987

Rasmussen J Vincente K (1989). Coping with human errors through system design: Implications for ecological interface design. *International Journal of Man Machine Studies* 31:517–534

Rasmussen N (1975). *Reactor Safety Study. An Assessment of Accident Risks in US Commercial Nuclear Power Plants*. Executive summary. WASH-1400 (NUREG-75014). Rockville, MD USA Nuclear Regulatory Commission

Ravetz J (2001). Safety in a globalising knowledge economy: An analysis by paradoxes. *Journal of Hazardous Materials* 86(1):1–16

Reason J (1987). The Chernobyl errors. *Bulletin of the British Psychological Society* 40:201–206

Reason J (1990). *Human Error*. Cambridge University Press, Cambridge

Reason J (1993). Managing the management risk: New approaches to organisational safety. In: Wilpert B Qvale T (Eds.). *Reliability and Safety in Hazardous Work Systems. Approaches to Analysis and Design. Interdisciplinary Study Group New Technology and Work (NetWork)*, pp. 7–22. Lawrence Erlbaum Associates-Publishers, Hove UK

Reason J (1997). *Manging the Risks of Organizational Accidents*. Ashgate, Aldershot

Reason J (2000). Human error: Models and management. *British Medical Journal* 320:768–770

Reason J (2013). *A Life in Error. From Little Slips to Big Disasters*. Ashgate, Farnham

Reason J Hobbs A (2003). *Managing Maintenance Error: A Practical Guide*. Ashgate Publishing Company, Burlington, VT

Reason J Hollnagel E Paries J (2006). *Revisiting the Swiss Cheese Model – Eurocontrol EEC Note No. 13/06*. European organisation for the safety of air navigation. Centre de Bois des Bordes

Redactie (1986). Het werkelijke gevaar van de menselijke fout (The real hazard of the human error). *Maandblad voor Arbeidsomstandigheden* 62(7/8):445

Redactie (1999). Arbeidsomstandigheden 75 jaar (Working conditions 75 years). *Arbeidsomstandigheden* 75(1):20–21

Redinger C Levine S (1998). Development and evaluation of The Michigan OSHMS assessment instrument. *AIHA Journal* 59(8):572–581

Reeves T Turner B (1972). A theory of organisation and behaviour in batch production factories. *Administrative Science Quarterly* 17(1):81–98

Reid D (1987). Social reform into practice: Labour Inspectors in France, 1892–1914. *Journal of Social History* 20(1):67–87

Reij W (1962). De betekenis van de ergonomie voor de ongevallenpreventie (The meaning of ergonomics for accident prevention). *De Veiligheid* 38(7):37–44

Reijnders L (1979). Drie visies op veiligheid, de deugdelijke machine en de ondeugdelijke mens (Three views on safety, the sound machine and the inadequate human being). *Risicobulletin* 1(1):5–7

Reiman T Oedewald P (2007). Assessment of complex sociotechnical systems – Theoretical issues concerning the use of organizational culture and organizational core task concepts. *Safety Science* 45(7):745–768

Reniers G Dullaert W Ale B Soudan K (2005a). The use of current risk analysis tools evaluated towards preventing external domino effects. *Journal of Loss Prevention in the Process Industries* 18:119–126

Reniers G Dullaet W Ale B Soudan K (2005b). Developing an external domino accident prevention. *Journal of Loss Prevention in the Process Industries* 18(3):127–138

Reniers G Khakzad N (2017). Revolutionizing Safety and Security in the Chemical and Process Industry: Applying the CHESS concept. *Journal of Integrated Security Science* (1):2–15

Rigby L Swain A (1971). In-flight target reporting – how many is a bunch? *Human Factors* 13(2):177–181

Rijpma J (1997). Complexity, tight coupling and reliability: Connecting normal accidents theory and high reliability theory. *Journal of Contingencies and Crisis Management* 5(1):15–23

Rijpma J (2003). From deadlock to dead end: The normal accidents-high reliability debate revisited. Book review essay. *Journal of Contingencies and Crisis Management* 11(1):37–45

Rio Declaration (1992). *Report of the United Nations Conference on Environment and Development*, Rio de Janeiro, 3–14 June. Annex I Rio declaration on environment and development

Rip A (1986). The mutual dependence of risk research and political context. *Science and Technology studies* 4(3–4):3–15

RIVM (2008). *The Quantification of Occupational Risk. The Development of a Risk Assessment Model and Software*. RIVM report 620801001, Bilthoven

Robens (1972). *Committee on Safety and Health at Work (1972). Report of the Committee 1970–1972, Chairman Lord Robens*. Her Majesty's Stationery Office, London

Roberts A Pritchard D (1982). Blast effect of unconfined vapour cloud explosions. *Journal of Occupational Accidents* 3:231–247

Roberts K (1988). Some characteristics of one type of high reliability organization. *Organization Science* 1(2):160–176

Roberts K (1989). New challenges in organizational research: High reliability organizations. *Industrial Crisis Quarterly* 3:111–125

Roberts K (1990). Some characteristics of one type of high reliability organisation. *Organization Science* 1(2):160–177

Roberts K Hulin C Rousseau D (1978). *Developing an Interdisciplinary Science of Organizations*. Jossey-Bass, San Francisco.

Roberts K Rousseau D (1989). Research in nearly failure-free, high reliability organizations: Having the bubble. *IEE Transactions on Engineering Management* 36(2):132–139

Robinson G (1982). Accidents and sociotechnical systems principles for design. *Accident Analysis & Prevention* 14(2):121–130

Robson L Clarke J Cullen K Bieecky A Severin C Bigelow P Irvin E Culyer A Mahood Q (2007). The effectiveness of occupational health and safety management system interventions: A systematic review. *Safety Science* 45:329–353

Rochlin G (1986). High reliability organisations and technical change. Some ethical problems and dilemma. *IEEE Technology and Society Magazine*, September:3–9

Rochlin G (1989). Informal organisational networking as a crisis avoidance strategy US naval flight operation. *Industrial Crisis Quarterly* 3:159–176

Rochlin G (1999). Safe operation as a social construct. *Ergonomics* 42(11):1549–1560

Rochlin G La Porte T Roberts K (1987). The self-designing high reliability organisation: Aircraft carrier flight operation at sea. *Naval War College Review* 40:76–90

Roland Holst H (1990). *System Safety Engineering and Management.* Wiley & Sons, New York

Roland Holst H (1902). *Kapitaal en Arbeid in Nederland. Bijdrage tot de Economische Geschiedenis der 19e Eeuw (Capital and Labour in The Netherlands, Contribution to the History of the '19th Century).* A Soep, Amsterdam. SUN reprint 1973, Nijmegen

Rolt L (1955). *Red for Danger, a History of Railway Accidents and Railway Safety Precautions.* Lane, Sutton Publishing Limited, London

Romein J Romein A (1973). *De Lage Landen bij zee. Een Geschiedenis van het Nederlandse Volk, 6e Druk (The Low Lands at Sea. A History of the Dutch People).* Querido, Amsterdam. ISBN 90 214 2029 5

Ronza A Félez S Darbra R Carol S Vílchez J Casal J (2003). Predicting the frequency of accidents in ports areas by developing event trees from historical analysis. *Journal of Loss Prevention in the Process Industries* 16:551–560

Roolvink B (1968). Rede van minister van Sociale zaken en Volksgezondheid, de heer B Roolvink, bij de opening van de 7[e] bedrijfsveiligheidsbeurs. (Presenation of Minister of Social Affairs and Public health, mr. B Roolvink, during the opening of the 7th Industrial Safety Exhibition). *De Veiligheid* 44(6):199–203

Roos A (1979). Humanisering van de arbeid, wat houdt het in (Humanisation of labour, what is it all about). *De Veiligheid* 55(7/8):384–387

Roper S (1899). *Roper's Engineer's Handy Book.* David McKay, Philadelphia

Rosen G (1976). *A History of Public Health,* 3rd edition. MD Publications, New York

Rosling H Rosling O Rönnlund (2018). *Factfulness, Ten Reasons We're Wron about the World and Why Things Are Beter Than You Think.* Flatiron Books, New York

Rosness R (2009). Classifying the cases. NeTWork proceedings special issue. *Safety Science* 47(6):899–901

RoSPA Royal Society for the Prevention of Accidents. RoSPA in early years www.rospa.com/history/index.htm. Site consulted, May 2008

Rowe W (1977). *An Anatomy of Risk.* John Wiley & Sons, New York

Rowe W (1988). *An Anatomy of Risk.* Robert Krieger Publishing Company, Malabar FL

Ruddick C (1957). The value of good records and how to use them. *NSC Transactions* 3:5–8.

Rundmo T (2000). Safety climate attitudes and risk perception Norsk Hydro. *Safety Science* 34:47–59

Saarela K Saari J Aaltonen M (1989). The effect of an informational safety campaign in the shipyard industry. *Journal of Occupational Accidents* 10:255–266

Saari J (1982). Summary theme accidents and progress in technology. *Journal of Occupational Accidents* 4:373–378

Saari J (1984). Accidents and disturbances in the flow of information. *Journal of Occupational Accidents* 6(1–3):91–105

Saari J (1998). Safety interventions: international perspectives. In: Feyer A Williamson A (Eds.). *Occupational Injury, Risk, Prevention and Intervention.* Taylor & Francis, London

Saari J Näsänen M (1989). Effect of feedback on industrial housekeeping and accidents shipyards: A long-term study at a shipyard. *International Journal of Industrial Ergonomics* 4:201–211

Sadee C Samuels D O'Brien T (1976). The characteristics of the explosion of cyclohexane at the Nypro Flixborough plant on 1st June 1974. *Journal of Occupational Accidents* 1:203–235

Sagan S (1993). *The Limits of Safety: Organizations, Accidents, and Nuclear Weapons.* Princeton University Press, Princeton, NJ.

Saleh J and Marais K (2006). Highlights from the early (and pre) history of reliability engineering. *Reliability Engineering & System Safety* 91:249–256

Salminen S Saari J Saarela K Räsänen T (1992). Fatal and non-fatal occupational accidents: Identical versus differential causation. *Safety Science* 15(2):109–118

Salzano E Cozzani V (2012). *Introduction External Hazard Factors in QRA Revista de Ingeniería*, pp. 50–56. Universidad de los Andes. Bogotá D.C

Schaack D van (1917). *Safeguards for the Prevention of Industrial Accidents.* The life insurance Co., Hartford.

Schein E (1972). *Organization Psychology.* Prentice-hall. Englewood Cliffs, NJ

Schein E (1992). *Organisational Culture and Leadership.* Jossey-Bass, San Francisco

Schein E (1996). Three cultures of management. The key to organisational learning. *Sloan Management Review* 38(1):9–20

Schein E (1999). *The Corporate Culture Survival Guide.* Jossey-Bass, San-Francisco

Schmitz P Swuste P Renier G Nunen K van (2020). Mechanical integrity of process installations: Barrier alarm management based on bowties. *Process Safety and Environmental Protection* 138:139–147

Schmitz P Swuste P Reniers G Nunen K van (2021). Predicting major accidents in the process industry based on the barrier status at scenario level: A practical approach. *Journal of Loss Prevention in the Process Industries* 71:104519

Schmitz P Reniers G Swuste P Nunen K van (2021). Predicting major hazard accidents in the process insustry based on organizational factors: A practical, qualitative approach. *Process Safety and Environmental Protection* 148:1268–1278

Schmitz P Reniers G Swuste P (2021a). Predicting major hazard accidents by monitoring their barrier system: A validation in retrospect. *Process Safety and Environmental Protection* 153:19–28

Schmitz P Reniers G Swuste P (2021b). Determinating a realistic ranking of the most dangerous process equipment of the ammonia production process. *Journal of Loss Prevention in the Process Industries* 70:104395

Scholte R (1993). Postdoctorale beroepsopleiding Veiligheid, Gezondheid en Welzijn in de arbeid (Post graduate course Safety, Health and Well-being at Work). *Nederlands Tijdschrift voor Geneeskunde* 137:1025

Schouten M Faas A (2007). *Achtergronden Dodelijke en Ernstige Arbeidsongevallen op basis van in 2006 afgesloten ongevalsonderzoeken (Backgrounds of lethal and serious occupational accidents, based upon accident reports, finished in 2006).* Arbeidsinspectie, kantoor Den Haag, groep monitoring en beleidsinformatie

Schupp S Hale A Pasman H Lemkovitz S (2006). Design support for systematic integration of risk reduction. *Safety Science* 44(1):37–54

Schwedtman F Emery J (1911). *Accident Prevention and Relief. An Investigation of the Subject in Europe with Special Attention to England and Germany, Together with Recommendations for Action in the United States of America.* National Association of Manufacturers of the USA, New York

Schwitters R (1991). *De Risico's van de Arbeid. Het Ontstaan van de Ongevallenwet 1901 in Sociologisch Perspectief (The Risks of Labour. The Start of the la won Industrial Injuries in Sociological Perspective).* Proefschrift Rijksuniversiteit Utrecht. Rechtswetenschappelijke reeks Wolters Noordhoff, Groningen

Sectie Veiligheidskunde (1983). *Verslag Symposium Veiligheidskundige Opleiding in Nederland (Report of the Symposium Safety Training in the Netherlands)*. THDelft, 26 januari
Senge P (1990). *The Fifth Discipline: The Art and Practice of the Learning Organization*. Doubleday, New York
SER (1997). *Sociaal Economische Raad, Ontwerp Arbobeleidsregels (Social Economical Council, Draft Occupational Policy Rules)*. SER, Den Haag
SER (2000). *Sociaal Economische Raad, De Winst van Waarden. Advies Over Maatschappelijk Ondernemen. Uitgebracht aan de Staatssecretaris van Economische Zaken (Social Economical Council, The Advantage of Values. Advice on Corporate Social Responibility)*. SER, Den Haag Publicatienummer 11, 15 december
Serdijn M (1962) Resultaten van ongevallenonderzoek (Results of accident investigation). *De Veiligheid* 38(7):25–28
Sertyesilisik B Tunstall A McLougllin J (2010). An investigation of lifting operations on UK construction sites. *Safety Science* 48(1):72–79
Shannon H (1980). The use of a model to record and store data on accidents. *Journal of Occupational Accidents* 3:57–65
Shannon H Manning D (1980). Differences between lost time and non-lost-time accidents. *Safety Science* 2:265–272
Shannon H Mayr J Haines T (1997). Overview relationship organizational and workplace factors and injury rates. *Safety Science* 26(3):201–217
Shannon H Robson L Guastello S (1999). Methodological criteria for evaluating occupational safety intervention research. *Safety Science* 31(2):161–179
Sharlin H (1989). Risk perception, changing the terms of the debate. *Journal of Hazardous Materials* 21(3):261–272
Shaw L Sichel H (1971). *Accident Proneness, Research in the Occurrence, Causation, and Prevention of Road Accidents*. Pergamon Press, Oxford
Sheeman J (1973). *Industrialization and Industrial Labour in Nineteenth-Century Europe*. Jon Wiley & Sons, New York
Shepherd A (1989). Training issues in information technology tasks. In: Bainbridge L Ruiz S (Eds.). *Developing Skills with Information Technology. Interdisciplinary Study Group New Technology and Work (NeTWork)*, pp. 191–201. Wiley, Chichester
Shewhart W (1931). *The Economic Control of Quality of Manufactured Product*. D. van Nostrand Company, Inc., New York
Shewhart W Deming W (1939). *Statistical Methods Quality Control*. The Graduate School, The Department of Agriculture, Washington
SHHFI (2008). Herbert Heinrich, http://shhofi.org/inductees/Bios/heinrich93.htm, consulted May 3rd 2008
Short J (1984). The social fabric at risk: Towards the social transformation of risk analysis. *American Sociological Review* 49(6):711–725
Shrader K (1990). The social construction of risk: Scientific method, anti-foundationalism and public decision making. *Risk Health Safety and Environment* 1(1):23–41
Shrivastava P (1992). *Bhopal Anatomy of a Crisis*. Paul Chapman Publishing Ltd, London
Shrivastava P Mitroff I Miller D Miglani A (1988). Understanding industrial crises. *Journal of Management Studies* 25(4):285–303
Shrivastava S Sonpar K Pazzaglia F (2009). Normal accident theory versus high reliability theory. *Human Relations* 62(9):1357–1390
Siccema E (1973). The environmental risk arising from the bulk storage of hazardous chemicals (in Dutch). *De Ingenieur* 85(24):502

Sievers F (1941). Individueele opvoeding van den arbeider tot veiligheid (Individual education of workers to safety) *De Veiligheid* 18(7):120–123

Silva E Nele M Melo F Konozcy L (2016). Underground parallel pipelines domino effect. *Journal of Loss Prevention in the Process Industries* 43:315–331

Simpson R (1996). Neither clear nor present: The social construction of safety and danger. *Sociological Forum* 11(3):549–562

Sinclair U (1906). *The Jungle*. Doubleday Page & Co, New York. *Heruitgave 1981 door* Bantam Books, New York

Singleton W (1960). An experimental investigation of speed controls for sewing machines. *Ergonomics* 3(4):365–375

Singleton W (1967a). Ergonomics in system design. *Ergonomics* 10(5):541–548

Singleton W (1967b). The system prototype and his design problems. *Ergonomics* 10(2):120–124

Singleton W (1969). Display design principles and procedures. *Ergonomics* 12(4):519–531

Singleton W (1971). The ergonomics of information presentation. *Applied Ergonomics* 2(4):213–220

Singleton W (1972). Techniques for determining the cause of error. *Applied Ergonomics* 3(3):126–131

Singleton W (1984). Future trends in accident research in European countries. *Journal of Occupational Accidents* 6:3–12

Sitter L de (1975). *Werkoverleg en Werkstrukturering in Zweden: Een Verslag van 500 Bedrijfsexperimenten op de Werkvloer (Work Consultation and Work Structuring in Sweden: A Report of 500 Business Experiments in the Workplace)*. Nederlandse Vereniging voor Management – NNE, no. 604, Den Haag

Skogdalen J Vinnem J (2010). Quantitative risk analysis offshore-Human and organizational factors. *Reliability Engineering and System Safety* 96:468–479.

Slob G (1961). *Enkele Onderwerpen uit de Bedrijfsveiligheid En – Hygiëne (Some Topics from Occupational Safety and Hygiene)*. Wolters, Groningen

Slovic P (1993). Trust: Perceived risk, trust and democracy. *Risk Analysis* 13(6):675–682

Slovic P (1998). The risk game. *Reliability Engineering & System Safety* 59(1):73–77

Slovic P (1999). Trust, emotion, sex politics, and science: Surveying the risk-assessment battlefield. *Risk Analysis* 19(4):689–701

Slovic P Finucane M Peters E MacGregor D (2004). Risk as analysis and risk as feelings: Some thoughts about affect, reason, risk, and rationality. *Risk Analysis* 24(2):311–322

Slovic P Fischhoff B Lichtenstein S (1984). Behavioural decision theory perspective on risk and safety. *Acta Psychol* 56:183–203

Sluis W (1983). Veiligheid tot welke prijs. Vrijheid verantwoordelijkheid en democratiering moeten de veiligheid tot stand brengen Safety at what costs. Freedom, responsibiliy and democratisation should effectuate safety). *De Veiligheid* 59(10):519–521; Eigen initiatief moet meer ruimte krijgen (Own initiative must get more room). *De Veiligheid* 59(11):581–582; Hoe werken veranderingen in op de feitelijke veiligheid (How do changes effect safety). *De Veiligheid* 59(12):606–607

Sluis W (1984). Humanisering van de arbeid bevordert sociaal beleid èn doelmatigheid in arbeidsorganisaties (Humanisation of work promotes social policy and efficiency in work organisations). *De Veiligheid* 60(5):267–269

Sluyterman K (2004). *Gedeelde zorg. Maatschappelijke Verantwoordelijkheid van Ondernemingen in Historisch Perspectief.* Oratie bij de Aanvaarding van het Ambt van Bijzondere Hoogleraar in de Faculteit der Letteren van de Universiteit Utrecht, 29 november (*Shared Care. Corporate Social Responsibility from a Historical Perspective.*

Inaugural lecture on the acceptance of the position of endowed professor in the Faculty of Arts of Utrecht University)

Smallhorn A (1967). The safe design of guillotines. *Design* 223:42–46

Smallman C (1996). Challenging the orthodoxy in risk management. *Safety Science* 22(1–3):245–262

Smit W (1971). Bedrijfsveiligheid & calamiteitenbeheersing (Occupational safety & disaster management). *De Veiligheid* 47(9):259–260

Smith M (1951). Obituaries. Dr. H. M. Vernon. *Nature* 167(4245):383–384

Snook S (2000). *Friendly Fire, the Accidental Shoot Down of US Black Hawks Over Northern Iraq*. Princeton University Press, Princeton NJ

Spaan E (1956). Veiligheid van man tot man (Safety from man to man). *De Veiligheid* 32:2–4, 25–26, 50–52, 101–106

Spaan E (1961). Effectief gebruik van veiligheidspropaganda platen (Effective use of safety propaganda posters). *De Veiligheid* 37(12):1–3

Spangenberg S (2010). Large construction projects and injury prevention. Doctoral dissertation. National Research centre for the Working Environment, Denmark & University of Aalborg, Denmark

Spangenberg S Baarts C Dyreborg J Jensen L Kines P Mikkelsen K (2003). Factors contributing to the difference in work related injury rates between Danish and Swedish Construction workers. *Safety Science* 41(6):517–530

Spangenberg S Mikkelsen K Kines P Dyreborg J Baats C (2002). The construction of the Øresund Link between Denmark and Sweden: the effect of a multi-faceted safety campaign. *Safety Science* 40(5):457–465

Spargo J (1906). *The Bitter Cry of Children*. Macmillan, New York

Sparreboom F (1947). De reizende tentoonstelling van het Veiligheidsmuseum (Travelling exhibition of the safety museum). *De Veiligheid* 23(1):1

Spies F (1958). Wie is verantwoordelijk voor de veiligheid in de onderneming? (Who is responsible for safety in a company?). *De Veiligheid* 34(1):34–36

Srinivasan R Natarajan S (2012). Developments in Inherent safety: A review of the progress during 2001–2011 and opportunities ahead. *Process Safety and Environmental Protection* 90:389–403

Stallen P (1980). De individuele beoordeling van risico's (The individual assessment of risks). *De Veiligheid* 56(1):3–7

Stallen P Tomas A (1985). *De beleving van industriële veiligheid in Rijnmond (The safety perception in Rijnmond)*. TNO Apeldoorn en Katholieke Universiteit, Nijmegen

Stallen P Vlek C (1980). De individuele beoordeling van risico's (The individual assessment of risks). *De Veiligheid* 56(2):67–73

Stankiewicz A Moulijn J (2000). Process intensification: Transforming chemical engineering, *Chemical Engineering Progress* 96(1):22–34

Stankiewicz A Moulijn J (2004). *Re-Engineering. The Chemical Processing Plant – Process Intensification*. Marcel Dekker, New York

Starr C (1969). Societal benefit versus technological risk. *Science* 165:1232–1238

Starren A Dijkman A Beek D Gallis R (2009). Improving safety at work for low skilled and high risk work. *Safety Science Monitor* 13(2) online tijdschrift

Stassen H (1981). Mens-machine systemen (Men-machine systems). *De Veiligheid* 57(9):391–396

Steffy L (2011). *Drowning in Oil, BP and the Reckless Pursuit of Profit*. McGraw-Hill, New York

Steijn W Kampen J van Beek D van der Groeneweg J Gelder P van (2020). An integration of human factors into Quantitative Risk analysis using Bayesian Belief Networks towards developing a 'QRA+'. *Safety Science* 122.

Steiner W (1939). Iets over de psychologie bij het opwekken tot medewerking ter bevordering der veiligheid (Something about the psychology during the call for coorporation to enhance safety). *De Veiligheid* 16(10):129–133

Stockdale E (1957). How do you investigate an accident? *NSC Transactions* 3:9–12

Stone RW (1931). Reviewed Work: Industrial Accident Prevention by H. W. Heinrich. *Social Service Review* 5(2):323–324. http://www.jstor.org/stable/30009710

Stout N (1987). Characteristics of work related injuries involving forklift trucks. *Journal of Safety Research* 18:179–190

Strien P van (1978). Humanisering van de arbeid en de kwaliteit van het bestaan (humanisation of labour and the quality of being). *Tijdschrift voor Sociale Geneeskunde* 56:682–689

Suchman E Scherzer A (1960a). *Specific Areas of Needed Research. Current Research in Childhood Accidents*, pp. 47–52. Association for the Aid of Crippled Children, New York, included in: Haddon W Suchman E Klein D (Eds.) (1964). *Accident Research, Methods and Approaches*. Harper & Row, New York

Suchman E Scherzer A (1960b). *Accident Proneness. Current Research in Childhood Accidents*, pp. 7–8. Association for the Aid of Crippled Children, New York, included in: Haddon W Suchman E Klein D (Eds.) (1964). *Accident Research, Methods and Approaches*. Harper & Row, New York

Sugiyama H Fischer U Hungerbuhler K (2008). Decision framework for chemical process design including different stages of environmental, health and safety assessment. *AIChE Journal* 54(4):1037–1053

Sulzer-Azaroff B (1987). The modification of occupational safety behaviour. *Journal of Occupational Accidents* 9:177–197

Sulzer-Azaroff B Santamaria C de (1980). Industrial safety hazard reduction. *Journal of Applied Behavioural Analysis* 13:287–295

Summers A (2003). Introduction to layers of protection analysis. *Journal of Hazardous Materials* 104:163–168

Suokas J (1985). On the reliability and validity of safety analysis. Thesis Tampere University

Suokas J (1988). The role of safety analysis in accident prevention. *Accident Analysis & Prevention* 20(1):67–85

Surry J (1969). *Industrial Accident Research. A Human Engineering Appraisal*. University of Toronto, Toronto

Svedung I Rasmussen J (1998). Organisational decision making and risk management under pressure from fast technological change. In: Hale A Baram M (Eds.). *Safety Management, the Challenge of Change. Interdisciplinary Study Group New Technology and Work (NeTWork)*, pp. 249–264. Pergamon, Oxford

Swain A (1964). Some problems in the measurement of Human performance in man-machine systems. *Human Factors* 6(6):687–700

Swain A (1973). Design of industrial jobs a worker can and will do. *Human Factors* 15(2):129–136

Swain A Guttmann H (1983). *Handbook of Human Reliability Analysis with Emphasis on Nuclear Power Plant Applications*. NUREG/CR-1278, US-NRC

Swuste P (1996). Occupational hazards, risk and solutions. Doctoral thesis Delft University of Technology

Swuste P (2008a) Editorial, WOS2006–1, occupational accident scenarios and accident analysis. *Safety Science* 46:151–154. doi: 10.1016/j.ssci.2007.07.001

Swuste P (2008b), Editorial, WOS2006 regulatory issues, safety climate, culture and management. *Safety Science* 46(3):345–348. doi: 10.1016/j.ssci.2007.07.002

Swuste P (2008c). Teachers and trainers of occupational safety courses, is certification necessary. *NVVK Info* 17(2):28–33

Swuste P (2008d). You will only see it if you understand it. *Human Factors* 18(4):438–453

Swuste P Albrechtsen E Hovden J (2012a). Editorial WOS2010, on the road to vision zero. *Safety Science* 50:1939–1940. doi: 10.1016/j.ssci.2012.01.005

Swuste P Drimmelen D van Burdorf L (1997b). Design analysis and solution generation. *Safety Science* 27(2–3):85–98

Swuste P Eijkemans G (2002). Occupational safety health, and hygiene in the Urban informal sector of Sub-Saharan Africa. *International Journal of Occupational and Environmental Health* 8(2):113–118

Swuste P Frijters A Guldenmund F (2012b). Is it possible to influence safety in the building sector? A literature review extending from 1980 until the present. *Safety Science* 50:1333–1343

Swuste P Goossens L Bakker F Schrover J (1997a). Evaluation of accident scenarios in a Dutch steel works using a hazard and operability study. *Safety Science* 26(12):63–74

Swuste P Groeneweg J Gulijk C van Zwaard W Lemkowitz S (2020c). The future of Safety Science. *Safety Science* 125:104593

Swuste P Groeneweg P Gulijk C van Zwaard W Lemkowitz S (2018). Safety management systems from Three Mile Island to Piper Alpha, a review in English and Dutch literature for the period 1979 to 1988. *Safety Science* 107:224–244 doi: 10.1016/j.ssci.2017.06.003

Paul a, Jop Groeneweg b, Coen van Gulijk c, Walter Zwaard d, Saul Lemkowitz

Swuste P Guldenmund F Hale A Heimplaetzer P Heming B Oortman Gerlings P (1993). *Evaluatie van een gedecentraliseerde veiligheidszorgsysteem in een geïntegreerde staalfabriek. Onderzoeksverslag (Evaluation of a decentrilised safety management system in an integrated steel works, Research report)*. vakgroep Veiligheidskunde, Technische Universiteit Delft

Swuste P Gulijk C van Groeneweg J Guldenmund F Zwaard W Lemkowitz S (2020a). Occupational safety and safety management between 1988 and 2010 Review of safety literature in English and Dutch language scientific literature. *Safety Science* 121:303–318. doi: 10.1016/j.ssci.2019.08.032

Swuste P Gulijk C van Groeneweg J Zwaard W Lemkowitz S Guldenmund F (2020b). From Clapham Junction to Macondo, Deepwater Horizon: Risk and safety management in high-tech-high-hazard sectors A review of English and Dutch literature: 1988–2010. *Safety Science* 121:249–282. doi: 10.1016/j.ssci.2019.08.031

Swuste P Gulijk C Zwaard W (2010a). Safety metaphors and theories, a review of the occupational safety literature of the US, UK, and the Netherlands, till the first part of the 20th century. *Safety Science* 48(8):1000–1018. doi: 10.1016/j.ssci.2010.01.020

Swuste P Gulijk C van Zwaard W Lemkowitz S Oostendorp Y Groeneweg J (2016). Developments in the safety science domain, in the fields of general and safety management between 1970–1979, the year of the near disaster Three Mile Island, a literature review. *Safety Science* 86:10–26. doi: 10.1016/j.ssci.2016.01.022

Swuste P Gulijk C van Zwaard W Lemkowitz S Oostendorp Y Groeneweg J (2019). Van veiligheid naar veiligheidskunde (From Safety to safety Science). Vakmedianet, Alphen

Swuste P Gulijk C van Zwaard Oostendorp W (2014). Occupational theories, models and metaphors in the three decades since World War II, In the United States, Britain and the Netherlands. *Safety Science* 62:16–27. doi: 10.1016/j.ssci.2013.07.015
Swuste P Hale A (1994). Databases on measures to prevent occupational exposure to toxic substances. *Applied Occupational and Environmental Hygiene* 9:57–61
Swuste P Hale A Guldenmund F (2002). Change in a steel works: Learning from failures and partial successes. In Wilpert B Fahlbruch B (Eds.). *System Safety: Challenges and Pitfalls of Intervention*, pp. 135–158. Elsevier, Amsterdam
Swuste P Hale A Zimmerman G (1997). Sharing workplace solutions by solution data banks. *Safety Science* 26(12):95–104
Swuste P Jongen M (2007). Behavioural Based safety werkt het? (Behavioural Based safety, does it work?). *Tijdschrift voor Toegepaste Arbowetenschap* 20(1 2):13–16
Swuste P Jongen M (2011). Resilience, wat wordt ermee bedoeld? *Tijdschrift voor Toegepaste Arbowetenschap* 24(2):68–70
Swuste P Jongen M (2013). Is taal een gevaar? (Is language a hazard?). Verslag van de CGC-NVVK-bijeenkomst 13 januari 2013. *Tijdschrift voor Toegepaste Arbowetenschap* 26(2):54–57
Swuste P Koukoulaki T Targoutzidis A (2010). Editorial WOS2008, prevention of occupational accidents in a changing environment. *Safety Science* 48:933–935. doi: 10.1016/j.ssci.2010.05.017
Swuste P Nunen K van Reniers G Khakzad N (2019) Domino effects in chemical factories and clusters: An historical perspective and discussion. *Process Safety and Environmental Protection* 124:18–30
Swuste P Nunen K van Reniers G Khakzad N (2020). Domino effects in chemical factories and clusters. In: Cozzani V Reniers G (Eds.). *Dynamic Risk Assessment and Management of Domino Effects and Cascading Events in the Process Industry*. Elsevier isbn: 9780081028384
Swuste P Reniers G (2017). Seveso inspections in the European Low Countries. History, implementation, and effectiveness of the European Seveso directives in Belgium and the Netherlands. *Journal of Loss Prevention in the Process Industries* 49:68–77. doi: 10.1016/j/jlp 2016.11.006
Swuste P Sillem S (2018). The quality of the post academic course 'management of safety, health and environment (MoSHE) of Delft University of Technology. *Safety Science* 102:26–37. http://dx.doi.org/10.1016/j.ssci.2017.09.026
Swuste P Theunissen J Schmitz P Reniers G Blokland P (2016). Process safety indicators: A review of literature. *Journal of Loss Prevention tin the Process Industries* 40:162–173
Swuste P Zwaard W Groeneweg J Guldenmund F (2019). Safety professionals in the Netherlands. *Safety Science* 114:79–88
Takala J (1993). Association between occupational hazards detected with log-linear statistical methods. *Safety Science* 17(1):13–28
Takala J (1999). Global estimation of fatal occupational accidents. *Epidemiology* 10:632–639
Taylor F (1911). *The Principles of Scientific Management*. Harper & Brothers, New York. An unabridged republication is published by Dover publications Inc., Minola NY in 1998
Taylor R (2007). Statistics of design error in the process industries. NeTWork proceedings Special issue. *Safety Science* 45(1–2):61–73
Templer J Archea J Cohen H (1985). Study of factors associated with risk of work related stairway falls. *Journal of Safety Research* 16:183–196

Ter Mors E Groeneweg J (2016). The potential of local community compensation of hosting facilities. Proceedings of the SPE International Conference and Exhibition on Health, Safety, Security, Environment and Social Responsibility in Stavanger, Norway, Ref. SPE 179228, Society of Petroleum Engineers, London

Thackrah C (1832). *The Effects of Arts, Trades and Professions and of Civic States and Habits of Living on Health and Longevity with Suggestions for Removal of Many of the Agents, which Produce Disease and Shorten the Duration of Life*, 2nd edition. Lorgman, London

THD (1978). *Technische Hogeschool Delft*. Universitair Onderwijs en Onderzoek in Veiligheid (University Education and Research in Safety), Deel 1, Eindverslag. Symposium bureau THDelft

Thomson J (2013). *Refinery Major Accident Losses 1972–2011*. Safety in Engineering Ltd. http://www.safetyinengineering.com/contact/

Tielemans E Heederik D Kromhout H Hemmen J van Meijster T Fransman W Spaan S (2004). Karakterisering van trends in blootstelling. Naar een gestructureerde benadering op het niveau van branches (Characterisation of exposure trends. Towards a structured approach on branch level). *Tijdschrift voor toegepaste Arbowetenschap* 17(4):74–81

Tinmannsvik R Hovden J (2003). Safety diagnosis criteria-development and testing. *Safety Science* 41(7):575–590

TNO (1983). *LPG A Study. Comparative Analysis of the Risks Inherent in the Storage, Transshipment, Transport and Use of LPG and Motor Spirit*. 00 General Report, TNO, Apeldoorn

Tolman C (1928). Safety and production. *Safety* 14(4):126–128, 154–155

Tombs S (1988). The management of safety in the process industry: A redefinition. *Journal of Loss Prevention in the process industry* 1:179–181

Török Z Ajtai N Turcu A Ozunu A (2011). Comparative consequence analysis of the BLEVE phenomena in the context on Land Use Planning; Case study: The Feyzin accident. *Process Safety and Environmental Protection* 89:1–7

Treffers G (1968). Toxicologische en dermatologische aspecten van verfwaren (Toxicological and dermatological aspects of paints). *De Veiligheid* 44(11):337–63

Trier A van (1975). De eis van maximale veiligheid als opgave voor de technicus (The requirement of maximum safety as a task for the technician). *De Veiligheid* 51(5):207; (6):253–260

Trist E (1981). *The Evolution of Socio-Technical Systems, a Conceptual Framework and an Action Research Program*. York University, Toronto

Trist E Bamforth K (1951). Some social and psychological consequences of the longwall method of coal-getting. *Human Relation* 4:3–38

Tuominen R Saari J (1982). A model for analysis of accidents and its application. *Journal of Occupational Accidents* 4(2–4):263–273

Turner B (1971). *Exploring the Industrial Subculture*. The MacMillan Press LTD, London

Turner B (1976). The organisational and inter-organisational development of disasters. *Administrative Science Quarterly* 21(3):378–397

Turner B (1978). *Man-Made Disasters*. Butterworth-Heinemann, Oxford

Turner B (1989). Accidents and non-random error propagation. *Risk Analysis* 9(4):437–444

Turner B (1994). Causes of disaster, sloppy management. *British Journal of Management* 5(3):215–219

Turner B Pidgeon N (1997). *Man-Made Disasters*, 2nd edition. Butterworth-Heineman, Oxford UK

Tversky A Kahneman D (1974). *Judgement under Uncertainty, Heuristics and Biases*. Macat International Ltd, London

US-NRC United States Nuclear Regularory Commission (2016). *Historical Review and Observations of Defence in Depth*. Brookhaven National Laboratory, Upton NY

Valk L van der (2007). *Werkgevers, Centrale Werkgevers Risico-Bank en de Uitvoering van de Ongevallenwet (1900–1940) (Employers, Central Employers' Risk Bank and the Introduction of the la won Industrial Accidents)*. Erasmus Universiteit Rotterdam, Rotterdam

Vaughan D (1996). *The Challenger Launch Decision. Risky Technology, Culture, and Deviance at NASA*. The University of Chicago Press, Chicago

Venart J (2004). Flixborough The explosion and its aftermath. *Process safety and Environmental Protection* 82(B2):105–127

Venema A Jettinghoff K Bloemhoff A Stam C (2007). *Monitor Arbeidsongevallen 2005 (Monitor Occupational Accidents 2005)*. TNO Kwaliteit van leven, Hoofddorp, Stichting Consument en Veiligheid, Amsterdam

Venkatasubramian R Zhao J Viswanathan S (2000). Intelligent systems for HAZOP analysis of complex process plants. *Computer and Chemical Engineering* 24:2291–2302

Vermeulen R Hartog J de Swuste P Kromhout H (2000). Trends in exposure to inhalable particles. *Annals of Occupational Hygiene* 44(5):343–354

Vernon H (1919). *The Influence of Work and of Ventilation on Output in Tinplate Manufacture*. Industrial Fatigue Board, report nr 1. HMSO, London

Vernon H (1920). *Fatigue and Efficiency in the Iron and Steel Industry*. Industrial Fatigue Board, report nr 5, HMSO, London

Vernon H (1936). *Accidents and their Prevention*. University Press, Cambridge

Vernooy A (1988). Naar een tijdschrift voor Arbowetenschap (Towards a Journal of applied Occupational Sciences). *Tijdschrift voor Arbowetenschap* 1(1):1–2 wetenschappelijk katern behorend bij Maandblad voor Arbeidsomstandigheden 1988:64(5)

Vesely W Goldberg F Roberts N Haasl D (1981). *Fault Tree Handbook*. System and Reliability Research Office of the Nuclear Regulatory Research US Nuclear Regulatory Commission. Washington D.C. NUREG-0492

Villard H (1913). *Workmen's Accident Insurance in Germany*, a series of articles, NY

Vinkenburg H (2010). *Verschillen Tussen Angelsaksen en Rijnlanders. Spanningen Rond Beheersing en Betrokkenheid (Differences bewteen Anglo-Saxons and Rhinelanders. Stress between Control and Committment)*. Sigma, april 2:4–9

Visser E Pijl Y Stolk R Neeleman J Rosmalen J (2007). Accident proneness does it exist? A review and meta-analysis. *Accident Analysis and Prevention* 39(3):556–564

Visser K (1998). Developments in HSE Management in Oil and Gas Exploration and Production. In: Hale A Baram M (Eds.). *Safety Management, the Challenge of Change*, pp. 43–66. Pergamon, Amsterdam

Vlek C (1990). *Beslissen Over Risico-Acceptatie. Een Psychologisch-Besliskundige Beschouwing Over Risicodefinities, Risicovergelijking en Beslissingsregels voor het Beoordelen van de Aanvaardbaarheid van Riskante Activiteiten (Decide on Risk Acceptance. A Psychological-Decision-Making Consideration about Risk Definitions, Risk Comparison and Decision Rules for Assessing the Acceptability of Risky Activities)*. Rijksuniversiteit Groningen, Groningen – Gezondheidsraad, Den Haag

Vlek C (2010a). Judicious management of uncertain risks: I Developments and criticisms of risk analysis and precautionary reasoning. *Journal of Risk Research* 13(4):517–543

Vlek C (2010b). Judicious management of uncertain risks: II Simple rules and more intricate models for precautionary decision making. *Journal of Risk Research* 13(4):545–569

Vlek C Stallen P (1979). Assessment of dangerous activities: A psychometric analysis. *De Ingenieur* 91:48 (in Dutch)

Vollenhoven P van (2008). Maatschappelijk debat over de essentiële veiligheidsrelatie tussen overheid en samenleving (Social debate about the essential safety relationship between government and society). *Tijdschrift voor toegepaste Arbowetenschap* 21(2):51–54

Vossnack E (1913). Het ongeval van de 'Titanic'en de waterdichte indeeling van grote mailboten (The accident of the Titanic and waterproof compartments of big mailboats). *De Ingenieur* 28(13):223–244

Vreeman R (1982). *De Kwaliteit van de Arbeid in de Nederlandse Industrie, Werknemersonderzoek Naar de Kwaliteit van de Arbeid (The Quality of Labour in the Dutch Industry, Workers' Research of the Quality of Labour)*. SUN, Nijmegen

Vuorio S (1982). Investigation of serious accidents. *Journal of Occupational Accidents* 4:281–290

Vuuren W van (1998). Organisational failure, an exploratory study in the steel industry and the medical domain. PhD thesis Institute for Business Engineering and Technology Application Technical University Eindhoven

Waart A de (1951). Veiligheid, hygiëne en bedrijfsgeneeskunde (Safety, hygiene and occupational medicine). Voordracht Commissie van overleg. *De Veiligheid* 27:24–28

Wachter J Ferguson L (2013). Fatality prevention. Findings from the 2012 forum. *Professional Safety* 58(7):41–49

Wagenaar W (1983). Menselijk falen (Human failure). *Nederlands Tijdschrift voor de Psychologie* 38:209–222

Wagenaar W (1998). People make accidents but organisations cause them. In: Feyer A Williamson A (Eds.). *Occupational Injury, Risk, Prevention and Intervention*, pp. 121–128. Taylor & Francis, London

Wagenaar W Groeneweg J (1987). Accidents at sea Multiple causes and impossible consequences International. *Journal of Man-Machine Studies* 27:587–598

Wagenaar W Groeneweg J Hudson P Reason J (1994). Promoting safety in the oil industry. The Ergonomics Society Lecture Presented at the Ergonomics Society Annual Conference, Edinburgh, 13–16 April. *Ergonomics* 37(12):1999–2013

Wallien J (1953). Opvoeding en veiligheid (Education and safety). *De Veiligheid* 29:193–195

Wansink J (1976). Risicobeheersingsmethodiek in de veiligheid (Risk control method in safety). *De Veiligheid* 52(10):377–386

Warner F (1981). Foreword by Warner in Griffith, R.F (1981), *Dealing with Risk. The Planning, Management and Acceptability of Technological Risk.* Manchester University Press, Manchester

WASH-740 (1957*). Theoretical Possibilities and Consequences of Major Accidents in Large Nuclear Power Plants. A Study of Possible Consequences of Certain Assumed Accidents.* March, United States Atomic Energy Commission, Washington, DC

WASH 1400 (1975). *Reactor Safety Study an Assessment of Accident Risks in US Commercial Nuclear Power Plants US, NUREG75/014.* Nuclear regulatory commission October Nuclear Regulatory Commission, Germantown

Waterson P (2011). World War II and other historical influences on the formation of the Ergonomics Society. *Ergonomics* 54(12):1111–1129

Waterson P Eason K (2009).'1966 and all that': Trends and developments in UK ergonomics during the 1960s. *Ergonomics* 52(11):1323–1341

Watson J (1913). Psychology as the behaviourist views it. *Psychological Review* 20:158–177

Webb W (1955). The illusive phenomena in accident proneness. *Public health report* 70(10):951–956

Weick K (1974). Middle range theories of social systems. *Behavioural Science* 19(6):357–367

Weick K (1976). Educational organisations as loosely coupled systems. *Administrative Science Quarterly* 21(1):1–19

Weick K (1979). *The Social Psychology of Organizing*. Addison-Wesley Publishing Company Readings, Massachusetts

Weick K (1987). Organisational culture as a source of high reliability. *California Man Review* 29(2):112–127

Weick K (1989a). Mental models of high reliability systems. *Industrial Crisis Quarterly* 3:127–142

Weick K (1989b). Theory construction as disciplined imagination. *The Academy of Management Review* 14(4):516–531

Weick K (1991). The non-traditional quality of organisational learning. *Organisation Science* 2(1):116–124

Weick K (1993). Collapse of sensemaking in organisations: The Mann Gulch disaster. *Administrative Science Quarterly* 38(4):628–652

Weick K (1995). *Sensemaking in Organisations*. Sage Publications Inc., Thousand Oaks CA

Weick K Roberts K (1993). Collective mind in organisations. *Administrative Science Quarterly* 38(3):357–381

Weick K Sutcliffe K (2001). *Manging the Unexpected. Resilience Performance in the Age of Uncertainty*, 2nd edition. John Wiley & Sons, Inc. (2007)

Welch J (1995). *Cycles of Learning: Observations of Jack Welch*. internet consulted 12 juli 2015

Welcker J (1978). *Heren en Arbeiders in de Vroege Nederlandse Arbeidersbeweging 1870–1914 (Gentlemen and Workers During the Early Dutch Labour Movement 1870–1914)*. Van Gennep, Amsterdam

Wellock T Budnitz R (2016). *WASH-1400 The Reactor Safety Study. The Introduction of Risk Assessment to the Regulation of Nuclear Reactors*. NUREG KM 0010. Office of Nuclear Reactor Regulation

West T (1908). *Accidents their Causes and Remedies. A Treatise of the Development of Care and Faithfulness to Aid the Safeguarding of Life and Property*. Beaver Printing Company, Publishes, Greenville, PA

Westerman F (2017). *In het Land van de Ja-Knikkers: Verhalen uit de Polder (In the Country of Yes-Nodders: Stories from the Polder)*. Querido, Amsterdam

Westerouen van Meeteren F (1893). *Handboek voor de Nijverheidshygiène, Deel I en II (Reference Book on Industrial Hygiene, Part I and II)*. Elsevier, Amsterdam

Westrum R (1988). *Organisational and inter-organisational thought*. Contribution to the World Bank Conference Safety control and risk management, October World Bank, Washington, DC

Westrum R (1993). Cultures with requisite imagination. In: Wise J Hopkin V Stager P (Eds.). *Verification and Validation of Complex Systems: Human Factors Issues*. NATO ASI Series. Springer Verlag, Berlin

Wetenschap en Samenleving (1978). *Boem in Rijnmond, Themanummer Risicoacceptatie (Boem in Rijnmond, Special Issue Risk Acceptance)*. Verbond van Wetenschappelijke Onderzoekers, Bond van Wetenschappelijke Arbeiders

Whitney A (1925) Safety for more and better adventures. *American Journal of Public Health* 15(3):223–226

Wijk L van (1977). Het ongevalsproces, een systeemmodel (The accident process, a system model). *De Veiligheid* 53(10):433–436

Wikipedia (2016). Ontsporingen 2015 oil rig fires 2015; 50 major engineering failures 1989–2001; major industrial accidents 1989–2006; List of aircraft accidents resulting in at least 50 fatalities (alleen commerciële ongevallen meegenomen – only commercial accidents); List of accidents and incidents involving commercial aircraft List of accidents and disasters by death toll, consulted Januari 2016

Wikipedia (2019). *The Chernobyl disaster*. https://en.wikipedia.org/wiki/Chernobyl_disaster consulted December 1st.

Wildavsky A (1988). *Searching for Safety*. Transaction Publishers, London

Wildavsky A Wildavsky A (2008). The concise encyclopedia of economics. *Risk and Safety*. Library of economics and liberty. Downloaded on 14 juli 2013 www.econlib.org/library/Enc/RiskandSafety.html

Wilde G (1982). The theory of risk homeostasis: Implications for safety and health. *Risk Analysis* 2(4):209–225

Wilde G (1986). Beyond the concept of risk homeostatis: Suggestions for research and application towards the prevention of accidents and lifestyle-related disease. *Accident Analysis & Prevention* 18(5):377–401

Willems J Kam J (1993). Risico-inventarisatie en – evaluatie moeten verder uit elkaar. *Arbeidsomstandigheden* 69(9):489–490

Willems R (2004). *Hier Werk je Veilig, of je Werkt hier Niet (Here You Work Safely, or Not at All)*. Shell Nederland, Den Haag

Williams S (1927). *The Manual of Industrial Safety*. Shaw Company, New York

Williamson A Feyer A Cairns D (1996). Industrial differences in accident causation. *Safety Science* 24(1):1–12

Wilpert B (2002). Foreword. In: Wilpert B Fahlbruch B (Eds.). *System Safety. Challenges and Pitfalls of Interventions. Interdisciplinary Study Group New Technology and Work (NeTWork)*, p vii-ix. Pregamon, Oxford

Wilpert B (2009). Impact of globalization on human work. NeTWork proceedings special issue. *Safety Science* 47(6):727–732

Winkel N (1936). beginselen van de beveiliging bij den arbeid in fabrieken (principles of occupational safety). *De Veiligheid* 13(8/9):114–123; 13(12):129–134

Winsemius W (1946). Psychologie en bedrijfsongevallen (Psychology of industrial accidents). *De Veiligheid* 22(14):117

Winsemius W (1951). De psychologie van het ongevalsgebeuren. Verhandeling van het Instituut voor Praeventieve Geneeskunde. Psychology of accidents, Dissertation of the Institute of Preventive Medicine. Kroese, Leiden

Winsemius W (1952). De psychologie van het ongevalsgebeuren (Psychology of accidents). *De Veiligheid* 28:17–20

Winsemius W (1958). Op weg naar een wetenschap der veiligheid (To set out for a science of safety). *Mens en Onderneming* 12:282

Winsemius W (1959). Op weg naar een wetenschap der veiligheid (To set out for a science of safety). *Mens en Onderneming* 13:24, 107, 165, 228, 291, 368

Winsemius W (1960). Op weg naar een wetenschap der veiligheid (To set out for a science of safety). *Mens en Onderneming* 14:13, 76

Winsemius W (1965). Some ergonomic aspects of safety. *Ergonomics* 8(2):151–162

Winsemius W (1969a). *Taakstructuren, Storingen en Ongevallen (Task Structures, Process Disturbances and Safety)*. Wolters Noordhoff, Groningen

Winsemius W (1969b). Werk, rommel en veiligheid (Work, debris and safety). *De Veiligheid* 45(3):73–78; (4):115–120

Winsemius W (1978). *Wonen als Feilbaar Systeem (Living as a Fallible System)*. Bijdrage aan het Symposium 'Universitair onderwijs en onderzoek in de veiligheid' (Contribution to the Symposium University education and research in safety), 11 en 12 oktober Technische Hogeschool Delft. Symposiumverlag deel I, p 265–274

Winsemius W (1980). Het ongevalsgebeuren (Accidents). In: Jonge G. de Rogmans W. Winsemius (Eds.). *Preventie van ongevallen bij kinderen (Prevention of children's' accidents)*. Stafleu's wetenschappelijke Uitgeversmaatschappij, Alphen aan den Rijn

Wolsink M (1985). Kernenergie en de structuur van de elektriciteitsvoorziening (Nuclear energy and the structure of the electricity supply). *Socialisme en Democratie* 42 58–64

Woo D Vincente K (2003). Sociotechnical systems, risk management, and public health: Comparing the North Battleford and Walkerton outbreaks. *Reliability Engineering & System Safety* 80(3):253–269

Woods D (1990a). Risk and human performance measuring the potential for disaster. *Reliability Engineering & System Safety* 29(3):387–405

Woods D (1990b). On taking human performance seriously in risk analysis. Comment on Doughtery. *Reliability Engineering & System Safety* 29(3):375–381

Wright R (1978). Wat is Loss management, 1–2 (What is Loss management 1–2). *De Veiligheid* 54(5):209–216; 54(6):297–302

WRR (2008). *Wetenschappelijke Raad voor het Regeringsbeleid. Onzekere-Veiligheid, Verantwoordelijkheden Rond Fysieke Veiligheid (Uncertain Safety, Responsibilities Regarding Physical Safety)*. Den Haag

Young A (1946) het voorkomen van ongevallen in een van de grootste Engelse machinefabrieken (The prevention of accidents in one of the biggest English machine factories). *De Veiligheid* 22(10):77–78; 22(12):93–95

Ziekemeyer M Nossent S (1990). Arbeidsomstandighedenbeleid in de rubberverwerkende industrie (Working condistions policy in the rubber manufacturing industry). *Tijdschrift voor Toegepaste Arbowetenschap* 3(2):29–35

Zielhuis R (1962). Medisch-hygiënische aspecten van de verwerking van gewapende kunststoffen (Medical and hygienic aspects of the production of armed plastics). *De Veiligheid* (3):25–27

Zielhuis R (1984). Normering van carcinogene stoffen in het beroep: Veel vragen, weinig antwoorden (Standardization of carcinogenic substances in the profession: Many questions, few answers). *Tijdschrift voor Sociale Gezondheidszorg* 62(1):9–15

Zohar D (1980a). Safety climate in industrial organisations. *Journal of Applied Psychology* 65(1):96–102

Zohar D (1980b). Het veiligheidsklimaat in bedrijfsorganisaties (Safety climate in business organisations). *De Veiligheid* 56(11):551–555

Zohar D (2008). Safety climate and beyond: A multi-level multi-climate framework. *Safety Science* 46(3):367–387

Zohar D (2010). Thirty years of safety climate research: Reflection and future directions. *Accident Analysis & Prevention* 41(5):1517–1522

Zuuren P van (1983). Veiligheidspsychologie (Safety psychology). *De Veiligheid* 59(5):265–266

Zwaailichten (2016). Overzicht van rampen en ernstige incidenten na 1945, consulted Zwaailichten.org Januari 2016

Zwaard W (1990). De menselijke fout: terug van weggeweest? (The human fault is back again?). *Maandblad voor Arbeidsomstandigheden* 66(1):6–11

Zwaard W (1993a). De groenteboer hoeft geen foutenboom te maken (The greengrocer does not have to make an fault tree). *Arbeidsomstandigheden* 69(1):7–10

Zwaard W (1993b). Risico's evalueren is tobben tussen denken en doen. Voor- en nazorg krijgen meer aandacht (Evaluating risks is worrying about thinking and doing. Pre- and aftercare receive more attention). *Arbeidsomstandigheden* 69(9):459–462

Zwaard W (1994). Tien misvattingen over risico-evaluatie. Relatieve ranking als instrument (Ten misconceptions about risk assessment. Relative ranking as an instrument). *Arbeidsomstandigheden* 71(1):17–20

Zwaard W (1995). *Risico-Inventarisatie en -Evaluatie (Risk Inventory and Evaluation)*. Kluwer, Deventer

Zwaard W (1996a). Nieuwe Arbobesluit daagt arbodeskundige uit. Verwijzen naar regeltjes kan niet meer (New Health and Safety Decree challenges health and safety expert. You can no longer refer to rules). *Arbeidsomstandigheden* 72(2):59–62

Zwaard W (1996b). Risico ranking is lastiger dan het lijkt. Rangorde is uitgangspunt, geen eindpunt (Risk ranking is more difficult than it seems. Ranking is the starting point, not the end point). *Arbeidsomstandigheden* 73(4):167–171

Zwaard W (1998). Ook nieuwe Arbowet werkt 'geen gezeik RI&E' in de hand (The new Occupational Health and Safety Act encourage a 'no bullshit RI&E'). *Arbeidsomstandigheden* 75(5):6–9

Zwaard W (1999). Gedrag hoort in de RIE (Behaviour belongs tot the RIE). *Arbeidsomstandigheden* 75(9):10–15

Zwaard W (2007). *Kroniek van de Nederlandse Veiligheid. Van Kinderarbeid en Wassend Water tot Ontploffend Vuurwerk. (Chronicle of Dutch Safety, from Child Labour and Rising Waters to Exploding Fireworks)*. Syntax Media, Arnhem

Zwaard W (2009). *Ontwikkeling in het Onvolmaakte. 40 Jaar Hogere Veiligheidskunde (40 Years of Higher Vocational Safety)*. Stichting PHOV, Utrecht

Zwaard W Burdorf A Bus J Christis J Groeneweg J Jongeneelen F Kromhout H Kwantes J Mossink J Passchier W Veld in 't R Vernooy F (1996). *Arbodeskundige (m/v) Zoekt Risico's. Risico-Inventarisatie en -Evaluatie Belicht Vanuit Verschillende Disciplines (Occupational Expert (m/f) is Looking for Risk. Risk Invertory and Evaluation by different Disciplines)*. Nederlands Instituut voor Arbeidsomstandigheden (NIA), Amsterdam

Zwaard W Goossens L (1997). Relative Ranking ais hulpmiddel voor risico-evaluatie (Relative risk ranking as an aid for risk evaluation). *Tijdschrift voor Toegepaste Arbowetenschap* 10(1):10–15

Zwaard W Groeneweg J (1999a). Risico- en kwaliteitsmanagement naderen elkaar (Risk and quality management. *Tripod versus Nederlandse Kwaliteit* 76(4):12–17

Zwaard W Groeneweg J (1999b). Gedrag verandert pas door organisatie (Behaviour only changes through organisations). *Arbeidsomstandigheden* 78(3):32–35

Zwaard W Passchier W (1995). Risicobepaling en Risicobeheersing (Risk Assessment and Risk Control). *Tijdschrift voor Toegepaste Arbowetenschap* 8(1):8–12

Zwaard W Veld R in 't (1994). Veiligheidskunde beweegt van gevaar naar risico. Risicoevaluatie en – evaluatie belicht (Safety science moves from hazard to risk. Risk evaluation and evaluation highlighted). *Arbeidsomstandigheden* 70(12):635–638

Zwam H van (1978). *Veilig Samenwerken. Veiligheid als Voorwaarde van Integraal Ondernemingsbeleid (Safe Working Together, Safety as a Condition of an Integral Company Policy)*. Van Gorcum, Assen

Zwam H van (1979a). Veiligheidsbeïnvloeding: De Werker en zijn Motieven (Safety Influence: The Workers and his Motives). *De Veiligheid* 55(1):11–16

Zwam H van (1979b). De regulering van veiligheidsgedrag (The regulation of safe behaviour). *De Veiligheid* 55(5):215–221

Zwanikken S Swuste P (2002). De veiligheid van afvalverbrandingsinstallaties (The safety of waste incineration plants). *Tijdschrift voor Toegepaste Arbowetenschap* 15(3):42–48

Author Index

Subject Index